HOLT
MATHEMATICS

8

HOLT, RINEHART AND WINSTON, PUBLISHERS
New York • Toronto • Mexico City • London • Sydney • Tokyo

AUTHORS

Eugene D. Nichols
Robert O. Lawton Distinguished Professor
   of Mathematics Education
Florida State University
Tallahassee, Florida

Paul A. Anderson
Former Elementary School Teacher
Instructor of Elementary Mathematics
University of Nevada
Las Vegas, Nevada

Francis M. Fennell
Associate Professor of Education
Director of Mathematics Clinic
Western Maryland College
Westminster, Maryland

Frances Flournoy
Professor of Elementary Education
University of Texas
Austin, Texas

Sylvia A. Hoffman
Resource Consultant in Mathematics
Illinois State Board of Education
State of Illinois

Robert Kalin
Professor of Mathematics Education
Florida State University
Tallahassee, Florida

John Schluep
Emeritus Professor of Mathematics
State University of New York
Adjunct in Mathematics
Hartwick College
Oneonta, New York

Leonard Simon
Former Mathematics Teacher and
   Assistant Director of Curriculum Development
New York City Board of Education
New York, New York

COMPUTER CONSULTANT

Frank M. Trunzo
Mathematics Curriculum Coordinator and
   Director of Lower School Computer Program
Mathematics Teacher
Germantown Academy
Fort Washington, Pennsylvania

Copyright © 1985 by
Holt, Rinehart and Winston, Publishers
All rights reserved
Printed in the United States of America

ISBN 0-03-064223-X

5 6 7 8 9 0 032 9 8 7 6 5 4 3

Art and photo credits appear on page 471.

Cover design by Carole Anson

Cover photos: TL, HRW photo by **Don Hunstein; TR**, Thomas Hopker/Woodfin Camp; **BL**, U.S. Naval Observatory/Photo Researchers; **BR**, HRW photo by **Marty Katz.**

TABLE OF CONTENTS

4 DECIMALS

5 OPERATIONS WITH DECIMALS

6 SYSTEMS OF MEASURE

7 OPERATIONS WITH FRACTIONS

8 GEOMETRY AND CONSTRUCTIONS

9 PROBLEM SOLVING AND PROPORTIONS

10 APPLICATIONS OF PERCENT

11 INTEGERS

12 REAL NUMBERS

NUMBERS AND
NUMBER THEORY

Highway interchanges provide entrance to and exit from metropolitan areas.

READING AND WRITING NUMBER NAMES

Los Angeles International Airport is one of the busiest airports in the world. One year, the number of people using the airport was 23,716,028. To read a number, think of the period names.

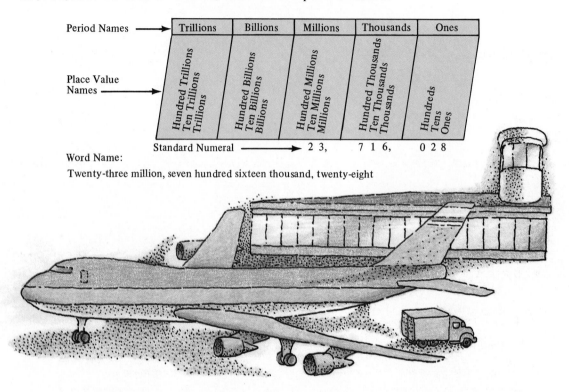

Period Names →	Trillions	Billions	Millions	Thousands	Ones
Place Value Names →	Hundred Trillions / Ten Trillions / Trillions	Hundred Billions / Ten Billions / Billions	Hundred Millions / Ten Millions / Millions	Hundred Thousands / Ten Thousands / Thousands	Hundreds / Tens / Ones

Standard Numeral → 2 3, 7 1 6, 0 2 8

Word Name:
Twenty-three million, seven hundred sixteen thousand, twenty-eight

Study and Learn

A. Read the numbers.

    1. 87,964   **2.** 564,192   **3.** 314,016,912   **4.** 974,184,312,147

B. Write standard numerals.

    5. 23 thousand   **6.** 4 million, 167 thousand

    7. Six million, seven hundred thousand, four hundred twenty-five

    8. Forty-five billion, six hundred million, twenty-four thousand

C. List in order from the least to the greatest.

    9. 81341070   81341700   81341007   82373710

2

D. Write standard numerals.

Examples $\frac{1}{2}$ thousand = 500 $6\frac{1}{2}$ thousand = 6,500

10. $14\frac{1}{2}$ thousand   **11.** $\frac{1}{2}$ million   **12.** $23\frac{1}{2}$ million   **13.** $\frac{1}{2}$ billion

Practice

Write standard numerals.

1. 742 thousand   **2.** 60 million   **3.** 724 billion

4. 386 trillion, 519 million   **5.** 32 billion   **6.** 688 trillion

7. Thirty-three million, six hundred fifteen thousand, forty

8. Seven billion, two hundred million, five hundred thousand

9. Sixty trillion, five million, seven hundred three

10. Ninety trillion, thirty million, sixty-four

List in order from the least to the greatest.

11. Invoice numbers:
81341678   96213419
81439201   81432901
82493719   96203419

12. Identification numbers:
205608339   205607999
204703712   205315629
204715111   205017832

Write standard numerals.

13. $11\frac{1}{2}$ million   **14.** $123\frac{1}{2}$ trillion   **15.** $46\frac{1}{2}$ billion   ★ **16.** 17.4 million

Solve.

17. If you spend three hundred dollars a day for forty years, you would spend four million, three hundred eighty-three thousand dollars. Write this amount as a standard numeral.

★ **18.** One year, the airport in Montreal had about $6\frac{3}{4}$ million passengers. Write the number of passengers as a standard numeral.

3

ROUNDING WHOLE NUMBERS

Today's attendance at the game was 74,759. Round the attendance to the nearest thousand and to the nearest ten thousand.

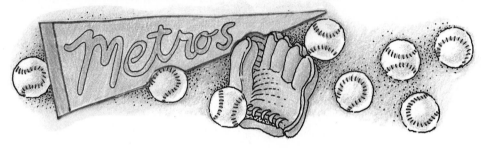

Round to	Number	Think	Write
nearest thousand	74,759	7 4,7 5 9 ↑ The digit to the right is 5 or greater.	74,759 ≐ 75,000 ↑ is approximately equal to
nearest ten thousand	74,759	7 4,7 5 9 ↑ The digit to the right is less than 5.	74,759 ≐ 70,000

Today's attendance rounded to the nearest thousand is 75,000 and rounded to the nearest ten thousand is 70,000.

Study and Learn

A. Round to the nearest thousand.

    1. 6,489   **2.** 37,916   **3.** 530,549   **4.** 1,789,769

B. Round to the nearest ten thousand.

    5. 54,320   **6.** 628,374   **7.** 2,635,000   **8.** 6,995,004

C. Round to the nearest hundred thousand.

    9. 498,372   **10.** 752,135   **11.** 5,841,000   **12.** 555,555,555

D. Round to the nearest million.

    13. 8,641,024   **14.** 34,952,000   **15.** 488,888,062

Practice

Here are the approximate seating capacities of some famous stadiums.

Wrigley Field: 37,741   San Diego: 48,460
Candlestick Park: 58,000   Metropolitan Stadium: 45,719
Three Rivers Stadium: 50,230   Municipal Stadium: 76,713

Round the seating capacities to the nearest thousand.

1. Wrigley Field   **2.** San Diego   **3.** Candlestick Park

4. Metropolitan   **5.** Three Rivers   **6.** Municipal

Round the seating capacities to the nearest ten thousand.

7. Wrigley Field   **8.** San Diego   **9.** Candlestick Park

10. Metropolitan   **11.** Three Rivers   **12.** Municipal

Round to the nearest hundred thousand.

13. 613,716   **14.** 893,423   **15.** 450,000   **16.** 965,708

17. 7,517,214   **18.** 2,684,916   **19.** 21,750,000

20. 609,253,899   **21.** 802,409,999   **22.** 5,234,569,000

Round to the nearest million.

23. 2,316,419   **24.** 5,623,900   **25.** 6,500,193

26. 26,319,814   **27.** 419,624,000   **28.** 374,815,000

29. 609,253,899   **30.** 802,409,999   **31.** 5,234,569,000

★ **32.** How many whole numbers when rounded to the nearest thousand round to 6,000?

Midchapter Review

Write standard numerals. *(2)*

1. Six billion, three hundred million, ten thousand, forty-five

2. Four trillion, nineteen thousand, two hundred six

Round. *(4)*

3. 3,714,016 to the nearest hundred thousand

4. 47,481,456 to the nearest million

5. 99,586,423 to the nearest ten thousand

6. 72,500 to the nearest thousand

Something Extra

Non-Routine Problems

Choose the box that completes the sequence.

COMPUTER—METHOD AND MEMORY

There are three steps in the computing process: Input, Compute, Output. The computing is done by the central processing unit (CPU). In most microcomputers, the CPU is contained in a single silicon chip called a microprocessor. This silicon chip has electrical connections to the input and output devices, to the random access memory (RAM) chips, and to the read only memory (ROM) chips.

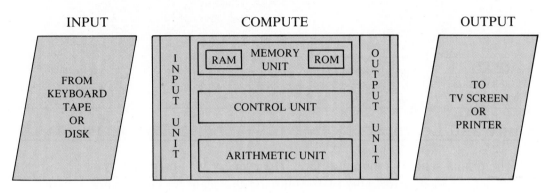

The key to the CPU is the control unit. It directs electrical current through the tiny electrical switches that make up the other four units. It does this by opening and closing switches millions of times per second. It controls the operation of the CPU.

We input information using the computer language BASIC which stands for Beginner's All-purpose Symbolic Instructional Code. The computer translates our commands into an on/off code that can be stored in the electrical switches of the memory. The memory is divided into two sections, RAM and ROM. RAM stores input in the on/off code. ROM stores switching instructions for the computer language. Here is the code for the BASIC command, **LET Y = 5**. We represent ons and offs in the switches with the binary digits 0 and 1.

L	E	T	Y	=	5
01001100	01000101	01010100	01011001	00111101	01100101

The arithmetic unit performs all computations and computer logic. When the control unit ''reads'' a command like **PRINT 3 + Y**, it opens and closes switches to get the number stored in Y from the RAM, then sends the computation to the arithmetic unit. After the arithmetic unit does the addition, the control sends the sum to the output unit. It is shown in base ten on the screen or printer.

Problem-Solving Skills

A Problem-Solving Method

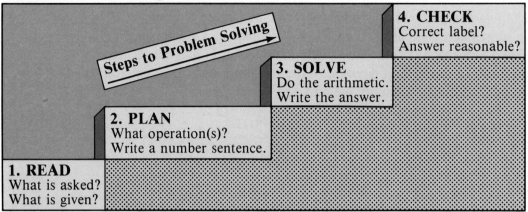

Steps to Problem Solving

4. CHECK
Correct label?
Answer reasonable?

3. SOLVE
Do the arithmetic.
Write the answer.

2. PLAN
What operation(s)?
Write a number sentence.

1. READ
What is asked?
What is given?

Study and Learn

A. The drama school is giving a play. The scenery was prepared by 12 actors. Each actor worked 16 hours on the scenery. How many hours were worked in all to prepare the scenery?

Step 1   READ the problem.

1. What is asked?   **2.** What is given?

Step 2   PLAN what to do.

3. What operation will you use?

4. Write a number sentence.

Step 3   SOLVE the problem.

5. $\begin{array}{r} 16 \\ \times\, 12 \\ \hline \end{array}$

Step 4   CHECK your answer. Reread the problem.

6. How many actors prepared the scenery?

7. How many hours did each actor work?

8. How many hours were worked in all?

9. Does it check?

B. Solve. Use the four problem-solving steps.

10. On Friday night 336 people attended the play. The auditorium has 24 seats in each row. How many rows did the 336 people fill?

Practice

Solve. Use the four problem-solving steps.

1. The script for the play was 36 pages, printed on one side. There were 24 copies of the script. How many sheets of paper were used to print the scripts?

2. Adult tickets cost $3.00 and children tickets cost $2.00 for the show. One night, 75 adult tickets and 112 children tickets were sold. How much money was collected at the ticket office that night?

3. The actors wrote the play in writing class. They used 20 periods of 45 minutes each to write the play. How many hours of class were used to write the play?

4. After the play refreshments were served. Cookies were sold for 35¢ per package and orange juice was 25¢ a container. Find the cost of 3 packages of cookies and 3 containers of orange juice.

5. The total cost of the costumes, scenery, tickets, and other items came to $100. A professional play recently cost $180,000. How many times as expensive was the professional play?

6. A special performance of the play was given for charity. Each ticket cost $5. The total amount collected was $1,000. How many people attended the performance?

7. The drama coach left his home at 6:45 pm and arrived at the school at 7:10 pm. The play started at 8:00 pm. How long after he left his home did the play begin?

★ 8. Three performances of the play were given. The average attendance at the three performances was 210. If there were 210 people at the first performance and 200 people at the second, how many people were present at the third performance?

★ 9. The auditorium stage is approximately 60 ft long and 20 ft wide. A section 40 ft by 20 ft was carpeted for the play. What part of the stage was carpeted?

DIVISIBILITY

There are 45 players in Mr. Kullen's basketball league. He wants to form teams with an equal number of players on each team. If each team has 5 players, will every player be on a team?

Since 45 ÷ 5 = 9, with *zero remainder*, 45 is *divisible* by 5. So, every player will be on a team.

▶ A number is divisible by a second number if the remainder is zero.

Study and Learn

A. Tell whether the first number is divisible by the second number.

1. 15; 5   **2.** 17; 3   **3.** 20; 10   **4.** 600; 1

▶ A number is divisible by:
  2 if the last digit is 0, 2, 4, 6, or 8.
  5 if the last digit is 0 or 5.
  10 if the last digit is 0.

B. Which numbers are divisible by 2? By 5? By 10?

5. 30   **6.** 62   **7.** 35   **8.** 40   **9.** 174

10. 285   **11.** 308   **12.** 400   **13.** 718   **14.** 68,405

▶ A number is divisible by:
  3 if the sum of the digits is divisible by 3.
  9 if the sum of the digits is divisible by 9.

C. Which numbers are divisible by 3? By 9?

15. 61   **16.** 93   **17.** 159   **18.** 207   **19.** 387

20. 427   **21.** 594   **22.** 675   **23.** 1,014   **24.** 7,614

▶ A number is divisible by 6 if it is divisible by both 2 and 3.

D. Which numbers are divisible by 6?

25. 36   **26.** 142   **27.** 908   **28.** 3,246   **29.** 8,490

Practice

Which numbers are divisible by 2?

1. 60   2. 455   3. 640   4. 935   5. 1,156

6. 1,400   7. 1,605   8. 12,015   9. 29,950   10. 46,662

Which numbers are divisible by 5?

11. 75   12. 501   13. 352   14. 680   15. 205

16. 315   17. 1,202   18. 5   19. 1   20. 5,435

Which numbers are divisible by 10?

21. 545   22. 6,455   23. 110   24. 17,320   25. 424

26. 4,000   27. 735   28. 820   29. 308   30. 10,005

Which numbers are divisible by 3?

31. 721   32. 903   33. 1,409   34. 5,013   35. 88,267

36. 683   37. 201   38. 1,333   39. 11,004   40. 771

Which numbers are divisible by 9?

41. 623   42. 409   43. 54,000   44. 818   45. 1,215

46. 3,000   47. 30,303   48. 991   49. 720   50. 1,116

Which numbers are divisible by 6?

51. 127   52. 309   53. 9,041   54. 5,013   55. 66,226

56. 72   57. 103   58. 192   59. 450   60. 518

Solve Problems

61. There are 89 baseball players. Each team has 9 players. Will every player be on a team?

★ 62. Give a rule for testing whether a number is divisible by 4.

FACTORS

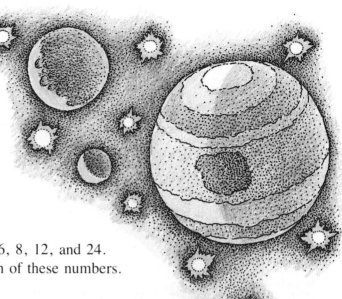

What are all the factors of 24?

Product = factor × factor
24 = 1 × 24
24 = 2 × 12
24 = 3 × 8
24 = 4 × 6

The factors of 24 are 1, 2, 3, 4, 6, 8, 12, and 24.
Notice that 24 is divisible by each of these numbers.

Study and Learn

A. List all the factors of each number.

1. 12   **2.** 20   **3.** 18   **4.** 23   **5.** 36

B. Study this table.

Number	Factors	Number of Factors
1	1	1
2	1, 2	2
3	1, 3	2
4	1, 2, 4	3
5	1, 5	2
6	1, 2, 3, 6	4
7	1, 7	2

A number that has exactly 2 different factors is called a **prime number.**

A number that has more than 2 factors is called a **composite number.**

The number 1 is neither prime nor composite.

6. Which numbers in the table are prime numbers?

7. Which numbers in the table are composite numbers?

C. Which numbers are prime numbers?

8. 15   **9.** 19   **10.** 21   **11.** 33   **12.** 47   **13.** 89

D. Which numbers are composite numbers?

14. 16   **15.** 11   **16.** 24   **17.** 37   **18.** 14   **19.** 111

E. When a number is written as a product of only prime numbers, we have the **prime factorization** of that number.

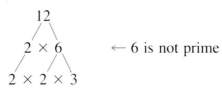

So, 12 = 2 × 2 × 3.

Write the prime factorization.

20. 8   **21.** 27   **22.** 40   **23.** 54   **24.** 81

Practice

List all the factors of each number.

1. 5   **2.** 19   **3.** 31   **4.** 6   **5.** 14

6. 27   **7.** 28   **8.** 32   **9.** 81   **10.** 30

11. 40   **12.** 48   **13.** 56   **14.** 60   **15.** 100

Tell whether each number is a prime number or a composite number.

16. 7   **17.** 11   **18.** 21   **19.** 23   **20.** 1

21. 32   **22.** 55   **23.** 36   **24.** 4   **25.** 9

26. 70   **27.** 21   **28.** 31   **29.** 85   **30.** 51

31. Copy and complete this chart through 100.

1 ② ③ 4 ⑤ 6 ⑦ 8 9 10
⑪ 12 13 14

Cross out 1. Beginning with 4, cross out all the numbers divisible by 2, by 3, by 5, and by 7. Circle all the remaining numbers. These numbers are the prime numbers less than 100. List them.

Write the prime factorization.

32. 22   **33.** 52   **34.** 45   **35.** 63   **36.** 91

SEQUENCES

A **sequence** is a group of numbers given in a specified order, according to some rule.   . . . means and so on.

Sequence: 5,   8,   11,   14, . . .
            ⟍add 3⟋   ⟍add 3⟋   ⟍add 3⟋

Rule: Add 3 to each number to get the next number in the sequence.
The next 3 numbers in the sequence are 17, 20, and 23.

Study and Learn

A. Find the rule for the sequence.

  1. 1, 4, 7, 10, 13, . . .   **2.** 1, 3, 9, 27, . . .

B. Find the number that fits the sequence.

  Example   Sequence: 1, 2, 4, 8, __?__, 32
                 The number 16 fits the sequence.

  3. 3, 9, 27, 81, __?__   **4.** 95, 85, 75, __?__, 55

C. The rule for a sequence can follow a pattern. Find the next 3 numbers in the sequence.

  Example   Sequence: 1,   3,   6,   10, . . .
                          ⟍add 2⟋   ⟍add 3⟋   ⟍add 4⟋

                 Rule: Add 2 to the first number, add 3 to the next
                        number, add 4 to the third number, and so on.
                 The next 3 numbers are 15, 21, and 28.

  5. 1, 3, 6, 10, 15, . . .   **6.** 2, 2, 4, 12, 48, . . .

D. Sometimes a sequence has 2 rules. Find the next 3 numbers in the sequence.

  Example   Sequence: 4,   5,   7,   8,   10, . . .
                          ⟍add 1⟋   ⟍add 2⟋   ⟍add 1⟋   ⟍add 2⟋

                 Rule: Add 1 to the first number, add 2 to the next
                        number, add 1 to the next number, and so on.
                 The next 3 numbers are 11, 13, and 14.

  7. 5, 10, 11, 16, 17, . . .   **8.** 2, 4, 5, 10, 11, . . .

Practice

Find the rule for the sequence.

1. 2, 7, 12, 17, 22, . . .

2. 3, 9, 15, 21, 27, . . .

3. 3, 6, 9, 12, . . .

4. 1, 4, 16, 64, . . .

Find the number that fits the sequence.

5. 2, 12, 22, 32, __?__

6. 41, 39, 37, __?__, 33

7. 0, 4, 8, 12, __?__

8. 48, 24, 12, 6, __?__

9. 5, 7, 11, 17, __?__

10. 0, 3, 7, __?__, 18

11. 1, 1, 2, 6, 24, __?__

12. 100, 98, 94, 88, 80, __?__

13. 6, 7, 5, 6, 4, __?__

14. 1, 3, 6, 8, 16, __?__

15. 1, 2, 6, 7, 21, __?__

16. 10, 20, 15, 25, 20, __?__

Find the next 3 numbers in the sequence.

17. 2, 4, 4, 8, 8, . . .

18. 10, 20, 25, 35, 40, . . .

19. 100, 98, 98, 96, 96, . . .

20. 89, 80, 71, 62, 53, . . .

★ Find the numbers that fit the sequence.

21. 60, __?__, __?__, 42, 36

22. 5, __?__, __?__, 24, 25

Something Extra
Non-Routine Problems

This grouping of numbers is called Pascal's triangle.
Discover the pattern and find rows 6 and 7 of Pascal's triangle.

```
row 1 ───────────→ 1
row 2 ───────────→ 1   1
row 3 ───────────→ 1   2   1
row 4 ───────────→ 1   3   3   1
row 5 ───────────→ 1   4   6   4   1
```

Problem-Solving Applications
Reading a Train Table

Train	Town A Lv	Town B Arr	Lv	Town C Arr	Lv	Town D Arr	Lv	Town E Arr
010	8:10 am	9:00	9:15	10:12	10:19	11:21	11:28	12:53 pm
020	8:45 am					→ 11:02	11:09	12:34 pm
030	9:15 am							→ 1:02 pm

1. How long does it take train 010 to go from Town A to Town D? [HINT: Look at the Arr column under Town D.]

2. At what towns does train 020 stop?

3. How long does it take train 010 to go from Town B to Town E?

4. How long does it take train 020 to go from Town D to Town E?

5. Ms. Parker takes Train 020 from Town A to Town D once a month. How long does the ride take?

6. How much longer does it take train 010 than Train 020 to go from Town A to Town D?

7. Mr. Parker lives in Town A and goes to Town E bimonthly on business. He takes the express train 030. How long does the ride take?

8. How much riding time is saved by taking the express train 030 from Town A to Town E rather than the local train 010?

Chapter Review

Write standard numerals. *(2)*

1. Three billion, five hundred million, eight thousand, fifty

2. Six hundred five trillion, twenty-two thousand, six

Round. *(4)*

3. 76,211 to the nearest thousand and to the nearest ten thousand

4. 7,536,917 to the nearest hundred thousand and to the nearest million

Complete. *(10)*

	Number	Divisible by					
		2	3	5	6	9	10
5.	100	✔	?	?	?	?	?
6.	16	?	?	?	?	?	?
7.	36	?	?	?	?	?	?
8.	60	?	?	?	?	?	?
9.	81	?	?	?	?	?	?
10.	95	?	?	?	?	?	?

List all the factors of each number. *(12)*

11. 15   12. 26   13. 41   14. 64   15. 8

Tell whether each number is a prime number or composite number. *(12)*

16. 5   17. 30   18. 13   19. 19   20. 49

Find the next 3 numbers in each sequence. *(14)*

21. 3, 13, 10, 20, 17, . . .   22. 3, 5, 9, 15, 23, . . .   23. 62, 58, 54, 50, 46, . . .

Solve.

24. There are 12 actors in a sewing
(8) class. Each worked for 6 periods of 30 minutes to make costumes for the play. How many hours did they work in all?

25. A train left Town A at 10:32 am
(16) and arrived in Town B at 12:15 pm that day. The trip usually takes 1 hour 24 minutes. How late was the train arriving in Town B?

Chapter Test

Write standard numerals. *(2)*

1. Two billion, six hundred million, twenty

2. Three hundred four trillion, seven hundred fourteen million

Round. *(4)*

3. 84,093 to the nearest thousand and to the nearest ten thousand

4. 8,647,938 to the nearest hundred thousand and to the nearest million

Complete. *(10)*

	Number	Divisible by					
		2	*3*	*5*	*6*	*9*	*10*
5.	300	✔	?	?	?	?	?
6.	12	?	?	?	?	?	?
7.	42	?	?	?	?	?	?
8.	45	?	?	?	?	?	?
9.	72	?	?	?	?	?	?
10.	500	?	?	?	?	?	?

List all the factors of each number. *(12)*

11. 16   **12.** 24   **13.** 7   **14.** 33   **15.** 40

Tell whether each number is a prime number or composite number. *(12)*

16. 3   **17.** 20   **18.** 7   **19.** 87   **20.** 100

Find the next 3 numbers in each sequence. *(14)*

21. 4, 14, 24, 34, 44, . . .   **22.** 4, 6, 10, 16, 24, . . .   **23.** 41, 37, 33, 29. 25. . . .

Solve.

24. There are 10 actors in an art
(8) class. Each worked for 4 periods of 30 minutes to make posters advertising the play. How many hours did they work in all?

25. A train left Town C at 9:47 am
(16) and arrived in Town D at 1:11 pm that day. The trip usually takes 3 hours 11 minutes. How late was the train arriving in Town D?

Skills Check

1. Add.

53,169
7,842
67,358
+ 9,014

A 136,383   B 137,383

C 137,833   D 138,383

2. Estimate the sum.

943 + 862

E 1,700   F 1,800

G 1,900   H 2,000

3. Subtract.

564,348
− 272,140

A 292,288   B 292,280

C 292,208   D 292,200

4. Estimate the difference.
8,543 − 2,601

E 5,000   F 6,000

G 7,000   H 12,000

5. Multiply.
7,604
× 17

A 129,468   B 129,268

C 125,268   D 60,832

6. Multiply.

564
× 703

E 396,492   F 395,492

G 394,592   H 41,172

7. Estimate the product.

36 × 81

A 3,200   B 2,400

C 240   D none of
          the above

8. Divide.
$6\overline{)43,164}$

E 7,094   F 7,184

G 7,189   H 7,194

9. Divide.
$34\overline{)2,584}$

A 760   B 706

C 85   D 76

10. Divide.

462 ÷ 42

E 111   F 101

G 11   H 10r2

11. Add.

3.016
24.512
6.79
+ 17.2

A 51.518   B 51.428

C 51.418   D 51.4

12. Add.

$ 42.21
60.39
35.65
+ 17.88

E $160.23   F $156.13

G $148.24   H $146.03

13. Subtract.
83.05
− 7.647

A 90.697   B 75.417

C 75.403   D 75.303

14. Subtract.
$ 421.65
− 96.21

E $425.64   F $402.75

G $350.68   H $325.44

15. Multiply.
1.346
× 8

A 1.0768   B 10.668

C 10.768   D 1,076.8

2 OPERATIONS WITH WHOLE NUMBERS

The telescope, located in Mona Kea, Hawaii, is 13,797 ft high.

FINDING SUMS BY MULTIPLYING

Consecutive numbers	1	2	3	4	5	6	7	8	9	. . .
Consecutive even numbers		2		4		6		8		. . .
Consecutive odd numbers	1		3		5		7		9	. . .

Sum
5
$\overbrace{(6)}$ × 3 = 18   Product
+ 7
———
18 ←

Sum
6
8
(10) × 5 = 50   Product
12
+ 14
———
50 ←

Sum of consecutive numbers = (middle number) × (number of addends).

Study and Learn

A. Find the sums by using multiplication. Check by addition.

1.   8
             9
        + 10
        ———

2.   29
            30
        + 31
        ———

3. 15 + 17 + 19

4. 22 + 24 + 26

B. Find the sums by using multiplication. Check by addition.

5.   10
            12
            14
            16
        + 18
        ———

6.   57
            58
            59
            60
        + 61
        ———

7. 4 + 6 + 8 + 10 + 12

8. 5 + 7 + 9 + 11 + 13

9. 30 + 31 + 32 + 33 + 34

Practice

Find the sums by using multiplication. Check by addition.

1.   9
            10
        + 11
        ———

2.   43
            45
        + 47
        ———

3.   122
           124
        + 126
        ———

4. 8 + 9 + 10 + 11 + 12

5. 77 + 79 + 81 + 83 + 85

STORAGE AND DISPLAY

Storage places in the computer's memory can be named with letters of the alphabet. These letters are used to tell the computer where to store data and then when to display it.

Here are some BASIC commands for storing and displaying data.

A. LET Y = 4   This tells the computer to store the number 4 in place Y. What command would you type to store 35 in X?

B. PRINT Y   This lets you see what is stored in place Y. What command would you type to see what is stored in place X?

C. PRINT Y + 7   This tells the computer to take the number stored in Y and add 7 to it. The computer will show 11 on the CRT.

D. LET Y = Y + 3   This tells the computer to take the number stored in Y and add 3 to it. What is the new value stored in Y?

E. Write the BASIC commands to store and display the following.

  1. 64 in place A   **2.** 38 in place M   **3.** 264 in place T

F. A computer program is a numbered list of commands. The control unit makes certain that the commands are followed in the order of their line numbers from lowest to highest.

Write the output for each of the following programs.

4.
```
10 LET Y = 5
20 PRINT Y
30 END
```

5.
```
10 LET A = 235
20 PRINT A + 35
30 END
```

6.
```
10 LET X = 25
20 LET X = X + 5
30 PRINT X
40 END
```

★ **7.**
```
10 LET X = 10
20 LET X = X + 15
30 LET X = X + 50
40 PRINT X
50 END
```

ADDING WHOLE NUMBERS

Marcia is on the class bowling team. Her scores for 3 games were
129, 136, and 132. What was her total score?

	Step 1	Step 2	Step 3
	ADD ONES	ADD TENS	ADD HUNDREDS

```
                 Step 1              Step 2              Step 3
               ADD ONES            ADD TENS          ADD HUNDREDS

                  1                   1                  1
   129          1 2 9               1 2 9              1 2 9
   136          1 3 6               1 3 6              1 3 6
 + 132        + 1 3 2             + 1 3 2            + 1 3 2
 _____        _____               9 7                3 9 7
                  7
         17 ones = 1 ten and 7 ones
```

Marcia's total score was 397.

Study and Learn

A. Complete.

1.	**2.**	**3.**	**4.**	**5.**
321	1,536	368	$ 23.37	55,821
+ 131	+ 2,107	429	36.84	9,839
		+ 148	+ 12.93	+ 67,092
2	43	5	4	2

B. To add numbers in horizontal form, write them in vertical form
and add.

Example   Add 1,243 + 13,286 + 9,259.

```
        1   1 1
            1,243
           13,286
         +  9,259
           _____
           23,788
```

Add.

6. 352 + 1,586 + 982

7. $123.56 + $59.07 + $368.92

C. Add.

8.	**9.**	**10.**	**11.**
5,624	$ 163.46	31,684	$ 1,784,819
+ 9,736	+ 89.49	3,714	68,720
		+ 23,932	+ 718,619

12. 6,007 + 9,486 + 701 + 650

13. 5,986 + 10,665 + 4,586 + 65,002

Practice

Add.

1. 844
  + 35

2. 1,600
  + 2,200

3. 4,634
  + 1,358

4. 5,201
  + 865

5. 44,360
  + 22,646

6. $ 364.58
  + 288.01

7. 34,624
  + 9,193

8. 567,896
  + 65,417

9. 1,421
  2,121
  + 1,215

10. $ 3,356
  4,107
  + 225

11. 3,384
  4,462
  5,734
  + 6,618

12. 36,582
  4,621
  59,007
  + 3,982

13. 4,613
  2,176
  6,345
  2,186
  + 1,794

14. 35,416
  8,902
  41,284
  32,369
  + 8,416

15. $ 634,019
  284,678
  72,516
  832,743
  + 62,713

16. $ 490,674
  7,785,109
  4,456,213
  489,113
  + 5,340,113

17. 234 + 511 + 1,724

18. $98.72 + $4.61 + $30.01

19. 23,304 + 546 + 4,279 + 9,806

20. 4,987 + 9,801 + 406 + 586

Solve Problems

21. The 5 members of the Lions bowling team scored 172, 195, 156, 127, and 169 in 1 game. What was their game total?

★ **22.** The sum of 7 addends is 137. 6 is added to each addend. What is the new sum?

Computer

A computer can be used as a calculator. Type each number without a comma. To add 1,421, 2,121, and 1,215, you type PRINT 1421 + 2121 + 1215, and press [RETURN] . The output is 4757.

H A N D S   O N

Type the input. Press [RETURN] after each line. Write the output.

1. 4,987 + 9,801 + 406 + 586

2. 567,896 + 65,417

3. 3,384 + 4,462 + 5,734 + 6,618

4. 4,613 + 8,752 + 19,201

SUBTRACTING WHOLE NUMBERS

There are 590 garbage cans along the city sanitation route. So far, 375 cans have been cleared. How many garbage cans remain to be cleared?

Step 1	Step 2	Step 3
RENAME TENS SUBTRACT ONES	SUBTRACT TENS	SUBTRACT HUNDREDS

$$
\begin{array}{r} 590 \\ -375 \\ \hline \end{array}
\qquad
\begin{array}{r} {}^{8\ 10}\\ 5\,\cancel{9}\,\cancel{0} \\ -3\,7\,5 \\ \hline 5 \end{array}
\qquad
\begin{array}{r} {}^{8\ 10}\\ 5\,\cancel{9}\,\cancel{0} \\ -3\,7\,5 \\ \hline 1\,5 \end{array}
\qquad
\begin{array}{r} {}^{8\ 10}\\ 5\,\cancel{9}\,\cancel{0} \\ -3\,7\,5 \\ \hline 2\,1\,5 \end{array}
$$

9 tens and 0 ones =
8 tens and 10 ones

Study and Learn

A. Subtract.

1. 780
   − 264

2. 7,936
   − 2,562

3. 89,837
   − 64,018

4. $ 653,682
   − 42,565

B. Sometimes you have to rename more than once to subtract.

Example   734 $\begin{array}{r}{}^{6}\,{}^{\overset{12}{\cancel{2}}}\,{}^{14}\\ \cancel{7}\,\cancel{3}\,\cancel{4} \\ -2\,5\,8 \\ \hline 4\,7\,6\end{array}$
            − 258

Subtract.

5. 634
   − 279

6. 4,316
   − 2,087

7. 56,342
   − 24,738

8. $ 841,356
   − 12,467

C. Subtraction with zeros should present no problem.

Example   6,000 $\begin{array}{r}{}^{5}\,{}^{9}\,{}^{9}\,{}^{10}\\ \cancel{6},\cancel{0}\,\cancel{0}\,\cancel{0} \\ -3,7\,5\,6 \\ \hline 2,2\,4\,4\end{array}$
            − 3,756

Subtract.

9. 800
   − 437

10. 3,000
   − 1,594

11. 6,002
   − 4,138

12. 50,000
   − 3,759

Practice

Subtract.

1.	7,687 − 4,194	**2.**	4,732 − 3,691	**3.**	74,824 − 61,290	**4.**	621,987 − 500,239
5.	8,436 − 6,147	**6.**	8,936 − 6,247	**7.**	34,836 − 13,159	**8.**	521,987 − 436,123
9.	5,204 − 1,328	**10.**	3,241 − 986	**11.**	73,405 − 14,376	**12.**	89,361 − 4,475
13.	$ 764,716 − 240,939	**14.**	$ 743,168 − 15,274	**15.**	$ 34,853 − 10,261	**16.**	$ 85,163 − 63,548
17.	500 − 324	**18.**	700 − 239	**19.**	6,000 − 3,451	**20.**	4,000 − 2,981
21.	6,004 − 3,213	**22.**	7,005 − 6,178	**23.**	50,000 − 27,616	**24.**	60,000 − 25,750

25. 800 − 247   **26.** 9,000 − 3,416   **27.** 10,053 − 2,594

28. 648,316 − 26,129   **29.** 2,816,347 − 954,819

Computer

You can use a simple program to subtract.

```
10 LET A = 1805
20 LET B = 371
30 PRINT A - B
40 END
```

This makes the computer

• store 1805 in A.

• store 371 in B.

• print the difference between A and B.

• stop the program.

H A N D S   O N

＋   Store the numbers you need in lines 10 and 20 to check your answers.
Press ⌈ RETURN ⌉ after each line, and after **RUN.**

ESTIMATING SUMS AND DIFFERENCES

The Yellow Bus Company had 672 passengers in one week. The second week, it had 548 passengers. The third week it had 450 passengers. Estimate how many passengers it had in all.

Estimate the sum by rounding each addend.

ACTUAL		ESTIMATE
672	⟶	700
548	⟶	500
+ 450	⟶	+ 500
		1,700

There were about 1,700 passengers.

Study and Learn

A. Estimate the sums.

Examples

361 is about   360
+ 23 is about + 20
                380

15,842 is about   15,800
+   756 is about +   800
                    16,600

1. 864 + 239	**2.** 1,207 + 6,641	**3.** 473 + 91	**4.** 7,111 + 326	**5.** 71,850 + 532

B. Estimate the differences.

Examples

4,949 is about   5,000
− 1,216 is about − 1,000
                  4,000

2,736 is about   2,700
−   287 is about −   300
                   2,400

6. 514 − 149	**7.** 8,279 − 4,630	**8.** 658 − 96	**9.** 3,339 − 602	**10.** 14,996 − 550

C. Estimate the sums.

11. 459 361 + 747	**12.** 4,516 8,290 + 927	**13.** 6,745 828 + 394	**14.** 64,078 3,953 + 17,224	**15.** 8,027 64 153 + 3,815

Practice

Estimate the sums.

1. 625 + 283	**2.** 371 + 59	**3.** 464 + 37	**4.** 3,219 + 6,540	**5.** 7,982 + 1,486
6. 5,195 + 644	**7.** 2,348 + 623	**8.** 4,591 + 74	**9.** 8,263 + 54	**10.** 14,889 + 63,241
11. 72,687 + 19,045	**12.** 53,622 + 8,829	**13.** 88,409 + 7,613	**14.** 41,720 + 231	**15.** $ 49,643 + 586
16. 456 713 + 949	**17.** 2,765 820 + 3,495	**18.** 907 5,634 + 723	**19.** 13,486 4,842 + 29,072	**20.** 3,268 40 312 + 652

Estimate the differences.

21. 748 − 394	**22.** 660 − 287	**23.** 882 − 25	**24.** 509 − 42	**25.** 5,072 − 2,458
26. 6,646 − 3,718	**27.** 8,643 − 399	**28.** 5,391 − 247	**29.** 8,416 − 99	**30.** 7,930 − 76
31. 36,212 − 14,084	**32.** 48,605 − 37,121	**33.** 45,116 − 4,470	**34.** 72,084 − 6,195	**35.** $ 24,192 − 901
36. 16,792 − 839	**37.** 52,817 − 13,680	**38.** 8,159 − 276	**39.** 7,518 − 7,485	★ **40.** 379,800 − 25,466

Solve Problems

41. A bus was driven 19,416 km one year, 24,961 km the second year, and 35,000 km the third year. Estimate the total distance driven.

42. One bus driver earned $9,345 and a second driver earned $4,745. Estimate how much more the first driver earned.

★ **43.** It is projected that in 1990 there will be 16,718,000 youths 10 to 14 years of age. This number is expected to increase to 20,153,000 by the year 2000. Estimate the increase.

Midchapter Review

Add. *(24)*

1.	2.	3.	4.	5.
2,346 + 497	$ 73,694 + 2,867	639 258 + 384	1,053 975 + 2,482	$ 13,867 29,045 + 6,690

Subtract. *(26)*

6.	7.	8.	9.	10.
897 − 645	$ 6,734 − 2,456	9,000 − 2,463	42,000 − 8,751	371,279 − 286,485

Estimate the sums. *(28)*

11.	12.	13.	14.	15.
6,438 + 7,126	36,416 + 9,319	$ 7,615 + 82	37,619 4,823 + 23,914	57 391 + 8,056

Estimate the differences. *(28)*

16.	17.	18.	19.	20.
8,341 − 2,942	7,016 − 914	40,041 − 9,562	5,218 − 76	$ 16,962 − 703

Something Extra

Non-Routine Problems

Which 2 shapes are mirror images?

1.
  a   b   c   d   e

2.
  a   b   c   d   e

COMPUTER GRAPHICS

Computer graphics allows you to draw shapes on the CRT.
Resolution refers to the size of the blocks being lit. The higher the
resolution, the smaller the blocks of lights, and the finer the detail.

A typical low-resolution graphics screen is made up of 40 columns
and 40 rows of blocks numbered 0 to 39. A typical high-resolution
screen has 280 columns numbered 0 to 279, and 160 rows numbered
0 to 159. Blocks are named according to their column and row
numbers. The block in the 6th column and 4th row is named 6,4.

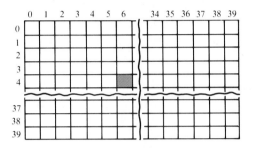
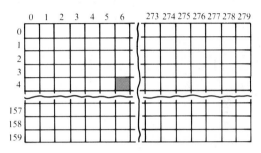

The programs below draw right triangles in the color blue (6). The
GR and HGR commands make the computer use the graphics mode.
Lines 30 and 70 cause the PLOT lines to use 0, then 1, then 2, and so on.
Which parts of the triangle do lines 40 through 60 plot?

LOW RESOLUTION

```
10 GR
20 COLOR = 6
30 FOR L = 0 TO 39 STEP 1
40 PLOT 0,L
50 PLOT L,39
60 PLOT L,L
70 NEXT L
```

HIGH RESOLUTION

```
10 HGR
20 HCOLOR = 6
30 FOR L = 0 TO 159 STEP 1
40 HPLOT 0,L
50 HPLOT L,159
60 HPLOT L,L
70 NEXT L
```

H A N D S   O N

To enter these programs, type the lines. Press [RETURN] after each
line. To run the programs, type **RUN** and press [RETURN] .

31

Problem-Solving Skills
Using Diagrams

A motorboat priced at $670 was on sale for $538.
How much was saved by buying during the sale?

Diagrams may help solve problems.

Original price

Sale price	} Savings

$$\begin{array}{r} \$\,670 \\ -\ 538 \\ \hline \$\,132 \end{array}$$

$132 was saved.

Study and Learn

A. Solve. Use the diagram.

1. Miss Greenberg bought a
   fishing boat on sale for $379.
   She saved $120.99. What was
   the original price of the boat?

Original price	
Sale price	Savings

2. Jane walked east from the
   campsite for 315 m. Olga
   walked west from the campsite
   for 405 m. How far apart
   were they?

   Olga ← 405 m △ 315 m → Jane

3. Juan cut a board that measured
   500 cm into 2 pieces. The
   smaller piece measured
   150 cm. What was the length
   of the larger piece?

   |← 500 cm →|
   | 150 cm } | ? |

B. Draw a diagram for each problem. Solve.

4. Steve pitched his tent 5 km due north of the ranger station.
   Dave pitched his tent 8 km due north of the ranger station.
   How far is Steve's tent from Dave's tent?

5. Mrs. Donaldsen is putting up 3 shelves in the living room.
   Each shelf is 5 cm thick. She wants 50 cm between each
   shelf. What will the distance be from the bottom of the
   bottom shelf to the top of the top shelf?

Practice

Draw a diagram for each problem. Solve.

1. Ms. Sterling bought a car stereo on sale for $188. The original price of the stereo was $375. How much did Ms. Sterling save by buying the stereo on sale?

2. Mr. Lietzke bought a television set on sale for $628.50. He saved $130.90 of the original price. What was the original price of the television set?

3. Paul and Linda drove toward each other from towns 26 mi apart. When they met, Paul had driven 17 mi. How far had Linda driven?

4. Frances cut a 45-cm board into two pieces. The shorter piece is 21 cm long. How long is the longer piece?

5. Pedro built a 60-cm shelf. He painted 35 cm on the left side of the shelf to indicate that he would store tools there. How long is the unpainted part of the shelf?

6. In Fred's room, there are 2 shelves on one wall. The first is 3 ft off the floor, the second is 3 ft above that. This last shelf is 2 ft from the ceiling. How high is the room?

7. The total length of a fence is 50 m. All but 14 m needs repair. How much fence needs repair?

8. The area of a vegetable garden is 108 m². Tomatoes and broccoli are planted in 65 m². How much space is left for cabbage?

9. Jane bought 300 cm of seam binding. She used 40 cm on a vest and 110 cm on a skirt. How much seam binding is left?

10. Joe used 175 mL of milk in a bowl of cereal and 250 mL to make a milk shake. How much milk did he use?

★ 11. A 24-foot pole is placed so that one half as much is below ground as above ground. How many feet of pole are above ground?

MULTIPLYING

A parking lot has 7 levels. There are 246 parking spaces on each
level. How many cars can be parked in the parking lot?

	Step 1	Step 2	Step 3
	MULTIPLY ONES	MULTIPLY TENS	MULTIPLY HUNDREDS

$$246 \times 7$$

Step 1:
$$\begin{array}{r} {}^{4}24\,6 \\ \times\ 7 \\ \hline 2 \end{array}$$

Step 2:
$$\begin{array}{r} {}^{3\ 4}24\,6 \\ \times\ 7 \\ \hline 22 \end{array}$$

Step 3:
$$\begin{array}{r} {}^{3\ 4}24\,6 \\ \times\ 7 \\ \hline 1{,}722 \end{array}$$

So, 1,722 cars can be parked in the parking lot

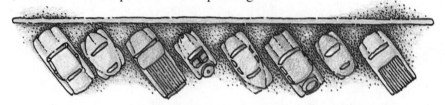

Study and Learn

A. Multiply.

1. 234	**2.** 1,312	**3.** 2,106	**4.** 14,081	**5.** $569.25
× 8	× 7	× 5	× 8	× 3

B. Complete.

6. Step 1	**7.** Step 2	**8.** Step 3
MULTIPLY BY ONES	MULTIPLY BY TENS	ADD

$$\begin{array}{r} 36 \\ \times 24 \\ \end{array}\qquad \begin{array}{r} 36 \\ \times 24 \\ \hline 144 \end{array}\qquad \begin{array}{r} 36 \\ \times 24 \\ \hline 144 \\ 720 \\ \hline \end{array}$$

← The zero may be left off.

C. Multiply.

9. 47	**10.** 673	**11.** 5,708	**12.** 32,411	**13.** $394.85
× 35	× 46	× 72	× 18	× 29

14. 8 × 371   **15.** 5 × 2,806   **16.** 6 × $5.42

17. 21 × 20   **18.** 46 × 85   **19.** 13 × 4,832

34

Practice

Multiply.

1. 42 ×2	**2.** 236 ×2	**3.** 107 ×8	**4.** 247 ×9	**5.** 364 ×7
6. 2,316 ×4	**7.** 3,215 ×6	**8.** 1,482 ×2	**9.** 56,822 ×4	**10.** $214.61 ×5
11. 22 ×34	**12.** 39 ×12	**13.** 46 ×37	**14.** 213 ×32	**15.** 324 ×16
16. 418 ×28	**17.** 672 ×53	**18.** 748 ×29	**19.** 6,133 ×24	**20.** $75.06 ×48
21. 6,834 ×76	**22.** 4,527 ×36	**23.** 97,543 ×88	**24.** 13,116 ×54	**25.** $358.96 ×21

26. 8 × 4,605   **27.** 92 × 27   ★ **28.** 54 × 8 × 61 × 7

Solve Problems

29. A survey showed that 256 cars used a parking lot each day. How many cars would use the parking lot in 5 days?

30. The cost of cleaning and lighting a parking lot in Suntown is $2,436 a month. What is the cost for 12 months?

MULTIPLYING LARGER NUMBERS

A space shuttle makes 125 trips a month. The space shuttle has 257 seats on it. How many people can travel on the space shuttle each month?

Step 1	Step 2	Step 3	Step 4
MULTIPLY BY ONES	MULTIPLY BY TENS	MULTIPLY BY HUNDREDS	ADD

$$
\begin{array}{r} 257 \\ \times\,125 \\ \hline 1285 \end{array}
\qquad
\begin{array}{r} 257 \\ \times\,125 \\ \hline 1285 \\ 514 \end{array}
\qquad
\begin{array}{r} 257 \\ \times\,125 \\ \hline 1285 \\ 514 \\ 257 \end{array}
\qquad
\begin{array}{r} 257 \\ \times\,125 \\ \hline 1\,285 \\ 5\,14 \\ 25\,7 \\ \hline 32{,}125 \end{array}
$$

So, 32,125 people can travel on the space shuttle each month.

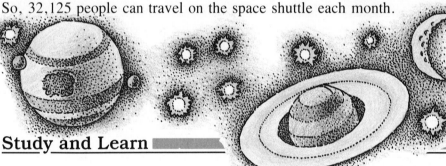

Study and Learn

A. Multiply.

1. $\begin{array}{r} 208 \\ \times\,764 \end{array}$
2. $\begin{array}{r} 1{,}192 \\ \times\,835 \end{array}$
3. $\begin{array}{r} 72{,}593 \\ \times\,219 \end{array}$
4. $\begin{array}{r} 2{,}356 \\ \times\,3{,}426 \end{array}$

5. $1{,}583 \times 19{,}037$
6. $4{,}176 \times 61{,}243$

B. Be careful of the zeros.

Examples

$$
\begin{array}{r} 238 \\ \times\,604 \\ \hline 952 \\ 14280 \\ \hline 143{,}752 \end{array}
\quad \longleftarrow 0 \times 238 = 0
$$

$$
\begin{array}{r} 2{,}601 \\ \times\,460 \\ \hline 156060 \\ 10404 \\ \hline 1{,}196{,}460 \end{array}
\quad \longleftarrow 0 \times 2{,}601 = 0
$$

Multiply.

7. $\begin{array}{r} 416 \\ \times\,408 \end{array}$
8. $\begin{array}{r} 2{,}609 \\ \times\,380 \end{array}$
9. $\begin{array}{r} 5{,}275 \\ \times\,1{,}004 \end{array}$
10. $\begin{array}{r} 7{,}230 \\ \times\,8{,}050 \end{array}$

36

Practice

Multiply.

1. 234 × 311	**2.** 642 × 236	**3.** 705 × 782	**4.** 318 × 945
5. 4,630 × 324	**6.** 5,724 × 678	**7.** 7,834 × 349	**8.** 9,076 × 414
9. 16,854 × 521	**10.** 2,508 × 1,366	**11.** 6,815 × 2,437	**12.** 32,184 × 5,183
13. 236 × 806	**14.** 1,342 × 280	**15.** 9,612 × 5,001	**16.** 3,152 × 3,010

17. 656 × 2,034

18. 1,830 × 5,926

19. 4,030 × 7,085

★ **20.** 90,009 × 34,341

Solve Problems

21. A trip on a space shuttle costs $836. There are 197 passengers on the shuttle. How much money will the space shuttle company take in?

Computer

The computer can multiply two numbers by using the symbol *.

```
10 LET A = 705
20 LET B = 782
30 PRINT A * B
```

This program makes the computer
store 705 in A.
store 782 in B.
print the product of A and B.

H A N D S   O N

Use the program to check your answers above. Modify lines 10 and 20 to change A and B. Press [RETURN] after each line and **RUN**.

DIVIDING BY A 1-DIGIT NUMBER

Disco Music Store ordered 456 records. Records are packaged 6 to a box. How many boxes of records did Disco Music receive?

Step 1
ESTIMATE.
How many 6's in 4? None.
How many 6's in 45? Try 7.

$$\begin{array}{r} 7 \\ 6\overline{)45\,6} \end{array}$$

Step 2
MULTIPLY.

$$\begin{array}{r} 7 \\ 6\overline{)456} \\ 42 \end{array}$$

Step 3
SUBTRACT. The estimate is correct. 42 is less than 45. Bring down the 6.

$$\begin{array}{r} 7 \\ 6\overline{)456} \\ 42\downarrow \\ \hline 36 \end{array}$$

Step 4
Repeat the steps.

$$\begin{array}{r} 76 \\ 6\overline{)456} \\ 42 \\ \hline 36 \\ 36 \\ \hline 0 \end{array}$$

Study and Learn

A. Complete.

1. $3\overline{)429}$ (1)

2. $2\overline{)\$948}$ (\$4)

3. $5\overline{)745}$ (1)

4. $2\overline{)816}$ (4)

5. $6\overline{)192}$ (3)

6. $4\overline{)\$344}$ (\$ 8)

7. $7\overline{)2,408}$ (3)

8. $8\overline{)3,456}$ (4)

B. Divide.

9. $4\overline{)3,224}$

10. $6\overline{)2,418}$

11. $3\overline{)5,121}$

12. $8\overline{)9,624}$

13. $3\overline{)164}$

14. $5\overline{)816}$

15. $4\overline{)314}$

16. $8\overline{)3,165}$

17. $4\overline{)848}$

18. $2\overline{)\$864}$

19. $3\overline{)9,969}$

20. $2\overline{)4,068}$

21. $3\overline{)951}$

22. $8\overline{)7,616}$

23. $6\overline{)4,874}$

24. $7\overline{)21,141}$

Practice

Divide.

1. $2\overline{)48}$ 2. $3\overline{)96}$ 3. $4\overline{)84}$

4. $3\overline{)76}$ 5. $3\overline{)82}$ 6. $6\overline{)89}$

7. $4\overline{)38}$ 8. $6\overline{)59}$ 9. $7\overline{)80}$

10. $2\overline{)420}$ 11. $3\overline{)963}$ 12. $4\overline{)884}$

13. $4\overline{)248}$ 14. $3\overline{)159}$ 15. $7\overline{)497}$

16. $6\overline{)846}$ 17. $3\overline{)516}$ 18. $4\overline{)\$568}$

19. $3\overline{)\$744}$ 20. $6\overline{)852}$ 21. $4\overline{)725}$

22. $3\overline{)609}$ 23. $4\overline{)412}$ 24. $5\overline{)105}$

25. $2\overline{)6,846}$ 26. $3\overline{)\$9,936}$ 27. $4\overline{)8,480}$

28. $3\overline{)8,196}$ 29. $4\overline{)9,220}$ 30. $4\overline{)3,284}$

31. $3\overline{)7,526}$ 32. $4\overline{)9,704}$ 33. $7\overline{)\$9,401}$

34. $6\overline{)\$6,006}$ 35. $4\overline{)4,008}$ 36. $3\overline{)1,509}$

37. $8\overline{)4,040}$ 38. $7\overline{)6,314}$ 39. $9\overline{)8,127}$

40. $4\overline{)31,284}$ 41. $6\overline{)29,642}$ 42. $7\overline{)12,346}$

43. $6\overline{)18,366}$ 44. $5\overline{)25,250}$ 45. $8\overline{)64,320}$

Solve Problems

46. 4 Model XYZ television sets were sold for a total of $1,556. What was the cost of each set?

★ 47. 4,644 cassettes are to be packed in boxes of 3 each. Then, the boxes are packed in cartons of 9 each. How many cartons will be filled?

DIVIDING BY LARGER NUMBERS

715 paperback books are placed in 31 boxes. Each box has the same number of books. How many books are in each box?

Step 1 Estimate.
How many 3 1's in 7 1?
Think: How many 3's in 7? Try 2.

Step 2 Multiply.

Step 3 Subtract. The estimate is correct. Bring down the 5.

Step 4 Repeat the steps.

```
        23 r 2
   31)715
       62↓
       95
       93
        2
```

There are 23 books in each box. There are 2 books left over.

Study and Learn

A. Divide.

   1. 36)864　　　**2.** 63)3,404　　　**3.** 52)$12,168　　　**4.** 312)13,416

B. Sometimes the first estimate is too large.

Example

Step 1 Estimate.
How many 3 4's in 6 2?
Think: how many 3's in 6? Try 2.

```
         2
   34)628
       68
        └── Too large
```

Step 2 Multiply.

Step 3 Subtract. The estimate is not correct. Try 1.
The new estimate is correct. Bring down the 8.

Step 4 Repeat the steps. Complete.

```
        18 r 16
   34)628
       34
       288
       272
        16
```

Divide.

   5. 49)8,869　　　**6.** 37)9,472　　　**7.** 29)1,830　　　**8.** 38)64,214

   9. 321)15,087　　　**10.** 716)499,052　　　**11.** 643)93,175

Practice

Divide.

1. $24\overline{)72}$ 2. $23\overline{)69}$ 3. $37\overline{)74}$ 4. $17\overline{)87}$ 5. $44\overline{)98}$

6. $63\overline{)378}$ 7. $21\overline{)168}$ 8. $43\overline{)349}$ 9. $54\overline{)268}$ 10. $82\overline{)616}$

11. $26\overline{)780}$ 12. $37\overline{)740}$ 13. $23\overline{)690}$ 14. $45\overline{)459}$ 15. $19\overline{)576}$

16. $41\overline{)902}$ 17. $33\overline{)\$858}$ 18. $36\overline{)475}$ 19. $16\overline{)388}$ 20. $27\overline{)820}$

21. $28\overline{)196}$ 22. $27\overline{)162}$ 23. $39\overline{)\$273}$ 24. $49\overline{)353}$ 25. $74\overline{)508}$

26. $29\overline{)762}$ 27. $17\overline{)394}$ 28. $27\overline{)675}$ 29. $31\overline{)296}$ 30. $63\overline{)450}$

31. $31\overline{)\$6,262}$ 32. $24\overline{)4,872}$ 33. $34\overline{)6,868}$ 34. $75\overline{)8,625}$

35. $43\overline{)5,117}$ 36. $21\overline{)9,093}$ 37. $34\overline{)7,316}$ 38. $52\overline{)9,407}$

39. $56\overline{)\$2,296}$ 40. $68\overline{)1,656}$ 41. $79\overline{)2,607}$ 42. $48\overline{)3,009}$

43. $47\overline{)52,781}$ 44. $67\overline{)54,605}$ 45. $38\overline{)22,867}$ 46. $87\overline{)436,200}$

47. $231\overline{)5,544}$ 48. $313\overline{)7,180}$ 49. $421\overline{)15,156}$ 50. $428\overline{)52,644}$

51. $712\overline{)242,080}$ 52. $541\overline{)124,971}$ ★ 53. $654\overline{)2,658,510}$ ★ 54. $207\overline{)1,657,863}$

Solve Problems

55. 24 cases of pencils cost $2,352. What was the cost of each case?

56. 24 cartons of paper cost a printer $672. What is the cost of a carton of paper?

Computer

```
10 LET A = 810
20 LET B = 27
30 PRINT A / B
```

Use this division program to check the answers above.

• Store 810 in A.

• Store 27 in B.

• Print the quotient.

A SHORTCUT IN DIVISION

Here is a short form you can use when the divisor is less than 10.

Short Form

Step 1
$$\begin{array}{r} 3 \\ 6\overline{)192} \\ \underline{18} \\ 1 \end{array}$$

$$6\overline{)19\,^{1}2}^{\,3}$$

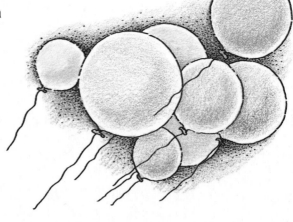

Step 2
$$\begin{array}{r} 32 \\ 6\overline{)192} \\ \underline{18} \\ 12 \\ \underline{12} \\ 0 \end{array}$$

$$6\overline{)19\,^{1}2}^{\,3\,2}$$

Practice

Divide. Use the short form.

1. $4\overline{)172}$ **2.** $6\overline{)384}$ **3.** $3\overline{)6,468}$ **4.** $9\overline{)4,338}$

5. $7\overline{)6,412}$ **6.** $8\overline{)42,319}$ **7.** $5\overline{)36,510}$ **8.** $7\overline{)342,612}$

9. $2\overline{)189}$ **10.** $3\overline{)4,526}$ **11.** $5\overline{)26,729}$ **12.** $6\overline{)723,601}$

13. $7\overline{)42,079}$ **14.** $8\overline{)561,602}$ **15.** $9\overline{)3,600,453}$

Something Extra
Non-Routine Problem

How many blocks are in the pile?

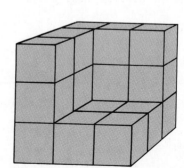

Skills Review

Write word names. *(2)*

1. 849,027   **2.** 5,361,004   **3.** 8,090,207,104

Write standard numerals. *(2)*

4. 942 billion, 671 thousand

5. 27 trillion, 86 million

6. Seventy trillion, three billion, six hundred two thousand

7. Eighty-nine billion, four hundred twelve million, eighteen

Round. *(4)*

8. 549,020 to the nearest thousand

9. 7,624,109 to the nearest ten thousand

10. 8,473,001 to the nearest hundred thousand

11. 625,501,784 to the nearest million

Complete. *(10)*

Number	Divisible by 2	3	5	6	9	10
12. 300	✔	?	?	?	?	?
13. 150	?	?	?	?	?	?
14. 62	?	?	?	?	?	?
15. 24	?	?	?	?	?	?
16. 72	?	?	?	?	?	?

List all the factors of each number. *(12)*

17. 22   **18.** 40   **19.** 200   **20.** 65   **21.** 72

Tell whether each number is a prime or a composite number. *(12)*

22. 61   **23.** 42   **24.** 19   **25.** 2   **26.** 106

Find the next 3 numbers in each sequence. *(14)*

27. 1, 5, 9, 13, . . .   **28.** 11, 13, 16, 20, . . .   **29.** 3, 6, 12, 24, . . .

ZEROS IN MULTIPLYING AND DIVIDING

In the multiplication examples below, compare the numbers of zeros in the factors with the number of zeros in the product.

$$4 \times 60 = 240 \qquad 40 \times 60 = 2,400 \qquad 40 \times 600 = 24,000$$

60 ← 1 zero	60 ← 1 zero	600 ← 2 zeros
× 4 ← + 0 zeros	× 40 ← + 1 zero	× 40 ← + 1 zero
240 ← 1 zero	2,400 ← 2 zeros	24,000 ← 3 zeros

The number of zeros in the product is equal to the *sum* of the numbers of zeros in the factors.

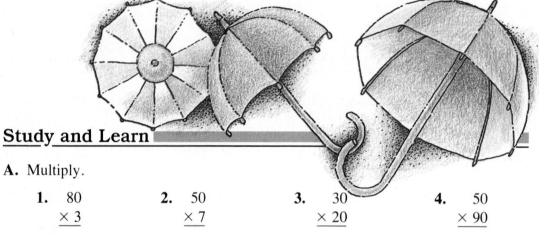

Study and Learn

A. Multiply.

1.	80	2.	50	3.	30	4.	50
	× 3		× 7		× 20		× 90

5.	800	6.	900	7.	600	8.	500
	× 40		× 80		× 700		× 300

9.	4,000	10.	4,000	11.	4,000	12.	50,000
	× 80		× 800		× 8,000		× 900

13. 80 × 40   **14.** 600 × 30   **15.** 700 × 400   **16.** 7,000 × 400

B. Divide.

Example $30 \overline{)600}$ with quotient 2 0   2 zeros − 1 zero = 1 zero

17. $40\overline{)160}$ **18.** $60\overline{)360}$ **19.** $70\overline{)140}$ **20.** $80\overline{)640}$

21. $40\overline{)1,600}$ **22.** $80\overline{)3,200}$ **23.** $60\overline{)3,600}$ **24.** $70\overline{)4,200}$

25. $400\overline{)8,000}$ **26.** $800\overline{)24,000}$ **27.** $700\overline{)210,000}$

Practice

Multiply.

1. 70
× 3

2. 40
× 7

3. 60
× 4

4. 90
× 2

5. 50
× 8

6. 30
× 20

7. 50
× 90

8. 40
× 10

9. 60
× 30

10. 70
× 40

11. 800
× 40

12. 900
× 80

13. 500
× 10

14. 300
× 20

15. 200
× 90

16. 600
× 700

17. 500
× 300

18. 400
× 700

19. 800
× 400

20. 700
× 700

21. 4,000
× 8

22. 4,000
× 80

23. 4,000
× 800

24. 4,000
× 8,000

25. 50,000
× 7

26. 50,000
× 70

27. 50,000
× 700

28. 50,000
× 7,000

29. 80 × 40

30. 600 × 30

31. 300 × 300

32. 20 × 7,000

33. 400 × 700

34. 7,000 × 400

35. 70,000 × 40

36. 800 × 40

Divide.

37. 40)160

38. 60)360

39. 20)140

40. 50)500

41. 70)140

42. 80)640

43. 30)180

44. 90)180

45. 4)1,200

46. 40)1,200

47. 40)12,000

48. 30)2,400

49. 60)3,600

50. 70)4,200

51. 50)4,500

52. 30)3,000

53. 400)16,000

54. 200)8,000

55. 400)160,000

56. 700)35,000

57. 800)24,000

58. 600)24,000

59. 300)30,000

60. 900)45,000

61. 43,000 ÷ 10

62. 43,000 ÷ 100

63. 43,000 ÷ 1,000

64. 7,000 ÷ 10

ESTIMATING PRODUCTS AND QUOTIENTS

A case of juice contains 24 cartons. A truck is loaded with 275 cases. About how many cartons of juice are on the truck?

Estimate by rounding each factor.

$$
\begin{array}{rll}
275 \text{ is about} & 300 \\
\times\,24 \text{ is about} & \times\,20 \\
\hline
& 6{,}000
\end{array}
$$

Study and Learn

A. Estimate the products. Complete.

1. $\begin{array}{r} 64 \\ \times\,31 \end{array} \longrightarrow \begin{array}{r} 60 \\ \times\,30 \end{array}$ 　2. $\begin{array}{r} \$58 \\ \times\,24 \end{array} \longrightarrow \begin{array}{r} \$60 \\ \times\,20 \end{array}$

3. $\begin{array}{r} 653 \\ \times\,56 \end{array} \longrightarrow \begin{array}{r} 700 \\ \times\,60 \end{array}$ 　4. $\begin{array}{r} 4{,}830 \\ \times\,750 \end{array} \longrightarrow \begin{array}{r} 5{,}000 \\ \times\,800 \end{array}$

B. Estimate the products.

5. $\begin{array}{r} 78 \\ \times\,51 \end{array}$ 　6. $\begin{array}{r} \$364 \\ \times\,79 \end{array}$ 　7. $\begin{array}{r} 4{,}389 \\ \times\,72 \end{array}$ 　8. $\begin{array}{r} \$6{,}431 \\ \times\,893 \end{array}$

C. Estimate quotients by rounding both the divisor and dividend.

Examples　　$18\overline{)562}$　　$20\overline{)600}$ ⟵ Estimate 30

　　　　$27\overline{)68}$　　$30\overline{)70}$ ⟵ Estimate 2

Estimate the quotients. Complete.

9. $24\overline{)62} \longrightarrow 20\overline{)60}$ 　　10. $37\overline{)\$821} \longrightarrow 40\overline{)800}$

11. $71\overline{)5{,}623} \longrightarrow 70\overline{)6{,}000}$ 　12. $253\overline{)71{,}684} \longrightarrow 300\overline{)70{,}000}$

D. Estimate the quotients.

13. $36\overline{)75}$ 　14. $63\overline{)420}$ 　15. $34\overline{)8{,}745}$ 　16. $125\overline{)65{,}120}$

Practice

Estimate the products.

1. 28 × 31	**2.** $ 36 × 59	**3.** 54 × 37	**4.** 21 × 45	**5.** 763 × 52
6. 681 × 86	**7.** 837 × 34	**8.** 450 × 126	**9.** 453 × 780	**10.** 786 × 431
11. 4,321 × 67	**12.** 6,512 × 74	**13.** 7,099 × 35	**14.** 8,312 × 410	**15.** 6,394 × 513
16. 43 × 46	**17.** 491 × 21	**18.** 803 × 194	**19.** 3,102 × 55	**20.** 7,732 × 187

Estimate the quotients.

21. 24⟌79 **22.** 36⟌$78 **23.** 32⟌85 **24.** 48⟌502

25. 42⟌$793 **26.** 25⟌912 **27.** 36⟌892 **28.** 91⟌873

29. 26⟌5,910 **30.** 12⟌$6,341 **31.** 37⟌2,134 **32.** 58⟌66,178

33. 321⟌76,430 **34.** 478⟌21,316 **35.** 651⟌31,416

Solve Problems

36. Some cans of soup are packed 48 to a box. Estimate the number of boxes needed to pack 25,418 cans of soup.

37. A box of doughnuts contains 12 doughnuts. A truck is loaded with 346 boxes of doughnuts. Estimate the number of doughnuts on the truck.

★ **38.** Oranges are packed 20 to a box. There are 362 dozen oranges to be packed. Estimate the number of boxes needed.

PROPERTIES OF OPERATIONS

Addition Properties	For all numbers a, b, and c	Examples
Commutative	$a + b = b + a$	$6 + 8 = 8 + 6$
Associative	$(a + b) + c = a + (b + c)$	$(2 + 4) + 7 = 2 + (4 + 7)$
Property of Zero	$a + 0 = a$ $0 + a = a$	$3 + 0 = 3$ $0 + 9 = 9$

Multiplication Properties	For all numbers a, b, and c	Examples
Commutative	$a \cdot b = b \cdot a$	$7 \times 3 = 3 \times 7$
Associative	$(a \cdot b) \cdot c = a \cdot (b \cdot c)$	$(2 \times 3) \times 4 = 2 \times (3 \times 4)$
Distributive Property of Multiplication over Addition	$a \cdot (b + c) = a \cdot b + a \cdot c$	$2 \times (3 + 4) = (2 \times 3) + (2 \times 4)$
Property of Zero	$a \cdot 0 = 0$ $0 \cdot a = 0$	$4 \times 0 = 0$ $0 \times 8 = 0$
Property of One	$a \cdot 1 = a$ $1 \cdot a = a$	$5 \times 1 = 5$ $1 \times 6 = 6$

Study and Learn

A. Name the property used.

   1. $8 + (2 + 3) = (8 + 2) + 3$ **2.** $14 \times 0 = 0$

   3. $4 \times (20 + 3) = (4 \times 20) + (4 \times 3)$ **4.** $45 + 27 = 27 + 45$

   5. $7 \times 10 = 10 \times 7$ **6.** $3 \times (2 \times 4) = (3 \times 2) \times 4$

   7. $0 + 23 = 23$ **8.** $1 \times 35 = 35$

B. Solve. Use the properties.

   9. $3 + n = 7 + 3$ **10.** $8 \times (9 \times 2) = (n \times 9) \times 2$

   11. $n \cdot 9 = 0$ **12.** $8 \cdot 3 = 3 \cdot n$

   13. $3 \times (6 + 4) = (3 \times 6) + (3 \times n)$ **14.** $n + 0 = 42$

Practice

Name the property used.

1. $21 \times (40 + 6) = (21 \times 40) + (21 \times 6)$ **2.** $57 \times 1 = 57$

3. $151 \times 23 = 23 \times 151$ **4.** $81 + (90 + 2) = (81 + 90) + 2$

5. $0 + 19 = 19$ **6.** $56 \times (71 \times 38) = (56 \times 71) \times 38$

7. $16 + 32 = 32 + 16$ **8.** $0 \times 75 = 0$

★ **9.** $3 + (17 + 4) = (17 + 4) + 3$ ★ **10.** $32 \times (80 \times 91) = 32 \times (91 \times 80)$

Solve. Use the properties.

11. $4 + n = 5 + 4$ **12.** $(5 + 6) + n = 5 + (6 + 8)$

13. $26 \cdot n = 0$ **14.** $9 \times (n + 6) = (9 \times 4) + (9 \times 6)$

15. $6 \cdot n = 3 \cdot 6$ **16.** $1 \cdot n = 18$

17. $3 \times (4 \times 9) = (3 \times n) \times 9$ ★ **18.** $(6 \times 23) + (7 \times 23) = n \times 23$

Something Extra
Calculator Activity

Study this pattern.

$1^2 = 1$

$2^2 = 1 + 2 + 1$

$3^2 = 1 + 2 + 3 + 2 + 1$

$4^2 = 1 + 2 + 3 + 4 + 3 + 2 + 1$

$5^2 = 1 + 2 + 3 + 4 + 5 + 4 + 3 + 2 + 1$

1. Check each equation above.

2. Continue the pattern for 6^2.

3. Check your answer on a calculator.

4. Continue the pattern for 7^2, 8^2, and 9^2. Check.

Problem-Solving Applications

Using Mathematics

		PLEASE BE SURE TO **DEDUCT** ANY PER CHECK CHARGES OR SERVICE CHARGES THAT MAY APPLY TO YOUR ACCOUNT					BALANCE	
CHECK NO.	DATE	CHECKS ISSUED TO OR DESCRIPTION OF DEPOSIT	(−) AMOUNT OF CHECK	√ T	(−) CHECK FEE (IF ANY)	(+) AMOUNT OF DEPOSIT	1,416	08
216	10-4	Gas and Electric Co.	163 28				163	28
							1,252	80
217	10-7	Shoe Co.	48 95					
218	10-13	Rent	400 00					
219	10-18	Telephone Co.	86 47					
220	10-22	XYZ Supermarket	53 29					

		PLEASE BE SURE TO **DEDUCT** ANY PER CHECK CHARGES OR SERVICE CHARGES THAT MAY APPLY TO YOUR ACCOUNT					BALANCE
CHECK NO.	DATE	CHECKS ISSUED TO OR DESCRIPTION OF DEPOSIT	(−) AMOUNT OF CHECK	√ T	(−) CHECK FEE (IF ANY)	(+) AMOUNT OF DEPOSIT	
221	10-25	Dentist	60 00				

1. What was the original balance in the checking account?

2. What was the amount of check #216?

3. How was the new balance found after the check for the Gas and Electric Company was written?

4. Find each new balance after checks #217–221 were written.

At the right is a record of the gross amount (salary) of a worker in a large city and the deductions taken.

Gross amount	$380.00
Soc Sec	25.46
Fed Tax	46.56
State Tax	21.57
City Tax	8.45
Disability Ins.	0.30
Net	

5. Add the deductions for social security, federal tax, state tax, city tax, and disability insurance.

6. Subtract your answer in #5 from the gross amount and compute the net take-home pay for the worker.

Chapter Review

Add. *(24)*

1.
```
  5,317
+ 7,879
```

2.
```
  81,756
+  7,975
```

3.
```
  314,437
  835,274
+  96,014
```

Subtract. *(26)*

4.
```
  8,135
- 5,916
```

5.
```
  37,156
-  8,529
```

6.
```
  30,000
- 17,494
```

Estimate the sums. *(28)*

7.
```
  1,661
+ 1,857
```

8.
```
  6,437
+   745
```

9.
```
  7,801
    528
+ 4,039
```

Estimate the differences. *(28)*

10.
```
  4,197
- 2,213
```

11.
```
  64,201
-  3,198
```

12.
```
  11,689
-    711
```

Multiply. *(34, 36)*

13.
```
  6,517
×    6
```

14.
```
  $736
×   68
```

15.
```
  6,286
×   408
```

Divide. *(38, 40)*

16. $4\overline{)848}$

17. $23\overline{)115}$

18. $43\overline{)\$13,588}$

Estimate the products and quotients. *(46)*

19.
```
  517
× 66
```

20.
```
  7,721
×    18
```

21.
```
  8,178
×   328
```

22. $63\overline{)6,245}$

23. $62\overline{)18,301}$

Solve. *(32, 50)*

24. Les bought 2 pieces of lumber each 12 ft long and 1 piece 8 ft long. What was the total length of the 3 pieces? Draw a diagram for the problem.

25. Ms. Rodriquez has a balance of $184.29 in her checkbook. She writes a check for $49.89 and another for $19.34. What is the new balance?

Chapter Test

Add *(24)*

1. 4,378
   + 7,516

2. 91,568
   + 9,625

3. 23,823
   714,278
   +   8,925

Subtract. *(26)*

4. 8,462
   − 1,839

5. 16,343
   − 7,619

6. 40,000
   − 16,257

Estimate the sums. *(28)*

7. 8,418
   + 4,765

8. 51,003
   + 7,114

9. 17,813
   28,902
   +   4,158

Estimate the differences. *(28)*

10. 56,713
    − 42,834

11. 19,621
    − 8,577

12. 68,001
    −   982

Multiply. *(34, 36)*

13. 6,417
    × 7

14. $589
    × 24

15. 4,845
    × 307

Divide. *(38, 40)*

16. 3)969

17. 21)147

18. 62)21,204

Estimate the products and quotients. *(46)*

19. 634
    × 17

20. 345
    × 836

21. 3,734
    × 529

22. 32)5,861

23. 59)17,500

Solve. *(32, 50)*

24. Charlotte drove 75 km in the morning and 116 km in the afternoon. How far did she drive in all? Draw a diagram for the problem.

25. Henry's salary is $225 a week. $15.08 is deducted for Social Security and $38.27 for taxes. What is his net take-home pay for the week?

Skills Check

1. Ms. Barbiero bought a television set for $399.50, a refrigerator for $289.99, and a stove for $249.49. What was the total cost?

   A $839.89   B $938.88

   C $938.89   D $938.98

2. Milton is on a diet. He lost 7 lb the first month, 8 lb the second month, gained 5 lb the third month, and lost 12 lb the fourth month. What was his net loss for the 4 months?

   E 32 lb   F 27 lb

   G 22 lb   H 8 lb

3. What is the sales tax on $0.49?

Price	Sales Tax
$0.30 to $0.42	3¢
$0.43 to $0.54	4¢
$0.55 to $0.67	5¢

   A 3¢   B 4¢

   C 5¢   D 49¢

4. John's pay envelope contained 6 twenties, 3 tens, 1 five, 4 ones, 1 quarter, 2 dimes, and 4 pennies. How much money did he receive?

   E $164.49   F $159. 49

   G $159.39   H $158.49

5. Brenda is 5 years older than Ryan. Ryan is 12 years old. How old is Brenda?

   A 25   B 17

   C 20   D 7

6. According to the thermometer, what is the temperature?

   E 30°F   F 40°F

   G 50°F   H 60°F

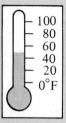

7. The Aurelios left their house at 8:35 am and returned home at 1:15 pm. How long were they gone?

   A 3 hours, 40 minutes

   B 4 hours, 40 minutes

   C 5 hours, 40 minutes

   D 7 hours, 20 minutes

8. According to the scale, what is the mass of the tomato?

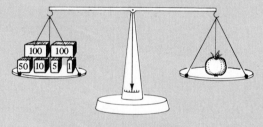

   E 300 g   F 298 g

   G 266 g   H 243 g

9. Sheila earns $896 a month. Estimate her salary for a year.

   A $8,000   B $9,000

   C $80,000   D $90,000

3 SOLVING EQUATIONS

Wind turbines are used to provide mechanical power.

DOING AND UNDOING

Think of a number. *Add* 2. From the sum *subtract* 2.
What is the result?

Suppose you
thought of 8:

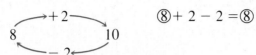

$(8) + 2 - 2 = (8)$

Adding and subtracting the same number undo each other.

Study and Learn

A. Start with any number. Represent the number by x.

   1. Add 5. What is the result?

   2. Subtract 5 from $x + 5$. What is the result?

   3. Complete. $x + 5 - 5 = \underline{\quad?\quad}$

B. What number should be subtracted so that the result is x?

   4. $x + 7$ **5.** $x + 4$ **6.** $x + \frac{3}{4}$ **7.** $x + 0.4$

C. Start with x.

   8. Subtract 3. What is the result?

   9. Add 3 to $x - 3$. What is the result?

   10. Complete. $x - 3 + 3 = \underline{\quad?\quad}$

D. What number should be added so that the result is x?

   11. $x - 2$ **12.** $x - 9$ **13.** $x - \frac{1}{2}$ **14.** $x - 0.4$

E. Multiplying and dividing undo each other.

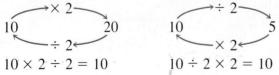

$$10 \times 2 \div 2 = 10 \qquad 10 \div 2 \times 2 = 10$$

Tell what to do so that the result is x.

   15. $3x$ **16.** $4x$ **17.** $\frac{x}{2}$ **18.** $\frac{x}{5}$

Practice

Tell what to do so that the result is x.

1. $x + 9$ **2.** $x + 12$ **3.** $x + 23$ **4.** $x + 1$

5. $x + \frac{1}{4}$ **6.** $x + \frac{5}{8}$ **7.** $x + 0.7$ **8.** $x + 0.06$

9. $x - 7$ **10.** $x - 8$ **11.** $x - 12$ **12.** $x - 14$

13. $x - \frac{1}{4}$ **14.** $x - \frac{2}{3}$ **15.** $x - 0.3$ **16.** $x - 0.016$

17. $2x$ **18.** $5x$ **19.** $8x$ **20.** $10x$

21. $6x$ **22.** $0.9x$ **23.** $0.15x$ **24.** $0.20x$

25. $\frac{x}{3}$ **26.** $\frac{x}{4}$ **27.** $\frac{x}{6}$ **28.** $\frac{x}{7}$

29. $\frac{x}{0.2}$ **30.** $\frac{x}{0.09}$ **31.** $\frac{x}{1.2}$ **32.** $\frac{x}{1.5}$

33. $x + 8$ **34.** $x - 2$ **35.** $\frac{x}{8}$ **36.** $7x$

37. $x + 0.2$ **38.** $x - 0.07$ **39.** $\frac{x}{0.8}$ **40.** $0.3x$

41. $x + \frac{7}{8}$ **42.** $x - \frac{3}{4}$ ★**43.** $\frac{3}{10}x$ ★**44.** $\frac{2}{3}x$

Something Extra
Non-Routine Problems

1. There are some books on a shelf. All are mathematics books except for two. All are story books except for two. All are science books except for two. How many books are on the shelf?

2. Helene has 8 more baseball cards than John has. If Helene gives John 6 baseball cards, John will have how many more baseball cards than Helene?

EQUIVALENT EQUATIONS

	Equation	Solution
	$x = 7$	
Adding 1 to both sides ⟶	$x + 1 = 8$	7
Adding 2 to both sides ⟶	$x + 2 = 9$	7
Adding 3 to both sides ⟶	$x + 3 = 10$	7

> **Equivalent equations have the same solution.**

▶ Adding the same number to both sides of an equation does not change the solution. This is called the *addition property for equations*.

Study and Learn

A. Which equations are equivalent?

    1. $x = 2$; $x + 3 = 2 + 3$; $x + 1 = 2$

    2. $y = 3$; $y + 4 = 7$; $y + 5 = 9$

B. Write 2 equivalent equations. Use the addition property for equations.

    3. $x = 7$ **4.** $b = 12$ **5.** $c = 4$ **6.** $d = 24$

Look at these equivalent equations.

	Equation	Solution
	$x + 3 = 12$	9
Subtracting 1 from both sides ⟶	$x + 2 = 11$	9
Subtracting 2 from both sides ⟶	$x + 1 = 10$	9
Subtracting 3 from both sides ⟶	$x = 9$	9

▶ Subtracting the same number from both sides of an equation does not change the solution. This is called the *subtraction property for equations*.

C. Write 2 equivalent equations. Use the subtraction property for equations.

    7. $x + 7 = 12$ **8.** $d + 8 = 14$ **9.** $n + 4 = 15$ **10.** $r + 6 = 10$

D. Which are pairs of equivalent equations?

    11. $r - 2 = 13 - 2$; $r = 13$ **12.** $z + 7 = 10$; $z + 7 + 7 = 10 - 7$

Practice

Write 2 equivalent equations. Use the addition property for equations.

1. $x = 3$ **2.** $x = 4$ **3.** $x = 7$ **4.** $x = 13$

Write 2 equivalent equations. Use the subtraction property for equations.

5. $x + 6 = 11$ **6.** $x + 7 = 15$ **7.** $x + 12 = 20$ **8.** $x + 14 = 30$

Which are pairs of equivalent equations?

9. $x + 1 = 5 + 1; x = 6$ **10.** $x + 8 = 2 + 8; x + 2 = 8$

11. $n + 2 = 18 + 2; n = 18$ **12.** $x - 1 = 19 - 1; x = 18$

13. $z = 14; z + 1 = 14 + 1$ **14.** $t + 1 = 8; t + 1 - 1 = 9 - 1$

15. $f + 4 = 7; f = 11$ **16.** $e - 9 = 22; e = 31$

17. $x + 3 - 3 = 9 - 3; x = 4$ **18.** $n - 8 + 8 = 11 + 8; n = 19$

★ **19.** $7 + y - 3 = 9; y = 12$ ★ **20.** $4 + y + 3 = 13; y = 6$

★ **21.** $15 = 6 + d - 9; d = 9$ ★ **22.** $11 - 8 = 8 - p; p = 5$

Something Extra
Calculator Activity

A special symbol is used for the product of consecutive numbers.
The following examples illustrate this new symbol.

$1! = 1$ Read 1! as 1 factorial.

$2! = 1 \times 2$, or 2   Read 2! as 2 factorial.

$3! = 1 \times 2 \times 3$, or 6   Read 3! as 3 factorial.

Use a calculator to compute each.

1. 6!   **2.** 7!   **3.** 8!   **4.** 9!   **5.** 10!   **6.** 11!

SOLVING EQUATIONS

Solve $x + 7 = 11$. Check your answer.

To get x alone, subtract
7 from both sides of the
equation.

$$x + 7 = 11$$
$$x + 7 - 7 = 11 - 7$$
$$x = 4$$

Solution is 4.

Check: $4 + 7 = 11$ True

Check

$x + 7$	11
$4 + 7$	11
11	

Study and Learn

A. What should be subtracted from both sides of the equation to get x alone?

    1. $x + 4 = 7$ **2.** $x + 5 = 12$ **3.** $x + 8 = 15$

B. Solve and check.

    4. $x + 4 = 7$ **5.** $x + 5 = 12$ **6.** $x + 8 = 15$

C. What should be added to both sides of the equation to get x alone?

    7. $x - 4 = 7$ **8.** $x - 5 = 12$ **9.** $x - 8 = 15$

D. Solve and check.

    Example $x - 4 = 12$

$$x - 4 = 12$$
$$x - 4 + 4 = 12 + 4$$
$$x = 16$$

Check

$x - 4$	12
$16 - 4$	12
12	

    10. $x - 4 = 7$ **11.** $x - 5 = 12$ **12.** $x - 8 = 15$

E. Solve and check.

    13. $b + 9 = 14$ **14.** $13 + n = 22$ **15.** $45 = t + 29$

    16. $m - 6 = 8$ **17.** $s - 32 = 14$ **18.** $28 = y - 11$

Practice

Solve and check.

1. $x + 3 = 8$

2. $n + 7 = 10$

3. $t + 5 = 12$

4. $z + 13 = 37$

5. $f + 18 = 45$

6. $e + 23 = 35$

7. $9 = x + 1$

8. $8 = c + 4$

9. $7 = b + 5$

10. $12 = d + 3$

11. $37 = x + 19$

12. $45 = r + 29$

13. $x + 17 = 123$

14. $y + 256 = 572$

15. $t + 173 = 426$

16. $323 = n + 58$

17. $199 = z + 47$

18. $387 = b + 241$

19. $a - 7 = 9$

20. $b - 14 = 17$

21. $d - 3 = 10$

22. $c - 21 = 45$

23. $x - 17 = 52$

24. $z - 25 = 41$

25. $37 = r - 17$

26. $39 = n - 10$

27. $12 = t - 7$

28. $8 = a - 5$

29. $12 = e - 9$

30. $14 = f - 3$

31. $b - 524 = 71$

32. $t - 36 = 649$

33. $1 = z - 417$

34. $112 = x - 42$

35. $413 = y - 58$

36. $x - 97 = 180$

★ **37.** $426 + b = 811$

★ **38.** $12 + f + 6 = 20$

★ **39.** $253 = 17 + z - 10$

61

Skills Review

Add. *(24)*

1. 543 + 652	**2.** 439 + 168	**3.** 4,511 + 2,689
4. 7,315 + 856	**5.** 36,618 + 27,179	**6.** 41,017 + 9,043

7. 8,036
1,920
+ 3,479

8. 12,579
3,008
73,861
+ 420

9. 868,413
93,016
234,526
+ 68,234

10. 3,416,812
9,236,014
167,312
43,091
+ 9,234,185

Subtract. *(26)*

11. 927 − 342	**12.** 3,519 − 1,736	**13.** 8,634 − 557	**14.** 63,295 − 48,039
15. 78,134 − 9,256	**16.** 427,631 − 275,988	**17.** 541,397 − 95,438	**18.** 9,831,245 − 760,098
19. 6,000 − 4,508	**20.** 37,000 − 18,124	**21.** 29,003 − 8,754	**22.** 316,004 − 94,625

Multiply. *(34, 36)*

23. 3,248 × 8	**24.** 516 × 23	**25.** 3,481 × 68	**26.** 15,088 × 41
27. 912 × 296	**28.** 4,168 × 257	**29.** 20,461 × 318	**30.** 7,239 × 1,451
31. 647 × 30	**32.** 821 × 509	**33.** 2,417 × 408	**34.** 1,560 × 7,003

Divide. *(38, 40)*

35. $2\overline{)868}$	**36.** $7\overline{)147}$	**37.** $8\overline{)336}$	**38.** $6\overline{)424}$
39. $9\overline{)3,319}$	**40.** $6\overline{)31,416}$	**41.** $15\overline{)495}$	**42.** $23\overline{)138}$
43. $37\overline{)727}$	**44.** $63\overline{)9,054}$	**45.** $41\overline{)17,356}$	**46.** $23\overline{)93,840}$
47. $156\overline{)312}$	**48.** $238\overline{)4,998}$	**49.** $852\overline{)7,668}$	**50.** $311\overline{)41,026}$
51. $526\overline{)15,387}$	**52.** $792\overline{)946,305}$	**53.** $407\overline{)235,517}$	**54.** $127\overline{)508,254}$

Midchapter Review

Which are pairs of equivalent equations? *(58)*

1. $x + 4 = 7 + 4; x = 28$

2. $x - 9 = 10 - 9; x = 1$

3. $5 + 2 = x + 2; x = 5$

4. $10 - 3 = x - 3; x = 7$

Solve and check. *(60)*

5. $x + 2 = 11$

6. $x + 4 = 16$

7. $x + 3 = 12$

8. $x + 41 = 70$

9. $x + 108 = 213$

10. $356 = x + 40$

11. $x - 3 = 7$

12. $x - 1 = 19$

13. $x - 11 = 20$

14. $x - 52 = 107$

15. $x - 39 = 21$

16. $129 = x - 57$

17. $x + 6 = 15$

18. $8 = b + 3$

19. $c - 3 = 4$

20. $249 = z - 128$

21. $b + 19 = 41$

22. $12 = n + 5$

23. $x - 19 = 34$

24. $212 = t - 15$

25. $13 = r + 7$

Something Extra
Non-Routine Problem

Fill in the empty cells so that the sum of each row, each column, and along each diagonal is 65. Use each number from 1 to 25 only once.

17	24	1	8	15
23	5			16
4		13		
10	12	19		
	18			9

EQUIVALENT EQUATIONS FOR MULTIPLICATION

	Equation	Solution
	$x = 5$	5
Multiplying both sides by 2 $\longrightarrow$	$2x = 10$	5
Multiplying both sides by 3 $\longrightarrow$	$3x = 15$	5
Multiplying both sides by 4 $\longrightarrow$	$4x = 20$	5

▶ Multiplying both sides of an equation by the same non-zero number does not change the solution. This is called the *multiplication property for equations*.

Study and Learn

A. Which equations are equivalent?

   1. $x = 2;$ $3 \cdot x = 3 \cdot 2;$ $5x = 15$

   2. $n = 3;$ $4 \cdot n = 12;$ $2 \cdot n = 6$

B. Write 2 equivalent equations. Use the multiplication property for equations.

   3. $x = 4$ **4.** $x = 6$ **5.** $x = 8$ **6.** $x = 0.7$

Look at these equivalent equations.

	Equation	Solution
	$4x = 16$	4
Dividing both sides by 2 $\longrightarrow$	$\frac{4x}{2} = \frac{16}{2}$ or $2x = 8$	4
Dividing both sides by 4 $\longrightarrow$	$\frac{4x}{4} = \frac{16}{4}$ or $x = 4$	4

▶ Dividing both sides of an equation by the same non-zero number does not change the solution. This is called the *division property for equations*.

C. Write 2 equivalent equations. Use the division property for equations.

   7. $4x = 8$ **8.** $6x = 30$ **9.** $8x = 24$ **10.** $9x = 18$

D. Which are pairs of equivalent equations?

   11. $x = 7; 3 \cdot x = 3 \cdot 7$ **12.** $\frac{x}{2} = 4; x = 8$

Practice

Which equations are equivalent?

1. $a = 2$ $3 \cdot a = 3 \cdot 2$ $3a = 6$ $3a = 9$

2. $x = 4$ $3x = 12$ $4x = 16$ $2 \cdot x = 2 \cdot 4$

3. $z = 6$ $\frac{z}{2} = 8$ $\frac{z}{2} = \frac{6}{2}$ $z = 4$

4. $c = 9$ $\frac{c}{3} = \frac{9}{3}$ $\frac{c}{3} = 27$ $\frac{c}{2} = 16$

Write 2 equivalent equations. Use the multiplication property for equations.

5. $n = 6$ 6. $r = 8$ 7. $x = 3$ 8. $z = 12$

Write 2 equivalent equations. Use the division property for equations.

9. $2b = 10$ 10. $8d = 32$ 11. $4n = 12$ 12. $15r = 30$

Which are pairs of equivalent equations?

13. $2z = 2 \cdot 8$; $z = 16$ 14. $3x = 15$; $x = 3$

15. $\frac{n}{4} = 20$; $n = 5$ 16. $\frac{t}{6} = 7$; $t = 42$

17. $6x = 42$; $x = 7$ 18. $\frac{x}{3} = 6$; $x = 2$

★ 19. $\frac{3x}{4} = 9$; $x = 3$ ★ 20. $\frac{2x}{3} = 8$; $x = 12$

Something Extra
Non-Routine Problems

1. The sum of 2 numbers is 70. Their difference is 18. What are the numbers?

2. The product of 2 numbers is 36. Their sum is 15. What are the numbers?

MORE SOLVING EQUATIONS

Solve $2x = 18$. Check your answer.
To get x alone, divide
both sides of the equation
by 2.

$$2x = 18$$
$$\frac{2x}{2} = \frac{18}{2}$$
$$x = 9$$

Solution is 9.

Check: $2 \cdot 9 = 18$ True

Check

$2x$	18
$2 \cdot 9$	18
18	

Study and Learn

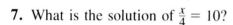

A. What should be done to both sides of the equation to get x alone?

    1. $3x = 48$ **2.** $24 = \frac{x}{3}$ **3.** $12x = 156$

B. Solve $\frac{x}{4} = 10$.

    4. What should be done to both sides of the equation to get x alone?

    5. Complete $\frac{x}{4} \cdot \underline{\ \ ?\ \ } = 10 \cdot \underline{\ \ ?\ \ }$

    6. Simplify $\frac{x}{4} \cdot 4 = 10 \cdot 4$. What equivalent equation do you get?

    7. What is the solution of $\frac{x}{4} = 10$?

C. Solve.

    8. $3x = 48$ **9.** $24 = \frac{x}{3}$ **10.** $12x = 156$ **11.** $\frac{x}{2} = 17$

D. Solve and check.

    Example $\frac{x}{2} = 16$ Check

$\frac{x}{2}$	16
$\frac{32}{2}$	16
16	

$$2 \cdot \frac{x}{2} = 16 \cdot 2$$
$$x = 32$$

    12. $\frac{x}{4} = 7$ **13.** $5x = 35$ **14.** $14x = 84$ **15.** $24 = \frac{x}{13}$

Practice

Solve and check.

1. $2x = 8$ **2.** $3y = 15$ **3.** $4z = 32$

4. $5r = 45$ **5.** $7n = 42$ **6.** $8a = 64$

7. $32 = 4c$ **8.** $16 = 2d$ **9.** $42 = 6k$

10. $18 = 3t$ **11.** $35 = 7z$ **12.** $72 = 9r$

13. $3n = 126$ **14.** $4t = 196$ **15.** $17t = 102$

16. $360 = 4a$ **17.** $483 = 7b$ **18.** $396 = 18c$

19. $\frac{d}{2} = 3$ **20.** $\frac{t}{4} = 1$ **21.** $\frac{r}{3} = 6$

22. $7 = \frac{s}{3}$ **23.** $9 = \frac{h}{5}$ **24.** $17 = \frac{r}{4}$

25. $15 = \frac{d}{5}$ **26.** $\frac{n}{3} = 71$ **27.** $\frac{t}{4} = 29$

28. $\frac{z}{3} = 7$ **29.** $\frac{x}{4} = 9$ **30.** $\frac{d}{7} = 8$

31. $\frac{x}{21} = 48$ **32.** $27 = \frac{c}{9}$ **33.** $301 = \frac{e}{4}$

★ **34.** $\frac{1}{2}x = 13$ ★ **35.** $\frac{3}{5}x = 15$ ★ **36.** $\frac{5}{6}x = 30$

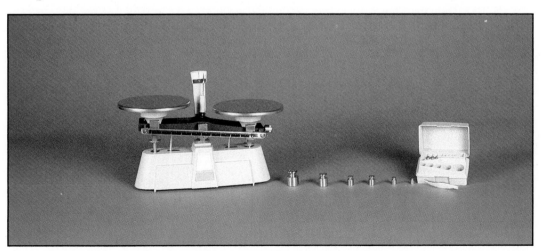

Problem-Solving Skills
Using Broken-Line Graphs

Broken-line graphs
are usually used
to show a trend or
to make comparisons.

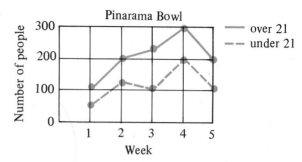

Study and Learn

A. Look at the broken-line graph above.

1. How many people over 21 used the bowling alley during the third week?

2. How many people under 21 used the bowling alley during the third week?

3. How many more people over 21 than under 21 used the bowling alley during the third week?

4. In general, do more people over 21 or under 21 use the bowling alley?

5. What is the general trend of people over 21 using the bowling alley? Is it up or down?

B. Look at the broken-line graph at the right. Sometimes 2 broken-line graphs may intersect.

6. How much were the record sales for the fourth week?

7. During which weeks were the record sales greater than the tape sales?

8. About how much more was taken in from the sale of records than from the sale of tapes during the second week?

9. What can you say about the trend in tape sales?

Practice

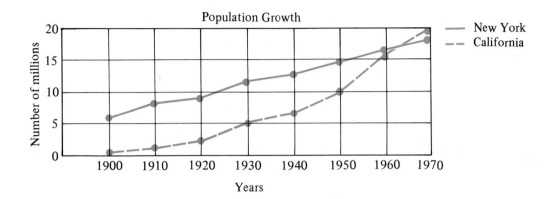

Population Growth

Number of millions

New York
California

Years

Answer the questions about the broken-line graph above.

1. The population growths of which two states are shown?

2. What was the approximate population of New York State in 1950?

3. What was the approximate population of California in 1950?

4. How much greater was the population of New York State than that of California in 1950?

5. What is the trend of population growth for New York State?

6. What is the trend of population growth for California?

7. Which state is increasing in population more quickly?

Answer the questions about the broken-line graph at the right.

8. What was Amy's score on the second test?

9. What was the class average on the fifth test?

10. How do Amy's test scores compare with the class average?

11. What is the trend for Amy's scores?

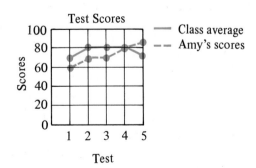

Test Scores

Scores

Class average
Amy's scores

Test

USING TWO EQUATION PROPERTIES

Sometimes it is necessary to use 2 equation properties to solve an equation.

Solve $2x + 3 = 13$. Check your answer.

Think: Get $2x$ alone by subtracting 3 from both sides. To get x alone, divide both sides of the equation by 2.

$$2x + 3 = 13$$
$$2x + 3 - 3 = 13 - 3$$
$$2x = 10$$
$$\frac{2x}{2} = \frac{10}{2}$$
$$x = 5$$

Check

$2x + 3$	13
$2 \cdot 5 + 3$	13
$10 + 3$	
13	

The solution of $2x + 3 = 13$ is 5.

Study and Learn

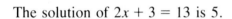

A. Solve $\frac{n}{2} - 4 = 7$. Check.

1. What should be done to get $\frac{n}{2}$ alone?

2. Complete. $\frac{n}{2} - 4 + \underline{\quad ? \quad} = 7 + \underline{\quad ? \quad}$

3. Simplify $\frac{n}{2} - 4 + 4 = 7 + 4$. What equivalent equation do you get?

4. Write an equivalent equation with n alone.

5. What is the solution of $\frac{n}{2} - 4 = 7$?

6. Check the solution in $\frac{n}{2} - 4 = 7$.

B. Solve and check.

7. $3a + 7 = 25$

8. $36 = 2b - 10$

9. $4c + 12 = 52$

10. $6 = \frac{d}{2} - 2$

11. $\frac{e}{4} + 3 = 19$

12. $\frac{f}{6} - 2 = 5$

Practice

Solve and check.

1. $3x + 4 = 19$

2. $4c + 7 = 31$

3. $7 = 3r + 1$

4. $2z + 19 = 51$

5. $3t + 18 = 57$

6. $9 = 2n + 1$

7. $59 = 3f + 8$

8. $62 = 8b + 6$

9. $421 = 10c + 11$

10. $3x - 1 = 8$

11. $4e - 7 = 21$

12. $8 = 3w - 16$

13. $7x - 21 = 35$

14. $9x - 27 = 72$

15. $18 = 2x - 20$

16. $21 = 2z - 9$

17. $43 = 8n - 13$

18. $500 = 9r - 103$

19. $\frac{x}{2} + 3 = 6$

20. $\frac{c}{4} + 7 = 10$

21. $\frac{e}{5} + 5 = 13$

22. $\frac{n}{5} + 7 = 12$

23. $\frac{z}{4} + 6 = 8$

24. $\frac{n}{8} + 1 = 4$

25. $\frac{m}{11} + 15 = 17$

26. $\frac{f}{9} + 18 = 25$

27. $15 = \frac{c}{11} + 8$

28. $\frac{z}{7} - 1 = 6$

29. $\frac{x}{3} - 2 = 3$

30. $\frac{r}{2} - 5 = 8$

31. $\frac{t}{5} - 7 = 2$

32. $\frac{n}{8} - 3 = 4$

33. $\frac{v}{50} - 32 = 20$

34. $\frac{w}{2} - 8 = 15$

35. $11 = \frac{z}{9} - 5$

36. $120 = \frac{t}{16} - 4$

37. $\frac{f}{2} + 3 = 12$

38. $5x - 3 = 12$

39. $8t + 1 = 17$

40. $40 = 3r - 2$

41. $100 = \frac{z}{7} - 85$

42. $4n + 5 = 29$

43. $\frac{x}{6} + 5 = 23$

44. $22x - 16 = 50$

45. $4x + 12 = 76$

★ **46.** $\frac{2x}{3} = 6$

★ **47.** $17 + x - 2 = 20$

★ **48.** $\frac{3}{4}x + 1 = 4$

Solve Problems

★ **49.** Bruce filled 12 boxes with the same number of bottles and had 3 bottles left over. He started out with 135 bottles. How many bottles were placed in each box?

★ **50.** Janet cut a piece of wood into 2 equal pieces for bookshelves. She cut 50 cm from one of the two pieces. It was then 150 cm long. How long was the original piece?

SOLVING INEQUALITIES

Solve $x + 1 < 7$. Replacements for x: 0, 1, 2, . . . , 10.

To help solve an inequality, solve the related equation.

$$x + 1 < 7 \longleftarrow \text{related equation} \longrightarrow x + 1 = 7$$
$$x + 1 - 1 = 7 - 1$$
$$x = 6$$

Try 6 in $x + 1 < 7$: $6 + 1 < 7$. False. So 6 is not a solution.
Try a number less than 6 in $x + 1 < 7$: $5 + 1 < 7$. True.
So, each number less than 6 is a solution.
Try a number greater than 6 in $x + 1 < 7$: $7 + 1 < 7$. False.
So, no number greater than 6 is a solution.
The solutions of the inequality $x + 1 < 7$ are 0, 1, 2, 3, 4, and 5.

Study and Learn

A. Solve $x - 4 > 3$. Replacements for x: 0, 1, 2, . . . , 10

1. Write the equation related to $x - 4 > 3$.

2. Solve $x - 4 = 3$.

3. Try a number greater than 7 in $x - 4 > 3$. Is the statement true?

4. Try a number less than 7 in $x - 4 > 3$. Is the statement true?

5. Give the solutions of $x - 4 > 3$.

B. Solve $2x > 8$. Replacements for x: 0, 1, 2, . . . , 10

6. Write the equation related to $2x > 8$.

7. Solve $2x = 8$.

8. Try a number less than 4 in $2x > 8$. Is the statement true?

9. Try a number greater than 4 in $2x > 8$. Is the statement true?

10. Give the solutions of $2x > 8$.

C. Solve. Replacements: 0, 1, 2, . . . , 10

11. $n + 1 > 3$ **12.** $2t < 6$ **13.** $2r - 1 > 7$

Practice

Solve. Replacements: 0, 1, 2, . . . , 10.

1. $x + 4 > 6$

2. $y + 1 < 3$

3. $z + 2 > 6$

4. $c - 4 > 1$

5. $d - 8 > 1$

6. $x - 1 < 5$

7. $3y < 9$

8. $3x > 12$

9. $2n < 20$

10. $x - 4 < 2$

11. $6r - 4 < 8$

12. $y + 4 < 10$

13. $2t - 1 < 9$

14. $3x < 9$

15. $4n - 8 < 24$

⋆**16.** $\frac{r}{2} > 2$

⋆**17.** $\frac{x}{3} + 1 > 3$

⋆**18.** $\frac{n}{2} - 1 < 5$

⋆**19.** $3t + 1 > 31$

⋆**20.** $4x - 2 < 2$

⋆**21.** $5x < 1$

Computer

You can use this program to solve inequalities for replacements
0 to 10. The **INPUT** command in line 30 shows a blinking light.
The computer is asking for information and waits for your answer.

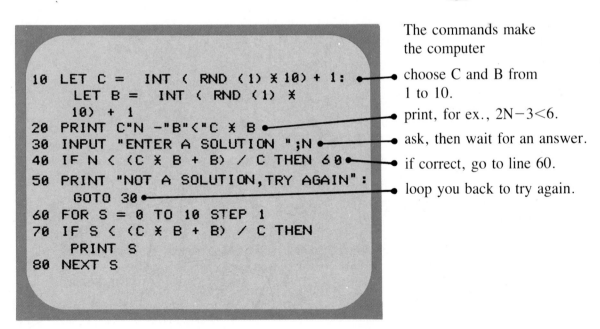

```
10 LET C =  INT ( RND (1) ✕ 10) + 1:
   LET B =  INT ( RND (1) ✕
   10) + 1
20 PRINT C"N -"B"<"C ✕ B
30 INPUT "ENTER A SOLUTION ";N
40 IF N < (C ✕ B + B) / C THEN 60
50 PRINT "NOT A SOLUTION,TRY AGAIN":
   GOTO 30
60 FOR S = 0 TO 10 STEP 1
70 IF S < (C ✕ B + B) / C THEN
   PRINT S
80 NEXT S
```

The commands make
the computer

choose C and B from
1 to 10.

print, for ex., 2N−3<6.

ask, then wait for an answer.

if correct, go to line 60.

loop you back to try again.

See the Computer Section at the back of the book for these commands.

Problem-Solving Applications
Reading a Telephone Bill

	State-Local Tax	Federal Tax	Charges Excluding Tax
Regular Monthly Charge	0|75	0|32	10|76
Local Usage	1|14	0|49	16|27
Toll Charges	0|66	0|46	15|27
Total Excluding Taxes			42|30
Taxes	2|55	1|27	3|82
Total of current charges			46|12

Solve.

1. What is the regular monthly charge excluding taxes?

2. How much is the total tax on the regular monthly charge?

3. How much more was the state-local tax than the federal tax on the local usage calls?

4. What would the total bill have been if there were no toll charges?

5. What would the total bill have been if there were no local calls?

6. What was the total cost of the local usage and toll calls?

7. During the month, 10 toll calls were made. About how much was the average cost per toll call including taxes?

8. The cost of local usage calls is divided among day and evening calls. If $11.23 was charged for 39 day calls, how much was charged for the evening calls?

9. Mr. Glinka kept a record of his toll call charges. The charges for 3 months were $18.56, $17.96, and $21.74. Estimate the average monthly charge for toll calls.

Chapter Review

Which are pairs of equivalent equations? *(58, 64)*

1. $x + 2 = 6 + 2$; $x = 2$

2. $x - 3 = 6 - 3$; $x = 6$

3. $2x = 22$; $x = 11$

4. $\frac{x}{5} = 10$; $x = 2$

Solve and check.

5. $x + 7 = 18$
(60)

6. $x + 21 = 43$
(60)

7. $z + 16 = 54$
(60)

8. $c - 3 = 8$
(60)

9. $t - 24 = 37$
(60)

10. $x - 75 = 97$
(60)

11. $42 = b + 3$
(60)

12. $49 = z - 7$
(60)

13. $456 = n - 108$
(60)

14. $4x = 24$
(66)

15. $8n = 72$
(66)

16. $3z = 27$
(66)

17. $\frac{x}{3} = 4$
(66)

18. $\frac{t}{5} = 206$
(66)

19. $\frac{e}{7} = 9$
(66)

20. $603 = 9c$
(66)

21. $21 = \frac{t}{2}$
(66)

22. $64 = 8y$
(66)

23. $2x + 5 = 11$
(70)

24. $3n - 7 = 11$
(70)

25. $31 = 6t + 7$
(70)

26. $\frac{r}{4} + 12 = 36$
(70)

27. $\frac{e}{5} + 15 = 85$
(70)

28. $6 = \frac{n}{2} - 4$
(70)

29. $\frac{x}{4} - 2 = 10$
(70)

30. $\frac{n}{6} + 4 = 22$
(70)

31. $54 = \frac{n}{2} - 8$
(70)

Use the graph to answer Exercise 32. *(68)*

32. In which game did Robin and Darrin score the same number of points?

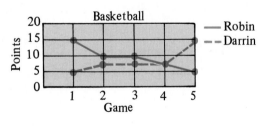

Solve. *(74)*

33. The Browns made 4 toll calls, which cost $8.60 in all. What was the average cost of each toll call?

Chapter Test

Which are pairs of equivalent equations? *(58, 64)*

1. $x + 3 = 7 + 3$; $x = 7$ **2.** $x + 4 = 12$; $x = 16$

3. $7x = 21$; $x = 3$ **4.** $\frac{x}{2} = 4$; $x = 8$

Solve and check.

5. $x + 5 = 13$ **6.** $x + 65 = 461$ **7.** $z + 14 = 65$
(60)   *(60)*   *(60)*

8. $x - 9 = 21$ **9.** $c - 15 = 85$ **10.** $e - 86 = 159$
(60)   *(60)*   *(60)*

11. $16 = z + 5$ **12.** $48 = n - 36$ **13.** $647 = b + 86$
(60)   *(60)*   *(60)*

14. $4x = 32$ **15.** $9z = 225$ **16.** $2n = 96$
(66)   *(66)*   *(66)*

17. $\frac{x}{4} = 7$ **18.** $\frac{n}{12} = 8$ **19.** $\frac{r}{3} = 24$
(66)   *(66)*   *(66)*

20. $18 = 6z$ **21.** $10 = \frac{r}{5}$ **22.** $14 = 2t$
(66)   *(66)*   *(66)*

23. $2t + 7 = 19$ **24.** $3z - 4 = 11$ **25.** $30 = 6t + 6$
(70)   *(70)*   *(70)*

26. $\frac{x}{4} - 3 = 7$ **27.** $\frac{r}{2} + 4 = 10$ **28.** $10 = \frac{n}{4} + 7$
(70)   *(70)*   *(70)*

29. $\frac{z}{3} - 4 = 16$ **30.** $\frac{s}{7} + 3 = 4$ **31.** $25 = \frac{n}{3} + 16$
(70)   *(70)*   *(70)*

Use the graph to answer
Exercise 32. *(68)*

32. How many more runs did Randy
score than Terry in the fourth
game?

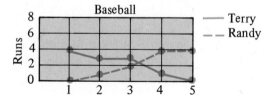

Solve. *(74)*

33. If 4 toll calls are made before 6 pm, the cost is $8.60. The same
4 toll calls made after 6 pm cost $6.80. How much more is the
cost of the calls before 6 pm?

Skills Check

1. What is the value of the underlined digit in 9,2**6**7,341?

A	B	C	D
7	700	7,000	70,000

2. Which shows three million, seven?

E	F	G	H
3,000,700	3,000,007	3,700	3,007

3. What is 741,683 rounded to the nearest hundred thousand?

A	B	C	D
742,000	741,700	740,000	700,000

4. Which is the smallest number?

E	F	G	H
790,832	791,382	790,382	1,790,382

5. What is the value of the underlined digit in 3.41**9**?

A	B	C	D
9	0.9	0.09	0.009

6. Which is the standard numeral for (1×1) + (4×0.1) + (1×0.01) + (3×0.001)?

E	F	G	H
1.00413	1.0413	1.413	14.13

7. Round 21.73 to the nearest whole number.

A	B	C	D
2,170	217	22	21

8. Which is the smallest number?

E	F	G	H
1.314	2	1.32	1.316

9. Give $\frac{8}{12}$ in simplest form.

A	B	C	D
$\frac{4}{6}$	$\frac{2}{3}$	$\frac{8}{12}$	$\frac{3}{4}$

10. Which number does not have the same value as $\frac{13}{4}$?

E	F	G	H
$3\frac{1}{4}$	$\frac{26}{8}$	$3\frac{3}{12}$	$3\frac{6}{8}$

11. Which is the greatest number?

A	B	C	D
$\frac{23}{50}$	$\frac{9}{20}$	$\frac{11}{25}$	$\frac{3}{10}$

12. What part of the rectangle is shaded?

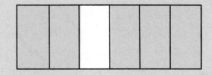

E	F	G	H
$\frac{1}{2}$	$\frac{3}{4}$	$\frac{5}{6}$	$\frac{7}{8}$

Chapter 4
DECIMALS

The Hoover Dam serves as a dam, an electrical power plant, and a reservoir.

READING AND WRITING DECIMALS

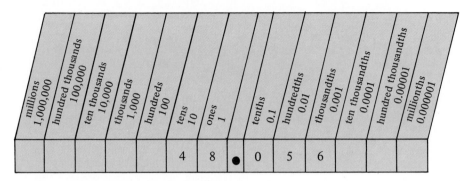

Standard numeral   Expanded numeral
48.056 = (4 x 10) + (8 x 1) + (0 x 0.1) + (5 x 0.01) + (6 x 0.001)

Read 48.056 as forty-eight and fifty-six *thousandths*.

⌐ Use the place of the last digit. ⌐

Study and Learn

A. Tell the place of the underlined digit.

   1. 0.<u>4</u>   **2.** 0.07<u>5</u>   **3.** 0.41312<u>8</u>   **4.** 6.9<u>4</u>12

B. Read.

   5. 0.4   **6.** 0.075   **7.** 0.413128   **8.** 6.9412

C. Write decimals.

   Example   six and thirty-four hundredths

               6.34 ⟵――――― place

   9. sixty-five thousandths   **10.** sixty-five ten-thousandths

   11. two and three hundred thousandths

   12. twenty-one and five millionths

D. What is the value of the underlined digit?

   Example   134.527<u>6</u>
                The digit 6 has the value 6 × 0.0001, *or* 0.0006.

   13. 37.56<u>9</u>   **14.** 175.004<u>3</u>   **15.** 25.00516<u>7</u>   **16.** 9<u>8</u>,762.4

E. Write expanded numerals.

Example   Standard numeral   Expanded numeral

48.056 = (4 × 10) + (8 × 1) + (0 × 0.1) + (5 × 0.01) + (6 × 0.001)

17. 3,876   **18.** 143,762   **19.** 0.241   **20.** 0.78602

F. Write standard numerals.

21. (9 × 100) + (6 × 10) + (0 × 1)

22. (9 × 0.1) + (0 × 0.01) + (4 × 0.001) + (6 × 0.0001) + (8 × 0.00001)

23. (2 × 10) + (0 × 1) + (5 × 0.1) + (2 × 0.01)

Practice

Write decimals.

1. three hundred thousandths   **2.** three hundred-thousandths

3. forty-nine ten-thousandths   **4.** ten and seven hundredths

5. thirty-six and eight hundred twenty-two millionths

What is the value of the underlined digit?

6. 0.<u>8</u>   **7.** 0.2<u>4</u>   **8.** 10,6<u>7</u>5   **9.** 1<u>7</u>,634   **10.** 73,543.2<u>1</u>

11. 0.12<u>4</u>   **12.** 0.2<u>2</u>22   **13.** 3.131<u>3</u>   **14.** 4.0000<u>9</u>   **15.** 0.00167<u>8</u>

Write expanded numerals.

16. 3,346   **17.** 725,798   **18.** 21.1576   **19.** 3.01608

Write standard numerals.

20. (5 × 1,000) + (3 × 100) + (4 × 10) + (8 × 1)

21. (0 × 0.1) + (2 × 0.01) + (0 × 0.001) + (9 × 0.0001)

22. (9 × 100) + (6 × 10) + (3 × 1) + (5 × 0.1) + (0 × 0.01) + (4 × 0.001)

COMPARING DECIMALS

Which town had a greater
rainfall on Monday?

0.3 = 0.30
    0.30 > 0.28
0.3 > 0.28

West town had the greater
rainfall.

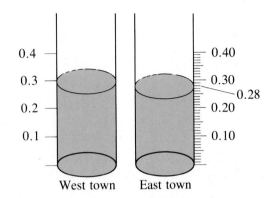

West town   East town

Study and Learn

A. Change to hundredths.

  1. 0.7   **2.** 0.6   **3.** 0.1   **4.** 3.0   **5.** 8.1

B. Change to thousandths.

  6. 0.5   **7.** 0.8   **8.** 0.32   **9.** 3.0   **10.** 4

C. Compare. Use >, <, and =.

  11. 0.9 ≡ 0.6   **12.** 0.07 ≡ 0.61   **13.** 1.350 ≡ 1.298   **14.** 0.9 ≡ 0.900

D. Change to hundredths and compare. Use >, <, and =.

  15. 0.8 ≡ 0.04   **16.**   0.4 ≡ 0.39   **17.** 0.03 ≡ 0.3   **18.** 4.24 ≡ 4.3

E. Compare. Use >, <, and =.

  19. 0.3 ≡ 0.139 **20.** 2.1 ≡ 2.009 **21.** 0.716 ≡ 0.85134 **22.** 0.64 ≡ 0.640

F. List in order from the greatest to the least.

   Example   0.07, 0.516, 0.0031
              Rewrite: 0.0700, 0.5160, 0.0031
           Then compare: 0.516 > 0.07 > 0.0031

  23. 0.06, 0.134, 0.9   **24.** 0.0037, 0.037, 0.37

  25. 0.07, 0.201, 0.13   **26.** 0.104, 0.03006, 0.4

Practice

Compare. Use >, <, and =.

1. 0.4 ≡ 0.6
2. 0.1 ≡ 0.7
3. 0.7 ≡ 0.3
4. 0.007 ≡ 0.30

5. 0.6 ≡ 0.09
6. 0.04 ≡ 0.10
7. 2.3 ≡ 2.229
8. 5.06 ≡ 5.6

9. 0.9 ≡ 0.09
10. 0.7 ≡ 0.70
11. 1.5 ≡ 1.37
12. 0.2 ≡ 1.02

13. 0.6 ≡ 0.140
14. 0.3 ≡ 0.400
15. 0.6 ≡ 0.590
16. 0.04 ≡ 0.400

17. 0.06 ≡ 0.500
18. 0.01 ≡ 0.009
19. 0.56 ≡ 0.59
20. 5.2 ≡ 5.090

21. 3.01 ≡ 3.01001
22. 27.02020 ≡ 27.2020
23. 0.09 ≡ 0.090000

List in order from the greatest to the least.

24. 0.014, 0.0435, 0.14
25. 0.0049, 0.049, 0.000049

26. 0.1, 0.09, 0.009
27. 0.064, 0.0064, 0.00064

Computer

Here is a program which uses **INPUT** commands to compare two numbers.

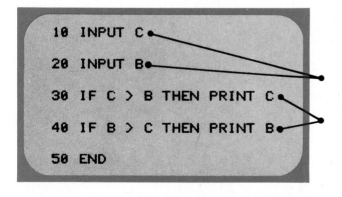

This program makes the computer

ask, then wait for a number to store in C, then in B.

compare the numbers and print the greater.

ROUNDING DECIMALS

This chart shows how to round decimals.

Round to	Number	Think	Write
nearest thousandth	0.0036	0.00 3 6 ↑ The ten-thousandths digit is 5 or greater.	$0.0036 \doteq 0.004$
nearest hundredth	4.2931	4.2 9 3 1 ↑ The thousandths digit is less than 5.	$4.2931 \doteq 4.29$
nearest tenth	3.647	3.6 4 7 ↑ The hundredths digit is less than 5.	$3.647 \doteq 3.6$
nearest whole number	7.506	7.5 0 6 ↑ The tenths digit is 5 or greater.	$7.506 \doteq 8$

Study and Learn

A. Round to the nearest thousandth.

   1. 0.4352   **2.** 0.7419   **3.** 2.31450   **4.** 3.006219

   5. 0.1694   **6.** 0.1695   **7.** 6.09904   **8.** 3.000219

B. Round to the nearest hundredth.

   9. 0.049   **10.** 0.054   **11.** 0.005   **12.** 0.6138

   13. 6.4183   **14.** 7.0061   **15.** 8.997   **16.** 3.002

C. Round to the nearest tenth.

   17. 9.68   **18.** 0.31   **19.** 1.0516   **20.** 3.449

   21. 2.506   **22.** 6.92   **23.** 2.90   **24.** 3.95

D. Round to the nearest whole number.

   25. 6.007   **26.** 7.3214   **27.** 2.63   **28.** 7.09

   29. 0.516   **30.** 0.432   **31.** 0.9   **32.** 0.499

Practice

Round to the nearest thousandth.

1. 1.5638   **2.** 2.3409   **3.** 0.4554   **4.** 6.0015

5. 8.0140   **6.** 8.0193   **7.** 7.0095   **8.** 8.9996

Round to the nearest hundredth.

9. 5.164   **10.** 0.307   **11.** 6.009   **12.** 7.095

13. 4.3641   **14.** 8.70253   **15.** 9.695   **16.** 4.997

Round to the nearest tenth.

17. 0.78   **18.** 3.64   **19.** 6.893   **20.** 6.031

21. 2.7163   **22.** 3.57631   **23.** 4.953   **24.** 6.0493

Round to the nearest whole number.

25. 0.8   **26.** 0.3   **27.** 3.64   **28.** 8.09

29. 5.934   **30.** 18.349   **31.** 14.382   **32.** 19.641

Complete.

	Number	Round to the nearest			
		thousandth	hundredth	tenth	whole number
33.	4.3165				
34.	0.7209				
35.	4.3652				
36.	2.0706				
37.	3.1495				

Solve.

38. A mechanic found the diameter of a pipe to be 1.7675 in. Round this to the nearest hundredth.

39. The mass of a steel bar is calculated to be 13.9680 kg. Round this to the nearest tenth.

Skills Review

Solve. *(60, 66, 70)*

1. $x + 7 = 13$

2. $x + 18 = 36$

3. $x - 5 = 9$

4. $x - 23 = 37$

5. $3 \cdot x = 27$

6. $4 \cdot x = 96$

7. $\frac{x}{2} = 4$

8. $\frac{x}{4} = 17$

9. $2x + 6 = 22$

10. $3x + 19 = 34$

11. $4x - 8 = 4$

12. $5x - 37 = 38$

Add. *(24)*

13. $27 + 532 + 96$

14. $2,153 + 705 + 6,927 + 458$

15.
```
   436
   819
   747
   864
+ 934
```

16.
```
  53,716
   8,734
  29,046
  38,654
+ 68,019
```

17.
```
  233,146
  416,897
  783,426
  825,713
+ 624,316
```

18.
```
$ 43,617
   7,816
  35,706
  19,816
+  7,613
```

Subtract. *(26)*

19.
```
  4,369
- 2,058
```

20.
```
  38,619
- 14,709
```

21.
```
  345,678
- 182,569
```

22.
```
  234,700
-  78,296
```

Multiply. *(34, 36)*

23.
```
428
× 9
```

24.
```
8,063
×   8
```

25.
```
 24
× 31
```

26.
```
 36
× 15
```

27.
```
 40
× 25
```

28.
```
 93
× 30
```

29.
```
613
× 43
```

30.
```
438
× 64
```

31.
```
7,450
×  36
```

32.
```
657
× 90
```

33.
```
 134
× 165
```

34.
```
43,167
×  803
```

Divide. *(38, 40)*

35. $7\overline{)2,149}$

36. $8\overline{)8,328}$

37. $6\overline{)42,315}$

38. $21\overline{)672}$

39. $32\overline{)768}$

40. $43\overline{)5,762}$

41. $37\overline{)16,872}$

42. $49\overline{)83,146}$

Midchapter Review

Write decimals. *(80)*

1. Thirty-one and six hundred four thousandths

2. Seventeen and four hundred thousand eight hundred millionths

What is the value of the underlined digit? *(80)*

3. 7.<u>4</u>
4. 0.65<u>1</u>
5. 2<u>3</u>.08
6. 1<u>9</u>,564

Compare. Use >, <, or =. *(82)*

7. 0.7 ≡ 0.9
8. 0.6 ≡ 0.60
9. 0.59 ≡ 0.6

Round. *(84)*

10. 2.1385 to the nearest thousandth

11. 4.834 to the nearest tenth

12. 1.361 to the nearest hundredth

13. 24.824 to the nearest whole number

14. 0.14 to the nearest whole number

15. 5.0672 to the nearest hundredth

Something Extra
Non-Routine Problems

Ann and Judy each have 30 birthday cards.
Ann wants to sell her cards at 2 for 5¢.
Judy wants to sell her cards at 3 for 5¢.
How much would they receive in all?

If they combine their cards and sell them
at 5 for 10¢, would they receive the same
total income? Check your guess.

Problem-Solving Skills

Selecting Equations

Millie works in a store that sells lumber, paint, and other items to fix up homes. Each section of a wood fence costs $11.99. Which equation should Millie use to find the cost of 8 sections?

$$8 \cdot c = 11.99 \qquad\qquad c = 8 \times 11.99 \qquad\qquad c - 8 = 11.99$$

PLAN   In the equation $8 \cdot c = 11.99$, the total cost is 11.99.

In the equation $c = 8 \times 11.99$, the total cost is c.

In the equation $c - 8 = 11.99$, the number of sections is subtracted from the total cost.

SOLVE   The correct equation is $c = 8 \times 11.99$ since each section costs 11.99 and Millie wants to find the cost of 8 sections.

Study and Learn

A. Select the equation to solve the problem.

1. The regular price of an entrance door with 6 panels is $159.99. During a sale the door sells for $149.50. How much is saved by buying the door during the sale?

    a. $s = 6 \times 159.99$ **b.** $s = 6 \times 149.50$

    c. $s = 159.99 - 149.50$ **d.** $s - 149.50 = 159.99$

D. Write expanded numerals. Use exponents.

Example $5,463 = 5 \times 10^3 + 4 \times 10^2 + 6 \times 10^1 + 3 \times 10^0$

17. 487   **18.** 5,685   **19.** 93,406   **20.** 403,617

Practice

Write in exponential notation.

1. 10,000,000,000   **2.** 100,000,000,000   **3.** 1,000,000,000,000

4. 10×10 **5.** $10 \times 10 \times 10 \times 10$ **6.** $10 \times 10 \times 10 \times 10 \times 10 \times 10 \times 10$

Multiply.

7. $10^2 \cdot 10^3$ **8.** $10^2 \cdot 10^1$ **9.** $10^3 \cdot 10^4$ **10.** $10^7 \cdot 10^5$

11. $10^1 \cdot 10^3$ **12.** $10^4 \cdot 10^3$ **13.** $10^4 \cdot 10^4$ **14.** $10^0 \cdot 10^6$

Divide.

15. $10^4 \div 10^1$ **16.** $10^5 \div 10^4$ **17.** $10^7 \div 10^4$ **18.** $10^{12} \div 10^3$

19. $10^9 \div 10^1$ **20.** $10^8 \div 10^6$ **21.** $10^9 \div 10^3$ **22.** $10^{12} \div 10^6$

23. $\dfrac{10^{16}}{10^9}$ **24.** $\dfrac{10^7}{10^4}$ **25.** $\dfrac{10^3}{10^2}$ **26.** $\dfrac{10^5}{10^5}$ **27.** $\dfrac{10^9}{10^3}$

Write expanded numerals. Use exponents.

28. 579   **29.** 7,718   **30.** 86,307   **31.** 506,982

Solve Problems

32. There are 1,000 library cards in a box. How many library cards are in 1,000 boxes?

33. Library cards are packed 1,000 cards to a box. How many boxes are needed for 10,000,000 cards?

34. How many $100 bills make $1,000,000?

35. There are 1,000,000 boxes in each of 100 warehouses. How many boxes are there in all?

SCIENTIFIC NOTATION

Mercury is the planet closest to the sun. It is about 60,000,000 km from the sun. Scientists use scientific notation to name such large numbers.

Standard Numeral	Scientific Notation
60,000,000	6×10^7

Number from 1 to 10   Power of 10

Study and Learn

A. Which are written in scientific notation?

1. 3×10^6 **2.** 14×10^2 **3.** 3×2 **4.** 9×10^{12} **5.** 4.7×10^6

B. Write in scientific notation.

6. 500   **7.** 40,000   **8.** 600,000   **9.** 7,000,000   **10.** 80,000,000

Write 360 in scientific notation.
Think: 360 is between 300 and 400.

$300 = 3 \times 10^2$ Number from 1 to 10
$360 = 3.6 \times 10^2$
$400 = 4 \times 10^2$ Power of 10

Shortcut: $360 = 3.6 \times 10^2$ ——— Move the decimal point 2 places
                                          to the left.
Need a number from 1 to 10

C. Write in scientific notation.

11. 7,400   **12.** 860,000

13. 3,800,000   **14.** 461,000,000,000

D. Write the standard numeral for 1.2×10^4.

The exponent is 4. Move the decimal point 4 places to right.
$1.2 \times 10^4 = 1.2000$
$\qquad\qquad = 12,000$

Write standard numerals.

15. 3.1×10^4 **16.** 6.2×10^6 **17.** 6×10^8 **18.** 7.31×10^9

Practice

Write in scientific notation.

1. 300   **2.** 6,000   **3.** 40,000   **4.** 800,000   **5.** 700,000,000

6. 240   **7.** 6,700   **8.** 31,000   **9.** 564,000   **10.** 3,120,000

11. 8,300,000   **12.** 16,000,000   **13.** 27,800,000

14. 6,400,000,000   **15.** 28,000,000,000,000   **16.** 716,000,000,000

Write standard numerals.

17. 3×10^3 **18.** 4×10^6 **19.** 6.1×10^3 **20.** 9.3×10^6

21. 3.12×10^3 **22.** 2.16×10^5 **23.** 3.49×10^6 **24.** 8.56×10^{10}

25. 7.62×10^{11} **26.** 7.216×10^4 **27.** 3.472×10^9 **28.** 4.1968×10^{12}

Write the numbers in scientific notation.

29. The earth is about 150,000,000 km from the sun.

30. Light travels about 6,000,000,000,000 mi in 1 year.

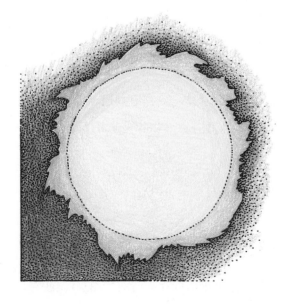

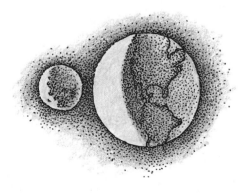

COMPUTING WITH SCIENTIFIC NOTATION

Light travels 5,870,000,000,000 mi in a year. How far does it travel
in 1,000,000 years?
Write each number in scientific notation. Then multiply.
$(5.87 \times 10^{12}) \times (1 \times 10^{6}) = 5.87 \times 1 \times 10^{12} \times 10^{6} = 5.87 \times 10^{18}$
Light travels 5.87×10^{18} mi in 1,000,000 years.

Study and Learn

A. Multiply. Use scientific notation.

    1. $93,000,000 \times 1,000,000$ **2.** $42,700,000,000 \times 1,000,000$

B. Multiply $30,000,000,000 \times 20,000$.

$$(3 \times 10^{10}) \times (2 \times 10^{4}) = 3 \times 2 \times 10^{10} \times 10^{4}$$
$$= 6 \times 10^{14}$$

Multiply. Write the answer in scientific notation.

    3. $30,000,000,000 \times 2,000,000$ **4.** $3,000,000 \times 3,000,000,000$

    5. $400,000 \times 200,000,000$ **6.** $2,000,000,000 \times 20,000,000$

C. Sometimes the answer must be rewritten to be in scientific
notation.
$$6,000,000 \times 4,000,000 = (6 \times 10^{6}) \times (4 \times 10^{6})$$
Think: 24 is not between 1 and 10. $= 24 \times 10^{12}$
Write 24 in scientific notation. $= 2.4 \times 10^{1} \times 10^{12}$
$$= 2.4 \times 10^{13}$$

Multiply. Write the answer in scientific notation.

    7. $400,000 \times 70,000$ **8.** $6,000,000 \times 30,000,000$

    9. $70,000,000 \times 300,000$ **10.** $200,000,000,000 \times 80,000,000$

D. Divide 9,000,000,000 by 3,000,000.

$$\frac{9,000,000,000}{3,000,000} = \frac{9 \times 10^{9}}{3 \times 10^{6}} = 3 \times 10^{3}$$

Divide. Write the answer in scientific notation.

    11. $\dfrac{4,000,000,000,000}{2,000,000}$ **12.** $\dfrac{36,000,000,000,000}{10,000,000,000}$

Practice

Multiply. Write the answer in scientific notation.

1. $3,000,000 \times 100,000$

2. $61,000,000 \times 10,000,000$

3. $4,000,000 \times 20,000$

4. $3,000,000 \times 3,000,000$

5. $20,000 \times 6,000,000$

6. $400,000 \times 40,000,000$

7. $200,000,000 \times 600,000$

8. $70,000,000 \times 400,000,000$

Divide. Write the answer in scientific notation.

9. $\dfrac{8,000,000,000}{2,000,000}$

10. $\dfrac{6,000,000,000,000}{3,000,000}$

11. $\dfrac{5,000,000,000,000}{1,000,000,000}$

12. $\dfrac{5,100,000,000,000,000}{1,000,000,000}$

13. $\dfrac{40,000,000,000}{200,000}$

14. $\dfrac{600,000,000,000,000,000,000}{2,000}$

15. $\dfrac{9,000,000}{3,000}$

16. $\dfrac{500,000,000}{50,000}$

Computer

A. The $\wedge$ key is the computer's exponent sign: $10\wedge6$ is 10^6. The computer displays number up to a billion as standard numerals. **PRINT (2 * 10**$\wedge$**2) * (4 * 10**$\wedge$ **3)**, gives the output **800000**.

B. To do $(3 \times 10^5) \times (2 \times 10^6)$ on the computer, you enter: **PRINT (3 * 10**$\wedge$**5) * (2 * 10**$\wedge$**6)**. The output is **6E+11**. $6 \times 10^{11} = 6E+11$. This is the computer's scientific notation.

C. What is the output of these PRINT statements?

1. PRINT (4 * 10$\wedge$**8) * (3 * 10**$\wedge$**4)**

2. PRINT (9 * 10$\wedge$**16) / (3 * 10**$\wedge$**10)**

H A N D S O N

Check your answers to the Practice. Remember to press RETURN .

Problem-Solving Applications

Better Buy

1. At Tannenbaum's Gift Shop, 2 coffee mugs sell for $1.79. At Grand Gifts, 3 coffee mugs sell for $2.65. Which is the better buy? [HINT: Find the cost of 1 coffee mug at each store.]

2. A store advertised 2 pens for $1.35 and a package of 3 of the same pens for $1.90. Which is the better buy?

3. A package of 6 coasters is marked $1.44 and a package of 8 of the same coasters is marked $1.76. Which is the better buy?

4. A box of 4 drinking glasses is marked $1.80 and a box of 6 of the same drinking glasses is marked $2.50. Which is the better buy?

5. A 12-pack of small pads is $1.09 and an 8-pack of the same small pads is $0.89. Which is the better buy?

6. At Santiago Gift Shop, picture frames are marked 2/$1.67 (2 for $1.67). At Superior Gifts, the same picture frames are marked 3/$2.20. Which is the better buy?

7. A package of 6 place mats is marked $8.00 and a package of 8 place mats is marked $10.00. Which is the better buy?

Chapter Review

Write decimals. *(80)*

1. Four hundred twenty-five millionths

2. Sixty-two thousandths

What is the value of the underlined digit? *(80)*

3. 280.51$\underline{1}$

4. 1$\underline{7}$,912.6

5. 40.623$\underline{4}$

6. 6.32214$\underline{6}$

Compare. Use $>$, $<$, and $=$. *(82)*

7. 0.73 $\equiv$ 0.730

8. 0.29 $\equiv$ 0.3

9. 0.1 $\equiv$ 0.009

Round. *(84)*

10. 4.329 to the nearest tenth

11. 2.359 to the nearest hundredth

12. 3.10459 to the nearest thousandth

13. 5.6214 to the nearest whole number

Write in exponential notation. *(90)*

14. 10,000

15. 10,000,000

Multiply or divide. *(90)*

16. $10^5 \cdot 10^3$

17. $10^4 \cdot 10^6$

18. $10^8 \div 10^8$

19. $\frac{10^7}{10^3}$

Write in scientific notation. *(92)*

20. 8,000,000

21. 4,900

Write standard numerals. *(92)*

22. 4×10^5

23. 3.7×10^6

Select the equation to solve the problem.

24. Ellen bought roof shingles for $5.99 a bundle. How much did she
(88) pay for 24 bundles?

    a. $24 \cdot c = 5.99$

    b. $c \div 24 = 5.99$

    c. $5.99 \div 24 = c$

    d. $24 \times 5.99 = c$

Solve.

25. Abner's Grocery is selling 4 cans of Colonel corn for $1.19.
(96) Westland Supermarket is selling 3 cans of Colonel corn for
$0.95. Which is the better buy?

Chapter Test

Write decimals. *(80)*

1. Eighty-six millionths   **2.** Thirty-two hundred-thousandths

What is the value of the underlined digit? *(80)*

3. 28.162<u>8</u>   **4.** <u>2</u>0,612.3   **5.** 10.624<u>5</u>8   **6.** 8.10923<u>7</u>

Compare. Use $>$, $<$, and $=$. *(82)*

7. 0.3 ▤ 0.7   **8.** 0.61 ▤ 0.610   **9.** 0.09 ▤ 0.1

Round. *(84)*

10. 3.148 to the nearest tenth   **11.** 2.158 to the nearest hundredth

12. 4.1487 to the nearest thousandth   **13.** 34.512 to the nearest whole number

Write in exponential notation. *(90)*

14. 1,000,000   **15.** 100,000,000

Multiply or divide. *(90)*

16. $10^4 \cdot 10^2$ **17.** $10^5 \cdot 10^7$ **18.** $10^6 \div 10^1$ **19.** $\dfrac{10^7}{10^7}$

Write in scientific notation. *(92)*

20. 600,000   **21.** 2,300,000

Write standard numerals. *(92)*

22. 3×10^4 **23.** 4.1×10^6

Select the equation to solve the problem.

24. Ira paid $16.50 for 12 ceiling panels and $32.95 for 4 cans of
(88) paint. How much did he spend?

    a. $c = 16.50 + 32.95$ **b.** $c = 32.95 - 16.50$

    c. $c = 12 \times 16.50 + 32.95$ **d.** $c = 16.50 + 4 \times 32.95$

Solve.

25. R&L Auto Store sells a package of 6 spark plugs for $5.34 and a
(96) package of 8 spark plugs for $7.04. Which is the better buy?

Skills Check

1. A radio that sells for $68.50 is on sale for $49.95. How much is saved by buying it on sale?

   A $118.45   B $18.65

   C $18.55   D $18.45

2. Henry works 35 hours a week. He earns $3.75 an hour. What does he earn each week?

   E $1,312.50   F $132.25

   G $131.25   H $130.25

3. Sam had $439.48 in his checking account. He made deposits of $120.14 and $96.50. What is his new balance?

   A $656.12   B $656.02

   C $222.84   D $210.64

4. The temperature is ⁻5°F outside. Which of the following temperatures is colder?

   E 15°F   F 10°F

   G ⁻4°F   H ⁻10°F

5. A roast was put in the oven at 4:15 pm. It takes $2\frac{1}{2}$ hours to cook. At what time will it be ready?

   A 6:16 pm   B 6:30 pm

   C 6:45 pm   D 7:45 pm

6. Ms. Jones ordered living room furniture October 2 and received it October 14. How long did she wait for her furniture?

   E 12 days   F 14 days

   G 31 days   H 377 days

7. Which unit would you use to tell the mass of an orange?

   A mg   B g

   C kg   D ton

8. Henry bowled five games and had scores of 145, 182, 175, 162, and 156. What was his average?

   E 175   F 168

   G 164   H 162

9. What is the perimeter of the right triangle?

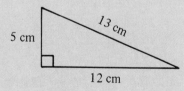

   A 780 cm   B 60 cm

   C 30 cm   D 25 cm

99

Chapter

5 OPERATIONS WITH DECIMALS

Brass sheet rolls are made of an alloy of copper and zinc.

ADDING DECIMALS

A team of 4 eighth-grade students ran in a relay race. Their times
were 12.4 seconds, 11.9 seconds, 11.8 seconds, and 12.1 seconds.
What was their total time for the relay race?

	Step 1	Step 2
12.4	12.4	12.4
11.9	11.9	11.9
11.8	11.8	11.8
+ 12.1	+ 12.1	+ 12.1
	48 2	48.2
	Add as with whole numbers.	Line up the decimal points.

Study and Learn

A. Add.

1.	**2.**	**3.**	**4.**
0.3	2.10	1.432	13.7194
0.2	3.41	0.416	24.8236
0.6	2.14	4.921	6.9139
+ 0.7	+ 4.20	+ 2.161	+ 2.6004

B. Add.

Example $12 + 3.4 + 7.29$

12.	—— or, change to hundredths ⟶	12.00
3.4	⟶	3.40
+ 7.29		+ 7.29
22.69		22.69

└── Line up the decimal points. ──┘

5. $4 + 2.7 + 6.008$ 6. $0.9 + 0.406 + 0.25$ 7. $1.1 + 6 + 3.006 + 0.91$

C. Add.

8.	**9.**	**10.**	**11.**
$ 6.41	$ 17.36	$ 0.31	$ 719.34
8.23	8.19	0.42	82.96
9.46	14.23	0.93	135.87
+ 9.24	+ 36.19	+ 0.88	+ 8.24

Practice

Add.

1. 2.4
   + 4.3

2. 6.04
   + 3.32

3. 41.74
   + 38.23

4. 31.238
   + 41.750

5. 0.96
   + 0.8

6. 2.8
   + 3.49

7. 34.065
   + 5.09

8. 7.638
   + 9.84

9. 0.6
    0.7
   + 0.4

10. 0.09
     0.04
   + 0.08

11. 0.348
     0.253
   + 0.617

12. 0.5367
     0.7204
   + 0.1689

13. 7.5
     8.3
   + 5.6

14. 17.3
     24.8
   + 3.9

15. 8.03
     4.17
   + 5.23

16. 12.05
     13.78
   + 24.86

17. 6.4163
     7.3187
     5.4128
   + 0.3517

18. 24.3168
     19.8904
      7.7146
   + 0.8238

19. 0.53016
     0.71023
     0.68424
   + 0.78346

20.   3.10967
     412.23687
       8.40816
   +   3.27912

21. 2.3 + 4.6

22. 7.04 + 6.39

23. 8.913 + 4.167

24. 3 + 1.4 + 0.56

25. 0.5 + 0.046 + 0.23

26. 12 + 3.5 + 0.069

27. $ 0.64
     0.79
     0.64
   + 0.98

28. $ 3.41
     0.89
     2.63
   + 7.84

29. $ 36.40
     8.68
     29.37
   + 16.79

30. $ 513.49
     89.84
     614.23
   + 83.45

★ **31.** 19 + 0.753 + 1.69 + 3.2 + 5.0006

★ **32.** 0.10569 + 25.6 + 3.992 + 104

Solve Problems

33. In a 400-m relay race, the times for the team members were 13.7, 12.9, 13.4, and 12.9 seconds. What was the total time for the team?

34. In three 100-m swimming races, a swimmer had times of 55.65, 55.49, and 56.02 seconds. What was the swimmer's total time for the 3 races?

SUBTRACTING DECIMALS

The average yearly rainfall in one city was 96.8 cm. The average yearly rainfall in another city was 174.9 cm. What is the difference in these rainfalls?

	Step 1	**Step 2**
174.9	174.9	174.9
− 96.8	− 96.8	− 96.8
	78 1	78.1
	Subtract as with whole numbers.	Line up the decimal points.

Study and Learn

A. Subtract.

1.	**2.**	**3.**	**4.**	**5.**
0.34	0.356	3.4	4.362	9.3468
− 0.17	− 0.170	− 2.7	− 2.574	− 3.1939

B. Subtract.

Example   6.8 − 3.57

6.8 ——— Change to hundredths ——→ 6.80
− 3.57   − 3.57
                                                  3.23

——— Line up the decimal points. ———

6. 0.4 − 0.07   **7.** 0.09 − 0.006   **8.** 0.59 − 0.412   **9.** 6.3 − 4.09

C. Decimals can be subtracted from whole numbers.

Example   7 − 2.4   7. ——— Change to tenths ——→ 7.0
                            − 2.4   − 2.4
                                                                         4.6

Subtract.

10. 6 − 3.1   **11.** 8 − 0.7   **12.** 58 − 3.47   **13.** 96 − 3.419

D. Subtract.

14.	**15.**	**16.**	**17.**	**18.**
$ 8.14	$ 9.00	$ 18.34	$ 40.83	$ 356.23
− 2.23	− 6.14	− 7.30	− 6.14	− 218.49

Practice

Subtract.

1. 36.8
   − 24.4

2. 526.8
   − 115.6

3. 3.7
   − 1.9

4. 241.6
   − 234.7

5. 24.29
   − 13.36

6. 41.68
   − 24.08

7. 34.74
   − 19.85

8. 124.04
   − 12.39

9. 0.008
   − 0.004

10. 0.063
    − 0.042

11. 0.746
    − 0.213

12. 8.348
    − 2.126

13. 0.621
    − 0.424

14. 0.831
    − 0.249

15. 8.340
    − 1.621

16. 9.356
    − 8.650

17. 0.0009
    − 0.0004

18. 0.0063
    − 0.0017

19. 0.0341
    − 0.0169

20. 0.4516
    − 0.2370

21. 2.0034
    − 1.0104

22. 6.3412
    − 4.1502

23. 7.3623
    − 4.0428

24. 9.7164
    − 2.3589

25. 0.8 − 0.3

26. 0.7 − 0.5

27. 0.07 − 0.03

28. 0.09 − 0.06

29. 0.008 − 0.006

30. 0.016 − 0.008

31. 0.934 − 0.824

32. 0.8136 − 0.4128

33. 0.6 − 0.24

34. 0.9 − 0.35

35. 0.07 − 0.008

36. 0.09 − 0.013

37. 0.46 − 0.234

38. 7.4 − 2.36

39. 8.9 − 2.47

40. 3.1 − 2.091

41. 6 − 0.4

42. 5 − 0.23

43. 8 − 1.16

44. 6 − 2.345

45. $ 7.12
    − 2.59

46. $ 8.00
    − 7.23

47. $ 40.72
    − 8.14

48. $ 752.24
    − 268.19

Solve Problems

49. One year Atlanta had 166 cm of rain. Jacksonville had 121.4 cm. How much more rain did Atlanta have?

50. One year Washington, D.C., had 128.3 cm of rain. San Juan had 130.2 cm. How much more rain did San Juan have?

ESTIMATING SUMS AND DIFFERENCES

Mario's 2 packed valises have masses of 9.3 kg and 6.5 kg. Estimate their sum and their difference to the nearest whole number.

Estimate by rounding.

Sum	Difference
9.3 = 9	9.3 = 9
+ 6.5 = + 7	− 6.5 = − 7
16	2

└──── Estimates ────┘

Study and Learn

A. Estimate to the nearest whole number.

1. 4.683	**2.** 7.823	**3.** 5.86	**4.** 6.7234
+ 0.71	6.41	− 3.456	− 3.5138
	+ 0.862		

B. Estimate to the nearest tenth.

Examples

Sum	Difference
3.623 = 3.6	4.86 = 4.9
0.29 = 0.3	− 1.374 = − 1.4
+ 5.386 = + 5.4	3.5
9.3	

└──── Estimates ────┘

5. 0.3	**6.** 7.428	**7.** 0.89	**8.** 4.8417
0.26	6.37	− 0.243	− 2.3689
+ 0.852	0.843		
	+ 1.609		

C. Estimate to the nearest dollar.

9. $ 8.23	**10.** $ 7.78	**11.** $ 7.47	**12.** $ 16.57
7.56	2.50	− 2.89	− 3.58
6.23	3.50		
+ 8.95	8.49		
	+ 6.38		

Practice

Estimate to the nearest whole number.

1.	0.89 + 2.083	2.	6.724 2.58 + 0.923	3.	8.316 2.492 + 3.542	4.	6.4139 2.8238 + 3.2500
5.	8.3 − 0.78	6.	4.9 − 0.87	7.	6.87 − 2.23	8.	8.7234 − 2.6009

Estimate to the nearest tenth.

9.	0.67 + 0.72	10.	0.835 0.27 + 0.85	11.	7.3 0.65 0.912 + 1.45	12.	6.412 2.375 3.46 + 8.9
13.	0.92 − 0.45	14.	0.68 − 0.2	15.	12.34 − 8.45	16.	6.319 − 2.4834

Estimate to the nearest dollar.

17.	$ 7.56 8.23 + 6.54	18.	$ 6.47 8.37 9.52 + 6.89	19.	$ 7.37 − 2.54	20.	$ 24.88 − 9.48

Solve Problems

21. A metal part is 3.04 cm thick. It is placed upon another part which is 1.89 cm thick. Estimate to the nearest whole number the total thickness of the two parts.

22. A wire is 2.47 m long. A second wire is 3.44 m long. Estimate to the nearest tenth the difference of the lengths.

23. Cristina bought a pair of shoes for $85.50, a coat for $79.00, and a pocketbook for $34.75. Estimate to the nearest dollar how much she spent.

24. The estimated cost of lumber for a woodwork project is $28.75. The estimated cost of finishing is $8.25. Estimate to the nearest dollar the total cost of the project.

Skills Review

Add. *(24)*

1.	2.	3.	4.
16,814	716,412	1,693,704	$ 71,604
29,312	89,374	286,803	86,416
8,768	421,821	4,416,214	9,789
+ 43,178	+ 86,764	+ 8,934,800	+ 11,689

Subtract. *(26)*

5.	6.	7.	8.
41,306	23,518	412,316	$ 9.00
− 29,800	− 8,916	− 89,427	− 4.87

Multiply. *(34, 36)*

9.	10.	11.	12.
4,136	4,789	3,418	5,689
× 7	× 28	× 204	× 860

Divide. *(38, 40)*

13. $7\overline{)147}$ **14.** $8\overline{)65,624}$ **15.** $9\overline{)64,483}$ **16.** $38\overline{)61,236}$

Compare. Use >, <, or = . *(82)*

17. 0.3 ▥ 0.19   **18.** 0.005 ▥ 0.04   **19.** 1.6 ▥ 0.28

20. 0.02 ▥ 0.2   **21.** 0.04 ▥ 0.040   **22.** 0.61 ▥ 0.6

Round. *(84)*

23. 23.062 to the nearest tenth

24. 1.631 to the nearest hundredth

25. 3.782 to the nearest whole number

26. 0.02753 to the nearest thousandth

Write in exponential notation. *(90)*

27. 10,000   **28.** 100   **29.** 100,000   **30.** 1,000,000

Write standard numerals. *(92)*

31. 3.5×10^4 **32.** 6×10^7 **33.** 5.2×10^5

Write in scientific notation. *(92)*

34. 800   **35.** 2,500   **36.** 90,000   **37.** 640,000,000

Midchapter Review

Add. *(102)*

1.
   8.6
   7.4
   6.5
$+ \, 8.2$

2. $3 + 2.1 + 0.056$

3. $17 + 0.158 + 5.3$

Subtract. *(104)*

4.
   8.4
$- \, 3.6$

5. $0.08 - 0.03$

6. $7 - 1.4$

Estimate to the nearest whole number. *(106)*

7.
  6.734
  2.04
$+ \, 4.351$

8.
  8.4
  9.3
$+ \, 6.4$

9.
  7.3
$- \, 0.78$

10.
  6.3419
$- \, 2.8124$

Estimate to the nearest tenth. *(106)*

11.
  0.7
  0.63
$+ \, 0.41$

12.
  0.841
  0.27
$+ \, 0.753$

13.
  0.83
$- \, 0.64$

14.
  17.56
$- \; 9.28$

Something Extra

Non-Routine Problems

Choose the number that belongs in the blank in the sequence.

1. 2.3, 3.4, 4.5, __?__ , 6.7, 7.8

   a. 5.5 **b.** 6.5 **c.** 5.6 **d.** 6.0

2. 2, 8, 18, __?__ , 50, 72, 98

   a. 24 **b.** 72 **c.** 28 **d.** 32

3. Assume that this statement is true.
If the road is covered with ice, it is slippery.
Which of the following **must** be true also?
  a. If the road is slippery, it is covered with ice.
  b. If the road is not covered with ice, it is not slippery.
  c. If the road is not slippery, it is not covered with ice.
  d. If the road is not covered with ice, it is slippery.

MULTIPLYING BY WHOLE NUMBERS

The thickness of a metal part for a machine is 1.2 cm. How high is a stack of 4 parts?

		Step 1	**Step 2**
1.2	THINK	1.2 or 1.2	1.2 ⟵ one decimal place
× 4		1.2 × 4	× 4
		1.2 4 8	4.8 ⟵ one decimal place
		+ 1.2	
		4.8	

Multiply as with whole numbers.

Determine where to place the decimal point.

So, a stack of 4 parts is 4.8 cm high.

Study and Learn

A. Multiply.

1. 3 × 0.2

2. 0.4 × 7

3. 3.2 × 4

4. 0.52 × 9

5. 0.13 × 12

Here is a shortcut for multiplying by powers of 10.

0.243	0.243	0.243
× 10	× 100	× 1,000
2.430	24.300	243.000

10 × 0.243 = 2.43   To multiply by 10, "move" the decimal point 1 place to the right.

100 × 0.243 = 24.3   To multiply by 100, "move" the decimal point 2 places to the right.

1,000 × 0.243 = 243   To multiply by 1,000, "move" the decimal point 3 places to the right.

B. Multiply by "moving" the decimal point.

6. 10 × 1.386   **7.** 100 × 1.386   **8.** 1,000 × 1.386   **9.** 100 × 2.34

C. Sometimes you need to insert zeros when you multiply by powers of 10.

Examples   100 × 0.6   1,000 × 0.6

           100 × 0.60 = 60   1,000 × 0.600 = 600

Multiply by "moving" the decimal point.

10. 100 × 0.9   **11.** 1,000 × 0.9   **12.** 1,000 × 0.08   **13.** 1,000 × 3.1

Practice

Multiply.

1. 2×0.4 **2.** 3×0.3 **3.** 6×0.1 **4.** 8×0.2

5.   0.7
$\underline{\times\ 4}$

6.   0.5
$\underline{\times\ 3}$

7.   0.9
$\underline{\times\ 6}$

8.   0.8
$\underline{\times\ 7}$

9.   3.2
$\underline{\times\ 3}$

10.   8.1
$\underline{\times\ 4}$

11.   9.3
$\underline{\times\ 5}$

12.   6.7
$\underline{\times\ 8}$

13.   24.3
$\underline{\times\ 6}$

14.   37.2
$\underline{\times\ 9}$

15. 124.3
$\underline{\times\ 53}$

16. 246.8
$\underline{\times\ 47}$

17. 3×0.03 **18.** 4×0.08 **19.** 5×0.05 **20.** 3×0.12

21.   0.06
$\underline{\times\ 2}$

22.   0.09
$\underline{\times\ 8}$

23.   0.14
$\underline{\times\ 3}$

24.   0.35
$\underline{\times\ 7}$

25.   2.31
$\underline{\times\ 2}$

26.   6.14
$\underline{\times\ 4}$

27.   3.04
$\underline{\times\ 37}$

28.   2.36
$\underline{\times\ 58}$

29. 4×0.002 **30.** 6×0.008 **31.** 3×0.012 **32.** 4×0.023

33. 10×0.63 **34.** 100×9.34 **35.** 100×0.8 **36.** $1{,}000 \times 2.75$

Solve Problems

37. The mass of a bar of steel is 12.06 kg. What is the mass of 8 of these bars?

38. Metal sheets are each 0.98 in. thick. How high is a stack of 32 of these sheets?

39. A Chinese unit of distance, li, is 630 yards. 3.5 li is how many yards?

MULTIPLYING DECIMALS

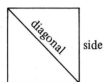

In a square, a side is approximately 0.7 times the length of a diagonal. How long is a side if a diagonal is 5.4 cm?

$$
\begin{array}{r}
5.4 \longleftarrow \text{ 1 decimal place} \\
\times 0.7 \longleftarrow \text{ + 1 decimal place} \\
\hline
3.78 \longleftarrow \text{ 2 decimal places}
\end{array}
$$

The number of decimal places in a product is the *sum* of the number of decimal places in the numbers multiplied.

Study and Learn

A. Multiply.

Example

$$
\begin{array}{r}
4.24 \longleftarrow \text{ 2 decimal places} \\
\times 1.3 \longleftarrow \text{ + 1 decimal place} \\
\hline
1\ 272 \\
4\ 24 \\
\hline
5.512 \longleftarrow \text{ 3 decimal places}
\end{array}
$$

Multiply.

1. 0.6×0.8

2. $\begin{array}{r} 2.34 \\ \times 0.4 \\ \hline \end{array}$

3. $\begin{array}{r} 6.213 \\ \times 0.7 \\ \hline \end{array}$

4. $\begin{array}{r} 1.23 \\ \times 0.34 \\ \hline \end{array}$

Sometimes it is necessary to insert zeros.

$$
\begin{array}{r} 0.64 \\ \times 0.007 \\ \hline \end{array}
\qquad
\begin{array}{r} 0.64 \\ \times 0.007 \\ \hline 448 \end{array}
\qquad
\begin{array}{r} 0.64 \\ \times 0.007 \\ \hline 0.00448 \end{array}
$$

THINK
$$
\begin{array}{r}
\text{2 decimal places} \\
\text{+ 3 decimal places} \\
\hline
\text{5 decimal places}
\end{array}
$$

Multiply as with whole numbers.

Place the decimal point. You need to insert 2 zeros.

B. Multiply.

5. $\begin{array}{r} 0.4 \\ \times 0.003 \\ \hline \end{array}$

6. $\begin{array}{r} 1.36 \\ \times 0.007 \\ \hline \end{array}$

7. $\begin{array}{r} 0.004 \\ \times 0.002 \\ \hline \end{array}$

8. $\begin{array}{r} 2.003 \\ \times 0.008 \\ \hline \end{array}$

9. $\begin{array}{r} 1.34 \\ \times 0.9 \\ \hline \end{array}$

10. $\begin{array}{r} 7.64 \\ \times 0.23 \\ \hline \end{array}$

11. $\begin{array}{r} 1.007 \\ \times 0.6 \\ \hline \end{array}$

12. $\begin{array}{r} 3.412 \\ \times 0.013 \\ \hline \end{array}$

Practice

Multiply.

1. $0.2 \times .04$ **2.** 0.6×0.4 **3.** 0.04×0.3 **4.** 0.1×0.01

5.
$$\begin{array}{r} 0.7 \\ \times\ 0.4 \\ \hline \end{array}$$

6.
$$\begin{array}{r} 0.1 \\ \times\ 0.1 \\ \hline \end{array}$$

7.
$$\begin{array}{r} 1.8 \\ \times\ 0.2 \\ \hline \end{array}$$

8.
$$\begin{array}{r} 34.1 \\ \times\ 0.6 \\ \hline \end{array}$$

9.
$$\begin{array}{r} 0.02 \\ \times\ 0.3 \\ \hline \end{array}$$

10.
$$\begin{array}{r} 0.08 \\ \times\ 0.4 \\ \hline \end{array}$$

11.
$$\begin{array}{r} 0.24 \\ \times\ 0.8 \\ \hline \end{array}$$

12.
$$\begin{array}{r} 1.34 \\ \times\ 0.4 \\ \hline \end{array}$$

13.
$$\begin{array}{r} 0.306 \\ \times\ 0.2 \\ \hline \end{array}$$

14.
$$\begin{array}{r} 1.412 \\ \times\ 0.8 \\ \hline \end{array}$$

15.
$$\begin{array}{r} 0.34 \\ \times\ 0.21 \\ \hline \end{array}$$

16.
$$\begin{array}{r} 0.34 \\ \times\ 0.36 \\ \hline \end{array}$$

17.
$$\begin{array}{r} 0.236 \\ \times\ 0.04 \\ \hline \end{array}$$

18.
$$\begin{array}{r} 0.156 \\ \times\ 0.34 \\ \hline \end{array}$$

19.
$$\begin{array}{r} 0.862 \\ \times\ 0.23 \\ \hline \end{array}$$

20.
$$\begin{array}{r} 2.3464 \\ \times\ 0.94 \\ \hline \end{array}$$

21.
$$\begin{array}{r} 0.2 \\ \times\ 0.003 \\ \hline \end{array}$$

22.
$$\begin{array}{r} 0.47 \\ \times\ 0.002 \\ \hline \end{array}$$

23.
$$\begin{array}{r} 0.007 \\ \times\ 0.003 \\ \hline \end{array}$$

24.
$$\begin{array}{r} 0.04 \\ \times\ 0.01 \\ \hline \end{array}$$

25.
$$\begin{array}{r} 3.46 \\ \times\ 0.003 \\ \hline \end{array}$$

26.
$$\begin{array}{r} 2.403 \\ \times\ 0.008 \\ \hline \end{array}$$

27.
$$\begin{array}{r} 16.47 \\ \times\ 1.36 \\ \hline \end{array}$$

28.
$$\begin{array}{r} 8.341 \\ \times\ 0.204 \\ \hline \end{array}$$

29.
$$\begin{array}{r} 0.5 \\ \times\ 0.7 \\ \hline \end{array}$$

30.
$$\begin{array}{r} 0.03 \\ \times\ 0.04 \\ \hline \end{array}$$

31.
$$\begin{array}{r} 0.6 \\ \times\ 0.04 \\ \hline \end{array}$$

32.
$$\begin{array}{r} 0.53 \\ \times\ 0.02 \\ \hline \end{array}$$

Solve Problems

33. The diagonal of a square is about 1.4 times the length of a side. How long is the diagonal of a square whose side is 8.3 cm? Round to the nearest whole number.

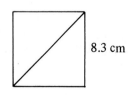

8.3 cm

34. A 1-m long steel bar has a mass of 0.78 kg. What is the mass of a steel bar which is 3.64 m long? Round the answer to the nearest tenth.

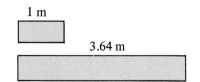

1 m

3.64 m

Problem-Solving Skills

Multi-Step Problems

Sometimes it is necessary to use more than 1 operation to solve a problem.

Ann called Steve in a nearby town. They talked on the phone for 24 minutes. The telephone rate is $0.95 for the first 2 minutes and $0.42 for each additional minute. How much did the call cost?

SOLVE **Step 1** The cost for the first 2 minutes is $0.95.

Step 2 The number of minutes not yet charged for is 22 minutes.
Find the cost of the next 22 minutes at $0.42 a minute.

$$\begin{array}{r} \$0.42 \\ \times\ 22 \\ \hline \$9.24 \end{array}$$

Step 3 Find the total cost.

$$\begin{array}{r} \$\ 0.95 \\ +\ 9.24 \\ \hline \$10.19 \end{array}$$ The call cost $10.19.

Study and Learn

A. Solve.

1. Ms. Valenti called Los Angeles person-to-person to speak to her mother. It cost $3.55 for the first 3 minutes and $0.38 for each additional minute. How much did a 10-minute call cost her?

2. Henry and Sue talked on the telephone for 40 minutes. The rate was 4 message units for the first 3 minutes and 1 message unit for each additional minute. Each message unit cost 8.2 cents. What did the call cost?

114

Practice

Solve.

1. Marcia called Henry in a nearby town. They spoke on the phone for 36 minutes. The rate is $0.54 for the first 2 minutes and $0.23 for each additional minute. How much did the call cost?

2. A lawyer called a client. They spoke for a half hour. The rate is $1.37 for the first 3 minutes and $0.23 for each additional minute. How much did the call cost?

3. The total cost of toll calls for 1 month was $16.16. Mrs. Perez made 1 toll call to New Jersey for $3.16. If the other 2 toll calls cost the same, how much did each cost?

4. The total cost of toll calls for one month was $35.07. Mr. Johnson's toll calls within his state totaled $22.16. He made another toll call to Nevada which cost $5.73. If the other 2 toll calls cost the same, how much did each cost?

5. Mr. Loo called a friend in the city and spoke for 25 minutes. The rate was 2 message units for the first 3 minutes and 2 message units for each additional minute. Each message unit costs 8.2 cents. What did the call cost?

6. Two friends spoke on the phone for 24 minutes. The rate was 6 message units for the first 2 minutes and 2 message units for each additional minute. Each message unit costs 8.2 cents. What did the call cost?

7. A direct-dial call of 20 minutes costs $1.35 for the first 2 minutes and $0.54 for each additional minute. The same call made person-to-person cost $3.29 for the first 3 minutes and $0.54 for each additional minute. How much is saved by a direct-dial call?

8. An operator-assisted call costs $2.15 for the first 3 minutes and $0.36 for each additional minute. A direct-dial call costs $0.52 for the first minute and $0.36 for each additional minute. How much is saved on a 10-minute call dialed direct?

9. A 30-minute call during the day costs $1.54 for the first 3 minutes and $0.27 for each additional minute. How much could be saved by calling at night for $5.46?

10. A 20-minute call at noon costs $0.54 for the first 2 minutes and $0.23 for each additional minute. How much is saved by calling after 11:00 pm when the same call costs $1.88?

DIVIDING DECIMALS BY WHOLE NUMBERS

A piece of trim is 0.8 m long. It is
cut into 4 equal parts. How long
is each part?

$$4\overline{)0.8} \quad \text{THINK} \quad \overset{2 \text{ tenths}}{4\overline{)8 \text{ tenths}}}, \text{ or } \overset{0.2}{4\overline{)0.8}}$$

To divide by a whole number, line up the decimal points in the
dividend and the quotient. Divide as with whole numbers.

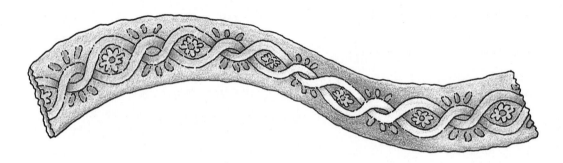

Study and Learn

A. Divide.

  1. $3\overline{)0.6}$ **2.** $4\overline{)1.2}$ **3.** $6\overline{)36.6}$ **4.** $7\overline{)13.3}$

B. Be careful with the zeros next to the decimal point.

   Example $\overset{0.02}{4\overline{)0.08}}$ $\overset{0.08}{4\overline{)0.32}}$ $\overset{0.002}{4\overline{)0.008}}$

                              Line up the decimal points.
                              Divide as with whole numbers.
                              Insert zeros if necessary.

Divide.

  5. $2\overline{)0.06}$ **6.** $6\overline{)0.42}$ **7.** $7\overline{)1.47}$ **8.** $8\overline{)23.04}$

  9. $3\overline{)0.009}$ **10.** $2\overline{)0.060}$ **11.** $5\overline{)0.500}$ **12.** $7\overline{)30.324}$

  13. $31\overline{)148.8}$ **14.** $23\overline{)56.35}$ **15.** $46\overline{)1.058}$ **16.** $79\overline{)158.79}$

  17. $27\overline{)13.689}$ **18.** $18\overline{)63.36}$ **19.** $23\overline{)10.373}$ **20.** $98\overline{)24.50}$

Practice

Divide.

1. 3)0.9   **2.** 3)0.3   **3.** 3)2.7   **4.** 7)3.5

5. 4)12.4   **6.** 4)24.8   **7.** 2)0.08   **8.** 6)0.36

9. 2)8.48   **10.** 4)1.64   **11.** 6)48.42   **12.** 3)48.45

13. 3)0.009   **14.** 2)0.014   **15.** 3)0.969   **16.** 3)0.159

17. 4)4.848   **18.** 2)8.124   **19.** 6)18.612   **20.** 3)42.012

21. 6)17.022   **22.** 4)1.624   **23.** 9)2.736   **24.** 4)34.136

25. 23)36.8   **26.** 41)98.4   **27.** 41)14.35   **28.** 59)18.88

29. 32)68.16   **30.** 18)60.12   **31.** 76)1,778.4   **32.** 27)1,379.7

33. 72)168.48   **34.** 86)145.34   **35.** 45)644.40   **36.** 56)113.68

37. 121)42.108   **38.** 142)59.214   **39.** 261)33.408   **40.** 343)29.498

41. 7)9.8   **42.** 9)0.117   **43.** 15)30.060   **44.** 92)100.28

45. 24)55.2   **46.** 86)318.2   **47.** 780)15.60   **48.** 317)177.52

Solve Problems

49. A piece of braid is 3.6 cm long. It is cut into 6 equal parts. How long is each part?

50. A bolt of fabric is 32.9 m long. It is cut into 47 equal parts. How long is each part?

51. A piece of seam binding is 97.2 cm long. It is cut into 12 equal pieces. How long is each piece?

DIVIDING BY DECIMALS

Jean has 3.5 m of wire for hanging pictures. She needs 0.5 m of wire for each picture. How many pieces of wire each 0.5 m long can be cut from the 3.5 m of wire?

$$0.5 \overline{)3.5} \quad \text{is the same as} \quad 5 \overline{)35}^{\,7}$$

Multiply by 10
Multiply by 10

Multiplying the divisor (0.5) and the dividend (3.5) by the same number does not change the quotient.

Study and Learn

A. Divide.

1. $0.2 \overline{)0.8}$ **2.** $0.3 \overline{)1.2}$ **3.** $0.9 \overline{)4.5}$ **4.** $0.6 \overline{)3.6}$

B. Sometimes you must multiply by a power of 10 to get a whole number divisor.

Examples $0.04 \overline{)0.16} \longrightarrow 4 \overline{)16}$ (multiplied by 100)
            $0.003 \overline{)2.16} \longrightarrow 3 \overline{)2,160}$ (multiplied by 1,000)

Divide.

5. $0.02 \overline{)0.12}$ **6.** $0.08 \overline{)0.32}$ **7.** $0.007 \overline{)0.021}$ **8.** $0.006 \overline{)0.126}$

> Here is a shortcut for multiplying by a power of 10.
>
> $0.5 \overline{)3.5}$ Use a carat (∧) to "move" the decimal point.
>
> Multiplying 0.5 and 3.5 by 10 results in "moving" the decimal point 1 place to the right.
> To multiply by 100, "move" 2 places to the right.
> To multiply by 1,000, "move" 3 places to the right.
>
> $0.04 \overline{)0.16}$ $0.003 \overline{)2.160}$ ←——Insert zero.

C. Divide.

9. $1.2 \overline{)0.144}$ **10.** $0.51 \overline{)0.1173}$ **11.** $0.004 \overline{)16}$ **12.** $\$0.23 \overline{)\$7.82}$

Practice

Divide.

1. $0.4\overline{)0.8}$ **2.** $0.2\overline{)0.6}$ **3.** $0.6\overline{)2.4}$ **4.** $0.5\overline{)3.0}$

5. $0.4\overline{)0.008}$ **6.** $1.2\overline{)3.6}$ **7.** $3.3\overline{)141.9}$ **8.** $0.9\overline{)27}$

9. $0.03\overline{)0.9}$ **10.** $0.01\overline{)0.1}$ **11.** $0.05\overline{)0.30}$ **12.** $0.08\overline{)0.72}$

13. $0.03\overline{)0.009}$ **14.** $0.09\overline{)0.072}$ **15.** $0.07\overline{)0.217}$ **16.** $0.04\overline{)0.288}$

17. $6.04\overline{)193.28}$ **18.** $5.07\overline{)16.224}$ **19.** $4.04\overline{)1292.8}$ **20.** $4.06\overline{)15.834}$

21. $0.003\overline{)0.009}$ **22.** $0.002\overline{)0.008}$ **23.** $0.003\overline{)0.06}$ **24.** $0.003\overline{)0.6}$

25. $0.006\overline{)1.602}$ **26.** $0.012\overline{)0.48}$ **27.** $0.046\overline{)23}$ **28.** $0.109\overline{)0.4469}$

29. $0.427\overline{)3.4587}$ **30.** $3.4\overline{)7.82}$ **31.** $0.1\overline{)2.6}$ **32.** $0.03\overline{)0.72}$

★ **33.** $\$0.12\overline{)\$3.00}$ ★ **34.** $\$0.47\overline{)\$9.87}$ ★ **35.** $\$0.36\overline{)\$15.12}$ ★ **36.** $\$0.24\overline{)\$93.60}$

★ **37.** $0.0006\overline{)3}$ ★ **38.** $0.0015\overline{)4.5}$ ★ **39.** $0.00025\overline{)0.8}$ ★ **40.** $0.00032\overline{)6.4}$

Computer

This computer program divides two numbers and prints the result.

```
10 INPUT A
20 INPUT B
30 PRINT A / B
```

The program tells the computer to
wait for you to put a number in A and B.
print A ÷ B.

A sample run:

```
RUN
■.6
■.2
3
```

Modify the program above so that it multiplies two numbers.

H A N D S O N

Run both programs several times using different values for A and B.
Remember to type RETURN .

ROUNDING QUOTIENTS

Sometimes quotients are rounded.

Divide 3.4 by 1.41. Round the quotient to the nearest tenth.

$1.41\overline{)3.4}$ $1.41\overline{)3.40}$

To find a quotient to the nearest tenth, carry the division to hundredths. Then round to the nearest tenth.

$$\overset{2.41 \doteq 2.4}{1.41\overline{)3.4000}}$$

Study and Learn

A. Find each quotient to the nearest tenth.

1. $0.3\overline{)0.5}$ **2.** $4\overline{)6.2}$ **3.** $1.9\overline{)3.4}$ **4.** $0.23\overline{)0.473}$

> To find a quotient to the nearest hundredth, carry the division to thousandths. Then round to the nearest hundredth.

$$0.08\overline{)0.2517} \quad 0.08\overline{)0.2517} \quad 0.08\overline{)0.25\,170} \quad \overset{3.146 \doteq 3.15}{0.08\overline{)0.25\,170}}$$

B. Find each quotient to the nearest hundredth.

5. $0.06\overline{)0.0257}$ **6.** $0.12\overline{)0.357}$ **7.** $13\overline{)0.946}$ **8.** $0.35\overline{)0.14897}$

Practice

Find each quotient to the nearest tenth.

1. $5\overline{)1.3}$ **2.** $7\overline{)2.6}$ **3.** $0.7\overline{)0.3}$ **4.** $0.9\overline{)0.6}$

5. $3.1\overline{)0.7}$ **6.** $4.2\overline{)5.34}$ **7.** $7.1\overline{)8.3}$ **8.** $0.34\overline{)0.789}$

Find each quotient to the nearest hundredth.

9. $0.04\overline{)0.3416}$ **10.** $0.23\overline{)0.516}$ **11.** $21\overline{)0.359}$ **12.** $15\overline{)0.746}$

13. $0.23\overline{)0.4768}$ **14.** $0.62\overline{)0.5049}$ **15.** $0.71\overline{)0.05617}$ **16.** $1.3\overline{)0.5163}$

ESTIMATING PRODUCTS AND QUOTIENTS

The heart pumps about 4.75 L of blood every minute. Estimate how many liters it pumps in 0.5 minutes.

To estimate a product with decimal factors:
    Step 1 Round each number. Keep the decimal points if needed.
    Step 2 Multiply.

	Step 1	**Step 2**
4.75	5	5
× 0.5	× 0.5	× 0.5
		2.5 So, it pumps about 2.5 L.

Study and Learn

A. Estimate. Complete.

1. $\begin{array}{r} 5.7 \rightarrow 6 \\ \times\,3 \rightarrow \times\,3 \end{array}$
    2. $\begin{array}{r} 3.5 \rightarrow 4 \\ \times\,6.4 \rightarrow \times\,6 \end{array}$
    3. $\begin{array}{r} 0.31 \rightarrow 0.3 \\ \times\,6.8 \rightarrow \times\,7 \end{array}$

Here is how to estimate a quotient.
    Step 1 Move the decimal points and place the decimal in the quotient as in dividing decimals.
    Step 2 Round the divisor if more than 1 digit.
    Step 3 Estimate the first digit in the quotient. Write 0's as needed.

Example	**Step 1**	**Step 2**	**Step 3**
$4.1\overline{)89.93}$	$4.1\overline{)89.9\,3}$	$40\overline{)899.3}$	$20.$ $40\overline{)899.3}$

B. Estimate.

4. $6\overline{)13.1}$
    5. $2.8\overline{)1.82}$
    6. $0.67\overline{)3.142}$
    7. $0.49\overline{)31.14}$

Practice

Estimate.

1. 4.8×6
    2. 3.1×8.2
    3. 8.6×5.9
    4. 0.21×4.1

5. $3\overline{)6.31}$
    6. $4.2\overline{)0.439}$
    7. $0.31\overline{)6.125}$
    ★ **8.** $5.7\overline{)0.483}$

121

EQUATIONS WITH DECIMALS

Solving equations with decimals is just like solving equations with whole numbers.

$$x + 0.7 = 0.9$$
$$x + 0.7 - 0.7 = 0.9 - 0.7$$
$$x = 0.2$$

$$3x = 1.2$$
$$\frac{3x}{3} = \frac{1.2}{3}$$
$$x = 0.4$$

Check
$x + 0.7$	0.9
$0.2 + 0.7$	0.9
0.9	

Check
$3x$	1.2
$3(0.4)$	1.2
1.2	

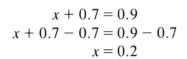

Study and Learn

A. Solve and check.

1. $x + 0.6 = 3.4$ **2.** $x + 3.7 = 5.9$ **3.** $x - 0.3 = 0.1$

4. $x - 0.9 = 1.7$ **5.** $5x = 0.25$ **6.** $0.4x = 2.4$

7. $\frac{x}{3} = 1.2$ **8.** $\frac{x}{0.2} = 0.2$ **9.** $\frac{x}{0.3} = 2.4$

B. Solve $2x + 0.3 = 1.5$. Complete.

10. $2x + 0.3 - \underline{\ ?\ } = 1.5 - \underline{\ ?\ }$

11. $2x = \underline{\ ?\ }$

12. $\frac{2x}{?} = \frac{1.2}{?}$

13. $x = \underline{\ ?\ }$

C. Solve and check.

14. $3x + 0.2 = 1.4$ **15.** $5x + 0.7 = 3.2$ **16.** $0.3x + 0.6 = 1.5$

17. $0.2x - 0.9 = 0.1$ **18.** $\frac{x}{4} + 0.3 = 0.6$ **19.** $\frac{x}{0.2} - 0.1 = 0.7$

Practice

Solve and check.

1. $x + 0.4 = 0.6$ 2. $x + 0.3 = 0.9$ 3. $x + 0.5 = 0.6$

4. $x + 0.5 = 2.3$ 5. $x + 0.7 = 3.1$ 6. $x + 0.9 = 2.3$

7. $x + 1.3 = 4.6$ 8. $x + 2.3 = 4.7$ 9. $x + 7.6 = 9.8$

10. $x - 0.4 = 0.3$ 11. $x - 0.7 = 0.2$ 12. $x - 0.1 = 0.4$

13. $x - 0.3 = 2.4$ 14. $x - 0.9 = 3.4$ 15. $x - 0.8 = 3.2$

16. $x - 1.4 = 3.9$ 17. $x - 2.3 = 7.4$ 18. $x - 14.7 = 32.9$

19. $2x = 0.8$ 20. $3x = 0.9$ 21. $4x = 0.8$

22. $3x = 2.4$ 23. $8x = 3.2$ 24. $4x = 0.16$

25. $0.4x = 0.8$ 26. $0.9x = 8.1$ 27. $0.6x = 42$

28. $\frac{x}{3} = 0.2$ 29. $\frac{x}{4} = 0.1$ 30. $\frac{x}{2} = 0.7$

31. $\frac{x}{0.3} = 4$ 32. $\frac{x}{0.4} = 0.9$ 33. $\frac{x}{2.4} = 0.7$

34. $7x + 1.6 = 3.0$ 35. $0.3x + 0.4 = 2.5$ 36. $0.6x + 1.2 = 3.6$

37. $2x - 0.7 = 0.3$ 38. $0.3x - 0.4 = 1.4$ 39. $0.5x - 1.6 = 2.9$

40. $\frac{x}{2} + 0.1 = 0.5$ 41. $\frac{x}{3} + 0.3 = 0.5$ 42. $\frac{x}{0.9} + 0.3 = 0.6$

43. $\frac{x}{3} - 0.1 = 0.4$ 44. $\frac{x}{4} - 1.3 = 0.2$ 45. $\frac{x}{0.5} - 0.8 = 0.2$

46. $x + 2.9 = 5.4$ 47. $3x - 0.1 = 2$ 48. $4x + 0.3 = 1.1$

49. $\frac{x}{1.3} = 6$ 50. $\frac{x}{5} + 0.3 = 2.8$ 51. $0.3x = 15$

52. The length of a display is 5.1 m. The length is 3 times the width. What is the width? [HINT: Use the equation $3x = 5.1$.]

Problem-Solving Applications

Career

Solve.

1. A plumber earns $11.41 an hour. How much does she earn in 35 hours? [HINT: Multiply.]

2. A plumber is paid bimonthly. His yearly salary is $21,840. How much is each paycheck?

3. A force of 45 kg is applied to a pipe wrench. A force 17 times as much is applied to the pipe. How much is the force applied to the pipe?

4. A piece of copper tubing is 6.32 m long. A pipe fitter cut 3 pieces, each 1.17 m, from the tubing. How long a piece of tubing remains?

5. The weight of 1 ft of a square steel bar is 11.95 lb per foot. Find the weight of 20 ft of the bar.

6. A plumber installed a new sink. He charged $17.50 to install the sink and $25.95 for the sink. What was the total charge?

★ 7. The outside diameter of a piece of copper tubing is 1.315 in. The thickness of the wall is 0.131 in. What is the length of the inside diameter?

★ 8. The inside diameter of a standard pipe is 26.62 mm. Its thickness is 3.38 mm. What is the length of the outside diameter?

Chapter Review

Add. *(102)*

1. 4.37
    1.68
    3.49
    + 8.63

2. $4 + 7.3 + 0.046$

3. $12 + 0.378 + 3.1$

Subtract. *(104)*

4. 8.406
    − 2.560

5. $0.6 - 0.38$

6. $8 - 1.24$

Multiply. *(110, 112)*

7. 3.14
    × 5

8. 0.26
    × 0.08

9. 0.01×0.468

Divide. *(116, 118)*

10. $4\overline{)0.8}$

11. $0.6\overline{)0.18}$

12. $6.3\overline{)7.812}$

13. Find the quotient to the nearest tenth and to the nearest hundredth. $0.06\overline{)0.428}$
(120)

Estimate to the nearest whole number. *(106, 121)*

14. 6.317
     4.936
     + 2.500

15. 8.3169
     − 2.9042

16. 3.6
     × 7

17. $3.1\overline{)6.176}$

18. Solve and check. $5x + 0.7 = 2.2$
(122)

Solve. *(114, 124)*

19. Two people spoke on the telephone for 25 minutes. The rate is $0.54 for the first 2 minutes and $0.23 for each additional minute. How much did the call cost?

20. A pipe fitter needs 4 pieces of copper tubing, each 0.95 m long. The 4 pieces are cut from a piece 5 m long. How long a piece of tubing remains?

Chapter Test

Add. *(102)*

1. 2.93
   3.46
+ 7.84

2. $6 + 4.9 + 0.024$

3. $9 + 0.234 + 11$

Subtract. *(104)*

4. 10.301
  − 1.876

5. $0.4 - 0.31$

6. $5 - 1.23$

Multiply. *(110, 112)*

7. 7.82
  × 3

8. 0.15
  × 0.09

9. 0.15×0.03

Divide. *(116, 118)*

10. $3\overline{)0.6}$

11. $0.4\overline{)0.20}$

12. $3.1\overline{)6.51}$

13. Find the quotient to the nearest tenth and to the nearest hundredth. $0.06\overline{)0.0255}$
(120)

Estimate to the nearest whole number. *(106, 121)*

14. 3.491
  5.764
+ 4.500

15. 7.4981
 − 3.6047

16. 5.8
 × 6.3

17. $3.2\overline{)6.512}$

18. Solve and check. $\frac{x}{0.2} - 0.6 = 0.4$
(122)

Solve. *(114, 124)*

19. Jeremy called Louise and they spoke for 18 minutes. The rate was 4 message units for the first 3 minutes and 2 message units for each additional minute. Each message unit cost 8.2¢. What did the call cost?

20. The weight of a square steel bar is 11.95 lb per foot. The bar is cut into 8 ft and 12 ft pieces. How much heavier is the larger piece?

Skills Check

1. Solve.

$$3x - 4 = 29$$

A	B	C	D
6	7	9	11

2. What is the value of *P* on the number line?

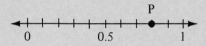

E	F	G	H
0.9	0.8	0.7	0.3

3. How many hours are there in 1,140 minutes?

A	B	C	D
23 hours	19 hours	15 hours	11 hours

4. How is the date September 26, 1979 written in numerical form?

E	F	G	H
9/26/79	26/9/79	9/79/26	10/26/79

5. Which percent has the same value as 0.3?

A	B	C	D
0.3%	3%	30%	300%

6. Which number has the same value as 53%?

E	F	G	H
0.53	53	5,300	0.0053

7. Which percent has the same value as $\frac{7}{10}$?

A	B	C	D
0.7%	7%	10%	70%

8. Which number does not have the same value as 60%?

E	F	G	H
$\frac{2}{5}$	$\frac{3}{5}$	$\frac{6}{10}$	$\frac{12}{20}$

9. Which percent of the region is shaded?

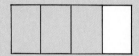

A	B	C	D
50%	60%	75%	80%

10. Which of the following decimals is equal to $\frac{5}{20}$?

E	F	G	H
0.25	0.46	0.85	0.92

11. What time is it?

A	B	C	D
12:31	6:00	6:31	12:29

12. Which time is the same as half-past one?

E	F	G	H
12:30	12:45	1:15	1:30

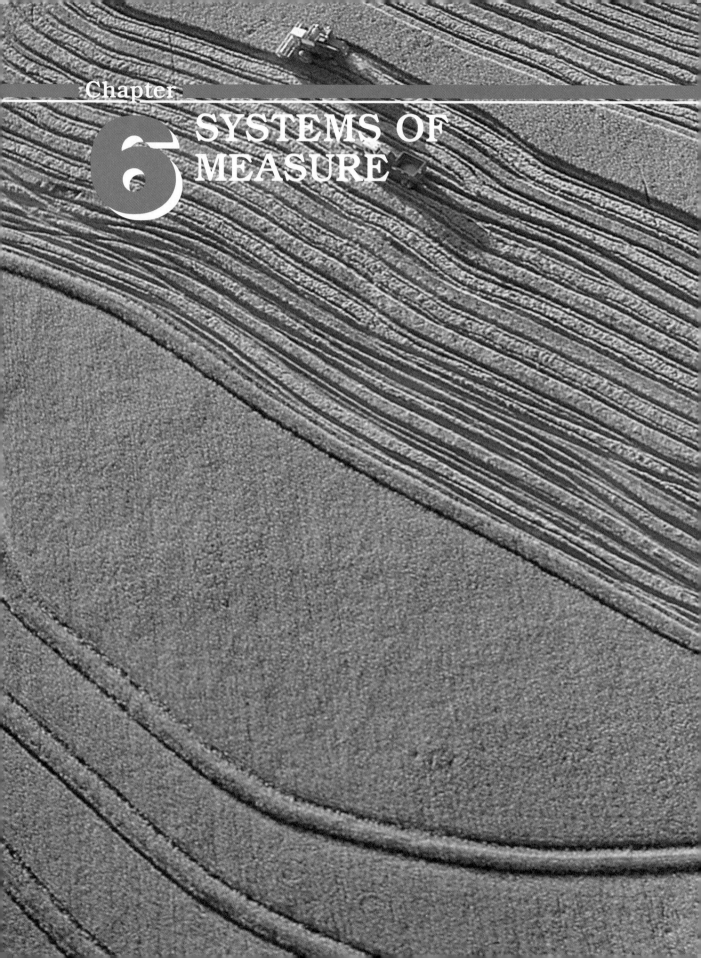

6

SYSTEMS OF MEASURE

Contour plowing and harvesting help prevent soil erosion.

LENGTH IN THE METRIC SYSTEM

The standard unit of length in the metric system is the meter. The other units are related to the meter. The prefix tells you its value. The most commonly used units are kilometer, meter, centimeter, and millimeter.

Unit of length	kilometer	hectometer	dekameter	meter	decimeter	centimeter	millimeter
Symbol	km	hm	dam	m	dm	cm	mm
Value	1,000 m	100 m	10 m	1 m	0.1 m	0.01 m	0.001 m
Prefix	kilo	hecto	deka		deci	centi	milli
Meaning	1,000	100	10		0.1	0.01	0.001

Study and Learn

A. Complete.

Example   4 km = ___?___ m   4 km = 4 × 1 km

$$= 4 \times 1,000 \text{ m} = 4,000 \text{ m}$$

1. 7 km = ___?___ m

2. 0.6 km = ___?___ m

3. 5 m = ___?___ cm

4. 0.04 m = ___?___ cm

5. 2 m = ___?___ mm

6. 30 m = ___?___ mm

> Note the following relationships.
> Since 1 km = 1,000 m, 0.001 km = 1 m.
> Since 1 cm = 0.01 m, 100 cm = 1 m.
> Since 1 mm = 0.001 m, 1,000 mm = 1 m.

B. Complete.

Example   200 cm = ___?___ m   200 cm = 200 × 1 cm

$$= 200 \times 0.01 \text{ m} = 2 \text{ m}$$

7. 500 cm = ___?___ m

8. 70 cm = ___?___ m

9. 6,000 mm = ___?___ m

10. 2,500 mm = ___?___ m

11. 9,000 m = ___?___ km

12. 800 m = ___?___ km

C. Estimate length in the metric system.

13. A baseball bat is about 1 m long. Mark off on the floor, with a piece of chalk, the distance you estimate is 1 m. Check your estimate with a meter stick.

Practice

Complete.

1. 5 km = __?__ m
2. 90 km = __?__ m
3. 0.8 km = __?__ m
4. 0.07 km = __?__ m
5. 26 km = __?__ m
6. 6 m = __?__ cm
7. 50 m = __?__ cm
8. 0.8 m = __?__ cm
9. 0.05 m = __?__ cm
10. 0.32 m = __?__ cm
11. 3 m = __?__ mm
12. 50 m = __?__ mm
13. 0.6 m = __?__ mm
14. 0.02 m = __?__ mm
15. 0.007 m = __?__ mm
16. 200 cm = __?__ m
17. 60 cm = __?__ m
18. 9 cm = __?__ m
19. 1,200 cm = __?__ m
20. 700 cm = __?__ m
21. 2,000 mm = __?__ m
22. 600 mm = __?__ m
23. 80 mm = __?__ m
24. 10,000 mm = __?__ m
25. 5,000 m = __?__ km
26. 300 m = __?__ km
27. 40 m = __?__ km
28. 6 m = __?__ km
★ 29. 50 dm = __?__ m
★ 30. 7 hm = __?__ m

Something Extra
Non-Routine Problems

Choose the box that completes the sequence.

1.   ?   a   b   c   d

2.   ?   a   b   c   d

131

PLACE VALUE AND THE METRIC SYSTEM

PLACE-VALUE CHART

Thousands	Hundreds	Tens	Ones	Tenths	Hundredths	Thousandths
1,000	100	10	1	0.1	0.01	0.001
km	hm	dam	m	dm	cm	mm
			2	3	4	5

The chart shows the relationship between place value and the metric system. The measure in the chart may be read in several ways:

$$2.345 \text{ m} = 234.5 \text{ cm} = 2,345 \text{ mm} = 0.002345 \text{ km}$$

The unit of metric measure determines the placement of the decimal point.

Study and Learn

A. Complete.

1. 0.678 m = __?__ cm   **2.** 0.678 m = __?__ km

3. 0.678 m = __?__ mm   **4.** 678 cm = __?__ m

You can change units by "moving" the decimal point.

3.4 cm = __?__ mm   Since 1 cm = 10 mm, multiply by 10.

3.4 cm = 34 mm   Move the decimal point 1 place to the right.

342 cm = __?__ m   Since 1 cm = 0.01 m, divide by 100.

342 cm = 3.42 m   Move the decimal point 2 places to the left.

B. Complete.

5. 5.21 m = __?__ cm   **6.** 3.2 m = __?__ mm   **7.** 2.4 km = __?__ m

8. 24 mm = __?__ cm   **9.** 432 mm = __?__ m   **10.** 37 m = __?__ km

Practice

Complete.

1. 4.1 cm = __?__ mm
2. 6.7 cm = __?__ mm
3. 2.34 cm = __?__ mm

4. 8.23 cm = __?__ mm
5. 7.42 m = __?__ cm
6. 8.32 m = __?__ cm

7. 6.2 m = __?__ cm
8. 9.1 m = __?__ cm
9. 4.67 km = __?__ m

10. 7.9 km = __?__ m
11. 2.7 m = __?__ mm
12. 3.64 m = __?__ mm

13. 24 mm = __?__ cm
14. 36 mm = __?__ cm
15. 6.7 cm = __?__ m

16. 16.4 cm = __?__ m
17. 416 mm = __?__ m
18. 2.13 mm = __?__ m

19. 63 m = __?__ km
20. 741.2 m = __?__ km
21. 9.3 cm = __?__ mm

22. 7.34 cm = __?__ mm
23. 12.41 m = __?__ cm
24. 4.364 m = __?__ cm

25. 8.34 km = __?__ m
26. 8.214 m = __?__ mm
★ 27. 9.4 km = __?__ mm

★ 28. 214.9 m = __?__ dam
★ 29. 4.1 dm = __?__ cm
★ 30. 3.9 hm = __?__ m

Solve Problems

31. Last year, City A reported 1.66 m of rain and City B reported 131.4 cm of rain. Which city had more rain?

32. One piece of electrical wire is 34 mm long and a second piece is 33.9 cm long. Which piece is longer?

● 33. Mary Myers can complete 23.5 units of work in 1 hour. How many units can she complete in 6 hours?

● 34. A wholesale dealer sold 7,540 toys at $1.35 each. What were his total sales?

● 35. A strip of metal is 420 cm long. How many 1.4 cm strips can be cut from it?

● 36. A strip of metal 134.4 cm long was cut into 32 equal parts. How long was each part?

ERROR OF MEASUREMENT

Materials: metric ruler

A. Measure each of the sides of
triangle *ABC* to the given
measure.

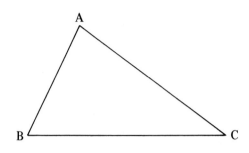

   1. To the nearest centimeter

   2. To the nearest millimeter

   3. Which measurement is more
      precise?

> ▶ The smaller the unit of measure, the more *precise* (closer to the
> actual length) is the measurement.

B. Which measurement is more precise?

   4. 3 cm or 8 mm   **5.** 14 m or 600 m   **6.** 4 mm or 7.4 mm

   7. 0.3 cm or 2 cm   **8.** 7 ft or 9 in.   **9.** 0.4 mi or 0.04 mi

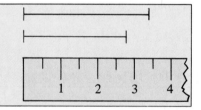

No measurement is ever exact. These
2 line segments are both measured to
be 3 cm to the nearest centimeter.
However, they are not the same
length.

> ▶ The **greatest possible error of measurement** is the greatest
> possible difference between the actual length and the measurement.
> The greatest possible error is one-half of the unit of measurement.

C. The measured length of $\overline{AB}$ is 3.4 cm.

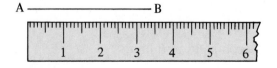

   10. What is the unit of measure
       used?

   11. What is $\frac{1}{2}$ of 0.1 cm?

   12. What is the greatest possible error of measurement?

D. Find the greatest possible error of measurement.

   13. 4.2 cm   **14.** 6 m   **15.** 8 mm   **16.** 6.04 km

Practice

Which measurement is more precise?

1. 5 cm or 5 mm

2. 6 m or 6 cm

3. 34 mm or 2 cm

4. 0.5 m or 5 m

5. 50 sec or 50 min

6. 16 mm or 0.16 m

7. 0.5 m or 0.5 mm

8. 5 cm or 50 mm

9. 7 mi or 1 ft

10. 0.3 in. or 0.03 in.

Find the greatest possible error of measurement.

11. 17 cm

12. 4 cm

13. 8 in.

14. 5 m

15. 2 cm

16. 8 m

17. 6.8 m

18. 18.5 km

19. 25.5 in.

20. 7.1 cm

21. 13.5 cm

22. 20.3 in.

23. 0.4 cm

24. 1.3 m

25. 2.8 in.

26. 0.003 cm

27. 0.3 in.

28. 0.90 m

29. 0.430 km

30. 6.05 cm

31. 8.6 m

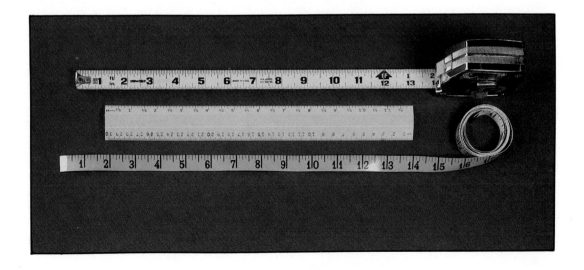

COMPUTER FLOWCHARTS

The flowchart below lists steps for playing a guessing game.
Use it to play the game with a friend.
The computer program makes the computer play the same guessing
game. Match the lines of the program with the steps in the flowchart.

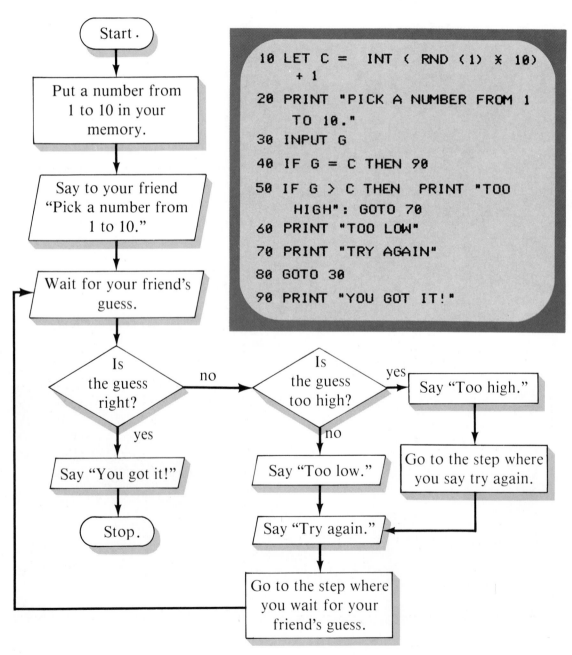

```
10 LET C =  INT ( RND (1) * 10)
   + 1
20 PRINT "PICK A NUMBER FROM 1
   TO 10."
30 INPUT G
40 IF G = C THEN 90
50 IF G > C THEN  PRINT "TOO
   HIGH": GOTO 70
60 PRINT "TOO LOW"
70 PRINT "TRY AGAIN"
80 GOTO 30
90 PRINT "YOU GOT IT!"
```

Midchapter Review

Complete. *(130, 132)*

1. 4,000 mm = __?__ m
2. 640 m = __?__ km
3. 32 cm = __?__ mm

4. 3.02 km = __?__ m
5. 7.3 cm = __?__ m
6. 6.59 m = __?__ mm

Which measurement is more precise? *(134)*

7. 7 cm or 7 m
8. 5 km or 8 m
9. 0.16 cm or 0.7 cm

Find the greatest possible error of measurement. *(134)*

10. 3.2 km
11. 4 mm
12. 6.05 cm
13. 0.2 m

Something Extra
Aid to Memory

Here is an interesting way to multiply.

Example   41 × 28

Take half of each number. Ignore the remainder. Continue until the result is 1.	Double each number.	Cross out the rows with an even number in the left column. Add the remaining numbers in the right column.

41	×	28	41	28
20		56	~~20~~	~~56~~
10		112	~~10~~	~~112~~
5		224	5	224
2		448	~~2~~	~~448~~
1		896	1	896
				1,148

Find the products. Use the method above.

1. 73 × 54
2. 88 × 27
3. 135 × 93
4. 602 × 321

Problem-Solving Skills

Choosing the Best Estimate

Some problems may be solved by rounding the numbers and then finding an estimate of the answer.

Study and Learn

A. Ms. Taxel is a traveling sales representative. One day, she drove 358 km using 76 L of gas. How many kilometers per liter (km/L) was this? Choose the best estimate.

300 km/L   5 km/L   50 km/L

Complete.

    1. 358 km is __?__ km to the nearest hundred.

    2. 76 L is __?__ L to the nearest ten.

    3. Think: 400 km ÷ 80 L = 5 km/L, so __?__ is the best estimate.

B. Ms. Taxel drove 354.9 km the first day, 274.6 km the second day, and 136.0 km the third day. How many kilometers did she travel in all? Choose the best estimate.

600 km   700 km   800 km

    4. 354.9 km is __?__ km to the nearest hundred.

    5. 274.6 km is __?__ km to the nearest hundred.

    6. 136.0 km is __?__ km to the nearest hundred.

    7. 400 + 300 + 100 = __?__ , so __?__ is the best estimate.

C. A liter of gasoline costs $0.29. What is the cost of 76.2 L? Choose the best estimate.

$14 $21 $24

    8. $0.29 is __?__ to the nearest ten cents.

    9. 76.2 L is __?__ L to the nearest ten liters.

    10. $0.30 × 80 L = __?__ , so __?__ is the best estimate.

Practice

Choose the best estimate.

1. On a car trip across the country, the driver kept a record of fuel used. One day she drove 223 mi and used 11 gal of gas. How many miles per gallon (mpg) was this?

   30 mpg   *25mpg*   *20 mpg*

2. The family travel record showed that one day they drove 341.8 km during the hours from 7:00 am to noon. What was the average number of kilometers per hour (km/h) driven during that period of time?

   80 km/h   60 km/h   50 km/h

3. Mr. Avery drove 614.8 km one day, 553.9 km the second day, and 424.7 km the third day. How many kilometers did he travel in all?

   1,500 km   1,600 km   1,700 km

4. A gallon of gas costs 101.8 cents. A company expense record showed 284.3 gal was used during a trip. What was the cost of the gasoline during the trip?

   $140 $210 $300

5. Sandy averaged 77.2 km/h for 7 hours. How far did she travel in that time?

   7 km   490 km   560 km

6. During the 5-day trip the total cost of food and motels was $487.50. What was the average cost for food and motels for each day?

   $9 $90 $100

7. A family drove 280.8 km using 66.5 L of gasoline. How many kilometers per liter was this?

   4 km/L   5 km/L   6 km/L

8. The distance between two cities is 3,142.8 km. A family has driven 1,874.9 km from one city toward the other. How much farther does the family need to drive to reach the other city?

   1,000 km   2,000 km   5,000 km

9. On a 9-day trip, the total cost of gasoline was $216.70. What was the average daily cost?

   $20 $200 $2,000

10. A liter of gasoline cost $0.27. A family used 1,256 L during a trip. What was the cost of gasoline for this trip?

    $2,400 $26.00 $300

SIGNIFICANT DIGITS

The length of the nail is 4.2 cm.
Unit of measure: 0.1 cm

Units in measurement: 42 because $0.1\overline{)4.2}$
Number of significant digits: 2

42 ⟵ 2 digits

▶ Significant digits are those digits in a measurement that give the number of times the unit is contained in the measurement.

Study and Learn

A. Complete.

	Measurement	Unit of Measure	Units in Measurement	Number of Significant Digits
1.	34 m	1 m		
2.	12 cm	1 cm		
3.	3 mm	1 mm		
4.	1.4 in.	0.1 in.		

B. When multiplying measures, the product should have the same number of significant digits as the measurements.

Example   What is the area of this metal strip?

Each measurement, 8.6 and 4.2, has 2 significant digits. So, the area should have 2 significant digits.

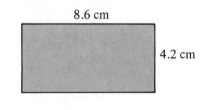

8.6 cm

4.2 cm

Area = 8.6 cm × 4.2 cm
     = 36.12 cm²

36.12 cm² rounded to 2 significant digits is 36 cm².

Find the area. Round to the correct number of significant digits.

5. Length: 4.7 m
    Width: 3.6 m

6. Length: 22.3 cm
    Width: 13.8 cm

Practice

Complete.

Measurement	Unit of measure	Units in measurement	Number of significant digits
1. 12 km	1 km		
2. 0.07 m	0.01 m		
3. 1,032 mi	1 mi		
4. 24 cm	1 cm		
5. 600 ft	100 ft		

Find the areas. Round to the correct number of significant digits.

6.

1.5 m

5.3 m

7. $l = 9.4$ mm
    $w = 7.6$ mm

8. $l = 8.6$ cm
    $w = 3.2$ cm

9. $l = 2.3$ ft
    $w = 1.8$ ft

★**10.** $l = 12.2$ km
    $w = 3.7$ km

Computer

This program computes the area of a rectangle.

```
10 INPUT "ENTER LENGTH ";L
20 INPUT "ENTER WIDTH ";W
30 INPUT "ENTER UNIT ";U$
40 PRINT "AREA = "L X W" SQUARE
   "U$
```

After the length and width are entered, the measurement of the unit is entered and stored in a word storage place, (U-string).

H A N D S   O N

You may use this program to check your answers to exercises **6–10** above. Type in the lines; then type **RUN**. Remember to press [RETURN] after each line. Round the output.

METRIC MEASURES OF MASS

In the metric system, the relationship between units of mass is based on 10. The most commonly used units are kilogram, gram, and milligram.

kilogram kg	hectogram hg	dekagram dag	gram g	decigram dg	centigram cg	milligram mg
1,000 g	100 g	10 g	1 g	0.1 g	0.01 g	0.001 g

$$1 \text{ kg} = 1,000 \text{ g} \qquad 1 \text{ g} = 1,000 \text{ mg}$$
$$1 \text{ g} = 0.001 \text{ kg} \qquad 1 \text{ mg} = 0.001 \text{ g}$$

Study and Learn

A. Complete.

Example
$$4.7 \text{ kg} = \underline{} \text{ g}$$
$$4.7 \text{ kg} = 4.7 \times \boxed{1 \text{ kg}}$$
$$= 4.7 \times \boxed{1,000 \text{ g}}$$
$$= 4,700 \text{ g}$$

Move the decimal point 3 places to the right.

1. $3.5 \text{ kg} = \underline{} \text{ g}$

2. $42 \text{ g} = \underline{} \text{ mg}$

3. $5.6 \text{ g} = \underline{} \text{ mg}$

4. $25 \text{ kg} = \underline{} \text{ g}$

B. Complete.

Example
$$35 \text{ mg} = \underline{} \text{ g}$$
$$35 \text{ mg} = 35 \times \boxed{1 \text{ mg}}$$
$$= 35 \times \boxed{0.001 \text{ g}}$$
$$= 0.035 \text{ g}$$

Move the decimal point 3 places to the left.

5. $125 \text{ mg} = \underline{} \text{ g}$

6. $50 \text{ g} = \underline{} \text{ kg}$

7. $225 \text{ g} = \underline{} \text{ kg}$

8. $5 \text{ mg} = \underline{} \text{ g}$

C. Another unit of mass is the ton (t). $1 \text{ t} = 1,000 \text{ kg}$
Complete.

9. $7 \text{ t} = \underline{} \text{ kg}$

10. $4,500 \text{ kg} = \underline{} \text{ t}$

142

D. Estimate mass in the metric system.

11. A paper clip has a mass of about 1 g. A book has a mass of about 1 kg. Estimate the masses of various objects in your classroom. Check your estimates by using a scale with metric weights.

Practice

Complete.

1. 3 kg = __?__ g **2.** 9 g = __?__ mg **3.** 17 g = __?__ mg

4. 4.3 kg = __?__ g **5.** 1.1 kg = __?__ g **6.** 9.02 g = __?__ mg

7. 650 g = __?__ mg **8.** 540 g = __?__ mg **9.** 91 kg = __?__ g

10. 2.6 kg = __?__ g **11.** 7.8 kg = __?__ g **12.** 0.1 g = __?__ mg

13. 100 mg = __?__ g **14.** 3,500 g = __?__ kg **15.** 50 g = __?__ kg

16. 0.1 g = __?__ kg **17.** 300 mg = __?__ g **18.** 1 mg = __?__ g

19. 25 mg = __?__ g **20.** 131 mg = __?__ g **21.** 7 g = __?__ kg

22. 500 g = __?__ kg **23.** 0.11 g = __?__ kg **24.** 200 mg = __?__ g

25. 3 t = __?__ kg **26.** 250 kg = __?__ t **27.** 30,000 kg = __?__ t

28. 7,000 kg = __?__ t **29.** 3.1 t = __?__ kg **30.** 4.56 t = __?__ kg

31. 100 kg = __?__ t **32.** 1,000 kg = __?__ t **33.** 0.5 t = __?__ kg

★ **34.** 325 mg = __?__ kg ★ **35.** 6 t = __?__ g ★ **36.** 0.03 t = __?__ mg

Solve Problems

37. A baby was 3.4 kg at birth. It gained 6.7 kg in 1 year. How heavy is the baby now?

38. A bag of potatoes is 5.1 kg. A bag of onions is 2.3 kg. How much heavier is the bag of potatoes?

METRIC MEASURES OF CAPACITY

The most commonly used units of capacity in the metric
system are the liter (L) and the milliliter (mL).

$$1 \text{ L} = 1,000 \text{ mL} \qquad 1 \text{ mL} = 0.001 \text{ L}$$

Study and Learn

A. Here is a method for relating units.

Examples

$3.7 \text{ L} = \underline{?} \text{ mL}$
$3.7 \text{ L} = 3.7 \times \boxed{1 \text{ L}}$
$\phantom{3.7 \text{ L}} = 3.7 \times \boxed{1,000 \text{ mL}}$
$\phantom{3.7 \text{ L}} = 3,700 \text{ mL}$

$73 \text{ mL} = \underline{?} \text{ L}$
$73 \text{ mL} = 73 \times \boxed{1 \text{ mL}}$
$\phantom{73 \text{ mL}} = 73 \times \boxed{0.001 \text{ L}}$
$\phantom{73 \text{ mL}} = 0.073 \text{ L}$

Complete.

1. $3 \text{ L} = \underline{?} \text{ mL}$

2. $8 \text{ L} = \underline{?} \text{ mL}$

3. $2 \text{ mL} = \underline{?} \text{ L}$

4. $15 \text{ mL} = \underline{?} \text{ L}$

5. $5.6 \text{ L} = \underline{?} \text{ mL}$

6. $12.8 \text{ L} = \underline{?} \text{ mL}$

7. $17 \text{ mL} = \underline{?} \text{ L}$

8. $384 \text{ mL} = \underline{?} \text{ L}$

B. Estimate capacity in the metric system.

9. A standard can of motor oil for an automobile
   contains about 1 L. Fill a large container
   with what you think is 1 L of water. Check
   your estimate by pouring the water into a
   graduated cylinder of 1 L.

Practice

Complete.

1. $4 \text{ L} = \underline{?} \text{ mL}$

2. $12 \text{ L} = \underline{?} \text{ mL}$

3. $4 \text{ mL} = \underline{?} \text{ L}$

4. $20 \text{ mL} = \underline{?} \text{ L}$

5. $8.7 \text{ L} = \underline{?} \text{ mL}$

6. $24.9 \text{ L} = \underline{?} \text{ mL}$

7. $27 \text{ mL} = \underline{?} \text{ L}$

8. $412 \text{ mL} = \underline{?} \text{ L}$

Skills Review

List all the factors of each number. *(12)*

1. 15 **2.** 30 **3.** 36 **4.** 21

For each number tell whether it is prime or composite. *(12)*

5. 16 **6.** 2 **7.** 13 **8.** 46

Add. *(24)*

9. $\begin{array}{r}16{,}254 \\ +\ 9{,}751\end{array}$	**10.** $\begin{array}{r}\$9{,}826 \\ +\ 7{,}249\end{array}$	**11.** $\begin{array}{r}383{,}109 \\ 265{,}355 \\ +\ 841{,}712\end{array}$	**12.** $\begin{array}{r}62{,}527 \\ 9{,}438 \\ +\ 116{,}019\end{array}$

Subtract. *(26)*

13. $\begin{array}{r}71{,}324 \\ -\ 9{,}836\end{array}$	**14.** $\begin{array}{r}90{,}000 \\ -\ 13{,}143\end{array}$	**15.** $\begin{array}{r}742{,}561 \\ -\ 400{,}698\end{array}$	**16.** $\begin{array}{r}165{,}329 \\ -\ 92{,}561\end{array}$

Multiply. *(34, 36)*

17. $\begin{array}{r}5{,}024 \\ \times\ 7\end{array}$	**18.** $\begin{array}{r}639 \\ \times\ 15\end{array}$	**19.** $\begin{array}{r}7{,}042 \\ \times\ 406\end{array}$	**20.** $\begin{array}{r}1{,}427 \\ \times\ 219\end{array}$

Divide. *(38, 40)*

21. $3\overline{)651}$ **22.** $27\overline{)297}$ **23.** $32\overline{)2{,}421}$ **24.** $135\overline{)3{,}645}$

Solve and check. *(60, 66, 70)*

25. $x + 7 = 15$ **26.** $x - 11 = 39$ **27.** $x + 64 = 108$

28. $4x = 40$ **29.** $12x = 72$ **30.** $\frac{x}{6} = 18$

31. $3x + 1 = 16$ **32.** $9x - 2 = 25$ **33.** $\frac{x}{7} + 4 = 13$

Round. *(84)*

34. 4.029 to the nearest tenth. **35.** 5.601 to the nearest whole number.

36. 23.056 to the nearest hundredth. **37.** 0.0015 to the nearest thousandth.

Multiply or divide. *(90)*

38. $10^6 \cdot 10^3$ **39.** $10^5 \cdot 10^5$ **40.** $10^3 \cdot 10^4$

41. $\frac{10^8}{10^2}$ **42.** $\frac{10^{18}}{10^9}$ **43.** $10^{10} \div 10^3$

CUSTOMARY UNITS OF LENGTH

A clock is 5 ft 3 in. high. Another clock is 64 in. high. Which is higher?

Change the height of the first clock to inches.

$$5 \text{ ft} = 5 \times (12 \text{ in.})$$
$$= 60 \text{ in.}$$
$$5 \text{ ft } 3 \text{ in.} = 60 \text{ in.} + 3 \text{ in.}$$
$$= 63 \text{ in.}$$

So, the second clock is higher.

Study and Learn

A. Complete. Use these relations between units of length.

1 foot (ft) = 12 inches (in.)   1 yard (yd) = 3 ft   1 yd = 36 in.
1 mile (mi) = 5,280 ft   1 mi = 1,760 yd

1. 2 ft = __?__ in.

2. 6 ft = __?__ in.

3. 3 yd = __?__ in.

4. 3 yd = __?__ ft

5. 2 mi = __?__ ft

6. 3 mi = __?__ yd

7. 48 in. = __?__ ft

8. 72 in. = __?__ ft

9. 72 in. = __?__ yd

10. 6 ft = __?__ yd

11. 12 ft = __?__ yd

12. 108 in. = __?__ yd

13. $\frac{1}{2}$ ft = __?__ in.

14. $\frac{1}{3}$ yd = __?__ in.

15. $2\frac{1}{2}$ mi = __?__ ft

16. 3 ft 4 in. = __?__ in.

17. 3 yd 1 ft = __?__ ft

18. 64 in. = __?__ ft __?__ in.

19. 41 in. = __?__ yd __?__ in.

Practice

Complete.

1. 3 ft = __?__ in.

2. 7 ft = __?__ in.

3. 9 ft = __?__ in.

4. 10 ft = __?__ in.

5. 2 yd = __?__ in.

6. 4 yd = __?__ in.

7. 2 yd = __?__ ft

8. 9 yd = __?__ ft

9. 3 mi = __?__ ft

10. 2 mi = __?__ yd

11. $\frac{1}{4}$ ft = __?__ in.

12. $\frac{1}{3}$ ft = __?__ in.

13. $\frac{1}{3}$ yd = __?__ ft

14. $\frac{1}{4}$ yd = __?__ in.

15. $\frac{1}{2}$ mi = __?__ ft

16. $\frac{1}{4}$ mi = __?__ ft

17. $\frac{1}{2}$ mi = __?__ yd

18. $\frac{1}{4}$ mi = __?__ yd

19. 24 in. = __?__ ft

20. 36 in. = __?__ ft

21. 144 in. = __?__ yd

22. 360 in. = __?__ yd

23. 9 ft = __?__ yd

24. 15 ft = __?__ yd

25. 6 ft 3 in. = __?__ in.

26. 7 ft 8 in. = __?__ in.

27. 7 yd 2 ft = __?__ ft

28. 6 yd 1 ft = __?__ ft

29. 26 in. = __?__ ft __?__ in.

30. 53 in. = __?__ ft __?__ in.

Solve Problems

31. A heavy-duty canvas costs $8/yd. Tom needs 72 in. to make a new carrier for his carpentry tools. How much will the canvas cost him?

32. The pilot of an airplane notes that the plane is flying at 33,000 ft. Approximately how many miles high is this?

33. Cesar bought a 6-foot spool of ribbon. He used 34 inches of ribbon for a package. How many inches of ribbon are left on the spool?

34. Mrs. Cohen needs 24 yards of wood molding for her house. The molding costs $1.25 per foot. How much will she have to pay for the molding?

CUSTOMARY UNITS OF WEIGHT AND CAPACITY

Pedro bought 2 lb 8 oz of meat. The meat cost $1.80/lb. How much did he pay for the meat?

$$2 \text{ lb at } \$1.80/\text{lb} \longrightarrow \$1.80 \times 2 = \$3.60$$
$$8 \text{ oz} \quad \text{Think} \quad 16 \text{ oz} = 1 \text{ lb,}$$
$$\text{so } 8 \text{ oz} = \tfrac{1}{2} \text{ lb} \longrightarrow \$1.80 \times \tfrac{1}{2} = \$0.90$$

$3.60 + $0.90 = $4.50
Pedro paid $4.50 for the meat.

Study and Learn

A. Complete. Use these relationships between units of weight.

   1 pound (lb) = 16 ounces (oz)   1 ton = 2,000 lb

1. 3 lb = __?__ oz   **2.** 2 tons = __?__ lb   **3.** $\frac{1}{2}$ ton = __?__ lb

4. $\frac{1}{4}$ lb = __?__ oz   **5.** 32 oz = __?__ lb   **6.** 10,000 lb = __?__ tons

B. Complete. Use the relationships between units of capacity.

   2 cups = 1 pint (pt)   1 quart (qt) = 2 pt   1 gallon (gal) = 4 qt
   1 pt = 16 fluid ounces (fl oz)   1 fl oz = 2 tablespoons (tbs)

7. 3 pt = __?__ cups   **8.** 2 pt = __?__ fl oz   **9.** 3 qt = __?__ pt

10. 2 gal = __?__ qt   **11.** 6 pt = __?__ qt   **12.** 3 fl oz = __?__ tbs

Practice

Complete.

1. 4 lb = __?__ oz   **2.** 6 tons = __?__ lb   **3.** 64 oz = __?__ lb

4. $\frac{1}{4}$ ton = __?__ lb   **5.** 70,000 lb = __?__ tons   **6.** 4 oz = __?__ lb

7. 4 pt = __?__ fl oz   **8.** 6 gal = __?__ qt   **9.** 4 cups = __?__ pt

10. 8 qt = __?__ gal   **11.** 9 pt = __?__ qt   **12.** 32 fl oz = __?__ pt

13. 5 fl oz = __?__ tbs   **14.** 12 qt = __?__ pt   **15.** 10 tbs = __?__ fl oz

CELSIUS TEMPERATURE

Temperature may be measured using the Celsius scale. Two fixed points on the Celsius scale are the boiling point of water and the freezing point of water. Normal body temperature is 37°C and normal room temperature is about 20°C.

Study and Learn

A. Look at the thermometer.

  1. What is the temperature at which water boils?

  2. What is the temperature at which water freezes?

  3. What is the temperature difference between the boiling and the freezing points of water?

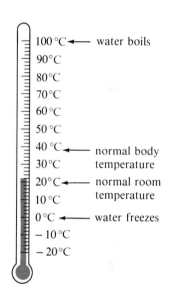

B. Comfortable, cold, or hot?

  Example   21°C is a normal room temperature. So, 21°C is comfortable.

  4. 4°C   **5.** 65°C   **6.** 19°C

C. Estimate the temperature for the following.

  7. Swimming in a lake   **8.** A refrigerated fruit

Practice

Comfortable, cold, or hot?

  1. 5°C   **2.** 94°C   **3.** 18°C   **4.** 39°C

  5. 99°C   **6.** ⁻8°C   **7.** 74°C   **8.** 61°C

Estimate the temperature for the following.

  9. Ice skating on a lake   **10.** Mowing a lawn

★ Cold, warm, or hot?

  11. Gold melts at 1,063°C.   **12.** Mercury freezes at ⁻38.87°C.

TIME ZONES

For each time zone you pass going west, the time changes to 1 hour earlier. For each time zone you pass going east, the time changes to 1 hour later.

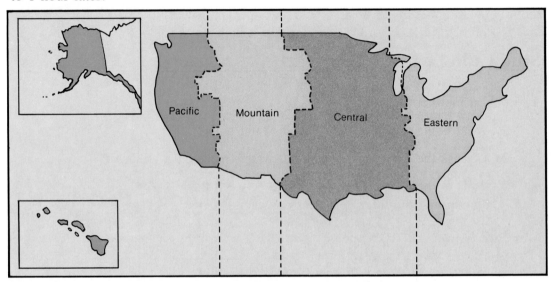

Study and Learn

A. Bill is in the Pacific time zone at 11:00 am. He calls Jennifer in the Eastern time zone, 3 time zones to the east. What time is it for Jennifer?

   1. Since the Eastern time zone is __?__ zones to the east, the time is 3 hours __?__ .

   2. 3 hours later than 11:00 am is __?__ .

   3. So, the time for Jennifer is __?__ .

B. It is 4:30 pm in the Eastern time zone. What time is it in each of these zones?

   4. 5 zones to the west   **5.** 8 zones to the east

> As you move across the International Date Line, in the Pacific Ocean, the date changes. If you are traveling eastward, the time stays the same but the date changes to one day earlier. If you are traveling westward, the time stays the same but the date changes to one day later.

C. On June 9th at 12:00 noon, Olga crossed the International Date Line going eastward.

    6. Did the time on her watch change?

    7. How did the date change?

    8. What time and date did it become when Olga crossed the International Date Line?

Practice

It is 10:30 am in the Pacific time zone. What time is it in each of these zones?

1. 6 zones to the east

2. 10 zones to the west

3. 4 zones to the west

4. 8 zones to the east

5. Mrs. Wong is traveling westward in an airplane. As her plane approaches the International Date Line, her watch says Wednesday, September 15, 3:00 pm. How should she adjust her watch on the other side of the line?

★ **6.** Mr. Goodman started traveling eastward in an airplane on April 27 at 9:00 am. His plane moved across 7 time zones and crossed the International Date Line. How did the time and date change for Mr. Goodman?

Something Extra
Non-Routine Problem

Daylight Savings Time was instituted to increase the amount of daylight available in the evenings. During World War II, the United States used Daylight Savings Time all year long so as to save fuel. Today it is only used from spring until fall and not used in all the states. In the spring, the clocks are turned ahead 1 hour. In the fall, they are turned back 1 hour (back to Standard Time).

If it is 6:00 am Daylight Savings Time in the Eastern time zone, what is the Standard Time in the Mountain time zone?

Problem-Solving Applications
Using a Table

The table shows the estimated pollution costs for one year.

ESTIMATED POLLUTION CONTROL COSTS

	Billions of dollars
Air pollution Public Private	0.8 23.9
Water pollution Federal State-local Private	0.2 12.6 25.3

Answer the questions. Use the table for Exercises 1–8.

1. What is the difference between estimated state-local and private costs for water pollution?

2. What is the estimated total amount to be spent on controlling air pollution?

3. What is the estimated total amount to be spent on controlling water pollution?

4. What is the total amount to be spent by the private sector in controlling air and water pollution?

5. How much more will the private sector spend on water pollution control than it will spend on air pollution control?

6. The estimated total cost for air pollution control is how many times the federal cost for water pollution control?

★ 7. Use 225,000,000 as an estimated population. How much per person will be spent for air pollution control by the public to the nearest dollar?

★ 8. Use 225,000,000 as an estimated population. How much per person is spent by the private sector for air pollution control to the nearest dollar?

● 9. A taxicab driver earned $16,345 one year and $11,228 the next year. Estimate the total amount earned in the two years.

● 10. A bus was driven 18,516 km in a year. A second bus was driven 7,824 km. Estimate the difference in the number of kilometers driven.

● 11. A case of milk contains 36 cartons. Estimate the number of cartons in 115 cases.

● 12. Pencils are packed 110 to a box. Estimate the number of boxes needed for 2,185 pencils.

Chapter Review

Complete. *(130, 132, 146)*

1. 400 mm = ___?___ m
2. 3.41 m = ___?___ cm
3. 270 m = ___?___ km

4. 4 ft 2 in. = ___?___ in.
5. 4 yd = ___?___ ft
6. $\frac{1}{2}$ mi = ___?___ ft

7. Which measurement is more precise? *(134)*   3 m or 3 mm

8. Find the greatest possible error of measurement. *(134)*   4.7 cm

Comfortable, cold, or hot? *(149)*

9. 37°C
10. 61°C

Complete. *(142, 144, 148)*

11. 1 kg = ___?___ g
12. 300 mg = ___?___ g
13. 1,600 g = ___?___ mg

14. 8 L = ___?___ mL
15. 0.3 L = ___?___ mL
16. 370 mL = ___?___ L

17. 6 lb = ___?___ oz
18. 3 tons = ___?___ lb
19. $\frac{1}{2}$ lb = ___?___ oz

20. 5 gal = ___?___ qt
21. 11 pt = ___?___ qt
22. 2 pt = ___?___ fl oz

Choose the best estimate. *(138)*

23. During a 10-day trip, the total cost of food and lodging was $681.45. What was the average cost for food and lodging per day?

$7 $70 $700

Solve. *(150, 152)*

24. It is 7:00 pm in the Central time zone. What time is it in the zone 6 zones to the east?

25. How much less oil per day was imported from Saudi Arabia in 1981 than in 1976?

Oil Imported from Saudi Arabia (thousand barrels per day)	
1971	128
1976	1,230
1981	1,125

Chapter Test

Complete. *(130, 132, 146)*

1. 600 cm = __?__ m   **2.** 93.5 m = __?__ mm   **3.** 550 m = __?__ km

4. 3 ft 8 in. = __?__ in.   **5.** 6 yd = __?__ ft   **6.** $\frac{1}{4}$ mi = __?__ ft

7. Which measurement is more precise? *(134)*   7 km or 9 m

8. Find the greatest possible error of measurement. *(134)*   2 cm

Comfortable, cold, or hot? *(149)*

9. 20°C   **10.** 2°C

Complete. *(142, 144, 148)*

11. 3 kg = __?__ g   **12.** 500 mg = __?__ g   **13.** 3,200 g = __?__ mg

14. 5 L = __?__ mL   **15.** 750 mL = __?__ L   **16.** 0.8 L = __?__ mL

17. 5 lb = __?__ oz   **18.** 4 tons = __?__ lb   **19.** $\frac{1}{4}$ lb = __?__ oz

20. 6 gal = __?__ qt   **21.** 7 pt = __?__ qt   **22.** 3 pt = __?__ fl oz

Choose the best estimate. *(138)*

23. A taxi driver drove 329 km one week and 573 km the second
week. How many more kilometers did he travel the second week?

        300 km   450 km   900 km

Solve. *(150, 152)*

24. It is 8:30 am in the Pacific time zone. What time is it in the zone
4 zones to the west?

25. How much more oil per day
was imported from Nigeria
in 1976 than in 1971?

Oil Imported from Nigeria (thousand barrels per day)	
1971	102
1976	1,025
1981	622

Skills Check

1. Stella bought a blouse for $15.95. How much change did she receive from a twenty-dollar bill?

   A $5.05   B $4.50

   C $4.05   D $3.05

2. Ms. Glinka earns $175 a week. How much does she earn in a year?

   E $9,100   F $9,000

   G $8,900   H $1,225

3. It was ⁻3°C at 9:00 am and it was 7° warmer at noon. What was the temperature at noon?

   A 10°C   B 4°C

   C ⁻4°C   D ⁻10°C

4. Linda practices the guitar 45 minutes a day Monday through Friday. How long does she practice during the 5 days?

   E 3 hours, 15 minutes

   F 3 hours, 45 minutes

   G 4 hours, 15 minutes

   H 5 hours

5. What is the perimeter of a square with one side 22 mm?

   A 484 mm   B 88 mm

   C 55 mm   D 44 mm

6. Which of the 3 foods is the best source of vitamin C?

Food	Vitamin C (mg)
apple	8
grapefruit	40
milk	2

   E apple   F milk

   G grapefruit   H both apple and milk

7. A chef used 1 lb 3 oz of flour for bread, 12 oz of flour for muffins and 1 lb of flour for rolls. How much flour was used altogether?

   A 6 lb 2 oz   B 4 lb 6 oz

   C 2 lb 15 oz   D 1 lb 7 oz

8. A salesperson earned a total of $31,185 in one year. Estimate the average weekly income.

   E $700   F $600

   G $500   H $400

Chapter

7 OPERATIONS WITH FRACTIONS

An industrial robot can be used in the welding of a car frame.

LEAST COMMON MULTIPLE

A *multiple* of 4 is a product of 4 and a counting number (not zero).
Since $4 \times 2 = 8$, 8 is a multiple of 4.

The steps below show you how to find the *least common multiple* of 4 and 6.

Step 1   List the multiples of 4: 4, 8, ⑫, 16, 20, ㉔, 28, 32, ㊱, 40, . . .

List the multiples of 6: 6, ⑫, 18, ㉔, 30, ㊱, 42, 48, 54, 60, . . .

Step 2   List the common multiples of 4 and 6: 12, 24, 36, . . .

Step 3   Least common multiple (LCM): 12

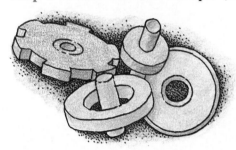

Study and Learn

A. A multiple of 3 is a product of 3 and a counting number (not zero). Since $3 \times 8 = 24$, 24 is a multiple of 3.

   1. Give the first 5 multiples of 3.

   2. Give the first 5 multiples of 8.

   3. Give the first 5 multiples of 12.

B. Find the least common multiple (LCM) of 8 and 12.

   4. List the multiples of 8.

   5. List the multiples of 12.

   6. What is the LCM of 8 and 12?

√ **C.** Find the LCM.

   7. 3, 8   **8.** 4, 10   **9.** 9, 15   **10.** 4, 6

  11. 6, 9   **12.** 3, 5   **13.** 4, 6, 9   **14.** 4, 8, 12

Practice

Give the first 5 multiples of each number.

1. 2   **2.** 4   **3.** 5   **4.** 7   **5.** 20

Find the LCM.

6. 3, 4   **7.** 2, 5   **8.** 7, 9

9. 8, 11   **10.** 3, 6   **11.** 4, 8

12. 2, 4   **13.** 10, 20   **14.** 6, 21

15. 10, 15   **16.** 9, 12   **17.** 14, 21

18. 2, 3, 4   **19.** 5, 6, 9   **20.** 3, 4, 6

★ **21.** 35, 42   ★ **22.** 8, 18, 20   ★ **23.** 9, 15, 12

Skills Review

Add. *(102)*

1.　6.19
　　　4.83
　　　+ 6.84

2.　3.01
　　　0.86
　　　+ 5.2

3.　6.7
　　　10.92
　　　+ 2.359

4. 3 + 6.4 + 0.009   **5.** 0.3 + 0.14 + 0.0019

Subtract. *(104)*

6. 7 − 4.6   **7.** 0.14 − 0.0016   **8.** 3.41 − 2

9. 15.86 − 9.832   **10.** 49 − 12.66   **11.** 98.3 − 21

Multiply. *(112)*

12. 3.41 × 2.6   **13.** 0.8 × 0.1   **14.** 0.08 × 0.04

15. 0.6 × 0.07   **16.** 5.3 × 0.72   **17.** 17.34 × 1.5

Divide. *(116, 118)*

18. $6\overline{)0.12}$ **19.** $0.7\overline{)3.43}$ **20.** $0.4\overline{)32}$

21. $0.9\overline{)345.6}$ **22.** $2.1\overline{)8.4}$ **23.** $0.53\overline{)1.325}$

EQUIVALENT FRACTIONS

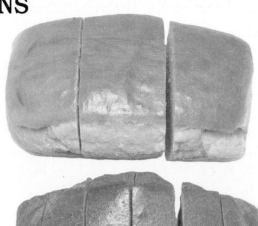

One loaf of bread is cut into 3 pieces. Another loaf of the same size is cut into 6 pieces. There is as much bread in 2 pieces of the first loaf as in 4 pieces of the second loaf.

$$\frac{2}{3} = \frac{4}{6}$$

$\frac{2}{3}$ and $\frac{4}{6}$ are *equivalent fractions*.

Note: $\frac{2}{3} = \frac{2 \cdot 2}{3 \cdot 2} = \frac{4}{6}$.

Study and Learn

A. Equivalent fractions are formed by multiplying the numerator and denominator by the same number.

$$\frac{1}{2} \qquad \frac{2}{4} \qquad \frac{3}{6} \qquad \frac{4}{8} \qquad \frac{5}{10} \qquad \frac{6}{12}$$

$$\qquad \frac{1 \cdot 2}{2 \cdot 2} \quad \frac{1 \cdot 3}{2 \cdot 3} \quad \frac{1 \cdot 4}{2 \cdot 4} \quad \frac{1 \cdot 5}{2 \cdot 5} \quad \frac{1 \cdot 6}{2 \cdot 6}$$

1. Find 3 equivalent fractions for $\frac{1}{3}$.

2. Find 3 equivalent fractions for $\frac{2}{5}$.

B. Sometimes you need to find an equivalent fraction with a given denominator.

Example   Find a fraction equivalent to $\frac{3}{4}$ with the denominator 20.

$$\frac{3}{4} = \frac{x}{20} \qquad \text{THINK} \qquad 4 \cdot ? = 20$$
$$\frac{3}{4} = \frac{3 \cdot 5}{4 \cdot 5} \qquad\qquad\quad 4 \cdot 5 = 20$$
$$\frac{3}{4} = \frac{15}{20}$$

Find equivalent fractions with the given denominators.

3. $\frac{1}{3} = \frac{x}{6}$ **4.** $\frac{3}{5} = \frac{x}{20}$ **5.** $\frac{1}{2} = \frac{x}{16}$ **6.** $\frac{5}{6} = \frac{x}{18}$

160

Practice

Find 3 equivalent fractions for each.

1. $\frac{1}{9}$
2. $\frac{4}{9}$
3. $\frac{5}{6}$
4. $\frac{3}{10}$
5. $\frac{5}{12}$

6. $\frac{1}{4}$
7. $\frac{5}{2}$
8. $\frac{3}{5}$
9. $\frac{1}{10}$
10. $\frac{3}{1}$

Find equivalent fractions with the given denominators.

11. $\frac{1}{2} = \frac{x}{16}$
12. $\frac{1}{2} = \frac{x}{8}$
13. $\frac{2}{3} = \frac{x}{12}$
14. $\frac{2}{3} = \frac{x}{15}$

15. $\frac{3}{4} = \frac{x}{8}$
16. $\frac{3}{2} = \frac{x}{24}$
17. $\frac{5}{8} = \frac{x}{40}$
18. $\frac{3}{1} = \frac{x}{40}$

19. $\frac{5}{6} = \frac{x}{12}$
20. $\frac{5}{6} = \frac{x}{36}$
21. $\frac{3}{10} = \frac{x}{20}$
22. $\frac{5}{8} = \frac{x}{24}$

23. $\frac{2}{3} = \frac{x}{9}$
24. $\frac{1}{2} = \frac{x}{10}$
25. $\frac{3}{4} = \frac{x}{16}$
26. $\frac{3}{10} = \frac{x}{100}$

27. $\frac{1}{5} = \frac{x}{10}$
28. $\frac{1}{3} = \frac{x}{9}$
29. $\frac{5}{6} = \frac{x}{24}$
30. $\frac{3}{4} = \frac{x}{20}$

31. $\frac{2}{5} = \frac{x}{40}$
32. $\frac{1}{2} = \frac{x}{24}$
33. $\frac{1}{3} = \frac{x}{18}$
34. $\frac{1}{4} = \frac{x}{100}$

35. $\frac{3}{8} = \frac{x}{16}$
36. $\frac{4}{5} = \frac{x}{20}$
37. $\frac{3}{5} = \frac{x}{100}$
38. $\frac{2}{3} = \frac{x}{30}$

39. $\frac{1}{4} = \frac{x}{28}$
40. $\frac{5}{1} = \frac{x}{3}$
41. $\frac{2}{9} = \frac{x}{27}$
42. $\frac{1}{6} = \frac{x}{42}$

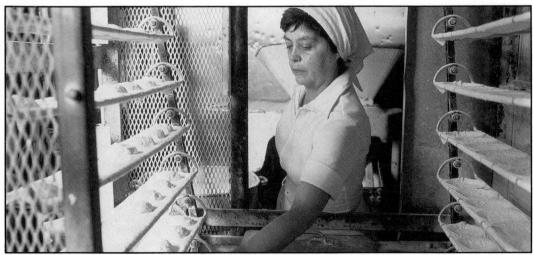

SIMPLIFYING FRACTIONS

Simplify $\frac{6}{10}$.

Step 1 Find the greatest common factor (GCF) of 6 and 10.

Method 1				Method 2	
Number	*Factors*	*GCF*	*Number*	*Prime factorization*	*GCF*
6	1, ②, 3, 6	2	6	②× 3	2
10	1, ②, 5, 10		10	②× 5	

Step 2 Divide the numerator and denominator by GCF.

$$\frac{6}{10} = \frac{6 \div 2}{10 \div 2} = \frac{3}{5} \qquad \frac{3}{5} \text{ is in simplest form.}$$

A fraction is in *simplest form* if the numerator and denominator have a GCF of 1.

Study and Learn

A. Find the GCF.

    1. 6, 15   **2.** 10, 12   **3.** 7, 11   **4.** 12, 18, 24

B. If the GCF of two or more numbers is 1, the numbers are relatively prime. Which numbers are relatively prime?

    5. 7, 13   **6.** 12, 15   **7.** 9, 16   **8.** 4, 5, 9

C. Which fractions are in simplest form?

    9. $\frac{3}{7}$ **10.** $\frac{3}{4}$ **11.** $\frac{9}{12}$ **12.** $\frac{6}{8}$ **13.** $\frac{3}{6}$

D. Simplify $\frac{6}{15}$. Use the GCF of 6 and 15.

    14. Complete. $\frac{6 \div 3}{15 \div 3} = \frac{?}{?}$ **15.** Is $\frac{2}{5}$ in simplest form?

E. Simplify.

    16. $\frac{2}{4}$ **17.** $\frac{2}{6}$ **18.** $\frac{5}{10}$ **19.** $\frac{6}{20}$ **20.** $\frac{8}{12}$

Practice

Find the GCF.

1. 15, 20   2. 20, 36   3. 18, 42

4. 8, 12, 20   5. 6, 15, 24   6. 14, 24, 48

Which numbers are relatively prime?

7. 3, 4   8. 6, 8   9. 5, 10

10. 4, 6, 10   11. 3, 5, 7   12. 4, 8, 12

Simplify.

13. $\frac{15}{30}$ 14. $\frac{2}{4}$ 15. $\frac{2}{6}$ 16. $\frac{6}{10}$ 17. $\frac{8}{24}$ 18. $\frac{8}{12}$

19. $\frac{5}{25}$ 20. $\frac{4}{16}$ 21. $\frac{15}{18}$ 22. $\frac{18}{20}$ 23. $\frac{5}{10}$ 24. $\frac{2}{12}$

25. $\frac{2}{8}$ 26. $\frac{4}{20}$ 27. $\frac{3}{4}$ 28. $\frac{3}{9}$ 29. $\frac{4}{8}$ 30. $\frac{2}{10}$

31. $\frac{3}{12}$ 32. $\frac{2}{15}$ 33. $\frac{3}{5}$ 34. $\frac{3}{8}$ 35. $\frac{8}{10}$ 36. $\frac{10}{12}$

37. $\frac{7}{9}$ 38. $\frac{14}{28}$ 39. $\frac{3}{6}$ 40. $\frac{4}{12}$ 41. $\frac{6}{9}$ 42. $\frac{3}{15}$

43. $\frac{4}{6}$ 44. $\frac{6}{12}$ 45. $\frac{10}{15}$ 46. $\frac{5}{15}$ 47. $\frac{8}{10}$ 48. $\frac{4}{10}$

49. $\frac{6}{15}$ 50. $\frac{3}{24}$ 51. $\frac{5}{6}$ 52. $\frac{9}{27}$ 53. $\frac{10}{12}$ 54. $\frac{10}{24}$

55. $\frac{6}{18}$ 56. $\frac{6}{24}$ 57. $\frac{14}{21}$ 58. $\frac{42}{48}$ 59. $\frac{28}{36}$ 60. $\frac{32}{40}$

61. 8 out of each 10 students buy lunch in school every day.
Simplify $\frac{8}{10}$.

62. 2 out of every 10 students do not buy lunch.
Simplify $\frac{2}{10}$.

COMPARING FRACTIONS

When comparing fractions with the same denominators, compare the numerators.

Example $\frac{7}{8} > \frac{6}{8}$ since $7 > 6$.

When comparing fractions with different denominators, first change them to equivalent fractions with the same denominators. Then compare numerators.

Example $\frac{7}{8} \equiv \frac{3}{4}$
$\frac{7}{8} \equiv \frac{6}{8}$
$\frac{7}{8} > \frac{6}{8}$ so $\frac{7}{8} > \frac{3}{4}$.

The pan with $\frac{7}{8}$ contains more lasagna.

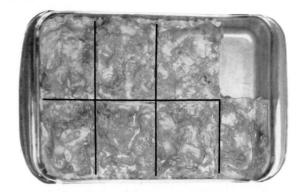

Study and Learn

A. Compare. Use $>$, $<$, or $=$.

1. $\frac{5}{16} \equiv \frac{3}{16}$ **2.** $\frac{7}{10} \equiv \frac{9}{10}$ **3.** $\frac{5}{6} \equiv \frac{2}{3}$ **4.** $\frac{2}{3} \equiv \frac{8}{12}$

B. Compare $\frac{3}{8}$ and $\frac{1}{6}$.

     5. What is the LCM of 8 and 6?

     6. Find equivalent fractions for $\frac{3}{8}$ and for $\frac{1}{6}$ using the LCM.

     7. Compare $\frac{9}{24} \equiv \frac{4}{24}$.

     8. Compare $\frac{3}{8} \equiv \frac{1}{6}$.

C. Compare. Use $>$, $<$, or $=$.

     9. $\frac{3}{4} \equiv \frac{1}{6}$ **10.** $\frac{3}{8} \equiv \frac{5}{12}$ **11.** $\frac{2}{3} \equiv \frac{10}{15}$ **12.** $\frac{3}{4} \equiv \frac{4}{5}$

Practice

Compare. Use $>$, $<$, or $=$.

1. $\frac{7}{8} \equiv \frac{1}{8}$ 2. $\frac{2}{5} \equiv \frac{3}{5}$ 3. $\frac{3}{10} \equiv \frac{9}{10}$ 4. $\frac{5}{6} \equiv \frac{1}{6}$

5. $\frac{5}{12} \equiv \frac{7}{12}$ 6. $\frac{7}{9} \equiv \frac{6}{9}$ 7. $\frac{8}{24} \equiv \frac{8}{24}$ 8. $\frac{3}{7} \equiv \frac{4}{7}$

9. $\frac{3}{8} \equiv \frac{3}{4}$ 10. $\frac{1}{3} \equiv \frac{5}{6}$ 11. $\frac{1}{2} \equiv \frac{3}{4}$ 12. $\frac{3}{5} \equiv \frac{7}{10}$

13. $\frac{1}{6} \equiv \frac{2}{12}$ 14. $\frac{3}{10} \equiv \frac{7}{20}$ 15. $\frac{1}{3} \equiv \frac{3}{9}$ 16. $\frac{3}{4} \equiv \frac{11}{12}$

17. $\frac{5}{6} \equiv \frac{23}{24}$ 18. $\frac{2}{5} \equiv \frac{7}{30}$ 19. $\frac{3}{7} \equiv \frac{6}{14}$ 20. $\frac{3}{8} \equiv \frac{7}{24}$

21. $\frac{1}{2} \equiv \frac{1}{3}$ 22. $\frac{2}{3} \equiv \frac{3}{4}$ 23. $\frac{3}{4} \equiv \frac{4}{5}$ 24. $\frac{1}{2} \equiv \frac{3}{3}$

25. $\frac{1}{3} \equiv \frac{2}{7}$ 26. $\frac{1}{2} \equiv \frac{3}{5}$ 27. $\frac{2}{3} \equiv \frac{4}{5}$ 28. $\frac{3}{4} \equiv \frac{7}{9}$

29. $\frac{3}{4} \equiv \frac{5}{6}$ 30. $\frac{3}{6} \equiv \frac{4}{8}$ 31. $\frac{3}{8} \equiv \frac{1}{12}$ 32. $\frac{7}{10} \equiv \frac{7}{15}$

33. $\frac{3}{4} \equiv \frac{7}{10}$ 34. $\frac{2}{8} \equiv \frac{5}{20}$ 35. $\frac{5}{12} \equiv \frac{7}{8}$ 36. $\frac{3}{8} \equiv \frac{5}{6}$

37. $\frac{7}{15} \equiv \frac{5}{6}$ 38. $\frac{3}{6} \equiv \frac{2}{4}$ 39. $\frac{1}{12} \equiv \frac{1}{16}$ 40. $\frac{7}{8} \equiv \frac{7}{10}$

41. $\frac{5}{6} \equiv \frac{2}{3}$ 42. $\frac{6}{7} \equiv \frac{5}{7}$ 43. $\frac{5}{12} \equiv \frac{3}{8}$ 44. $\frac{2}{3} \equiv \frac{4}{5}$

45. $\frac{1}{2} \equiv \frac{14}{28}$ 46. $\frac{3}{100} \equiv \frac{21}{100}$ 47. $\frac{2}{10} \equiv \frac{3}{100}$ 48. $\frac{3}{7} \equiv \frac{4}{9}$

Solve Problems

49. A brass rod has a diameter which is $\frac{3}{4}$ in. A second brass rod has a diameter of $\frac{1}{2}$ in. Which brass rod has the greater diameter?

50. Two pieces of lumber are each 1 yd long. $\frac{3}{4}$ yd is cut from one and $\frac{7}{8}$ yd from the second. Which piece remaining is longer?

165

MIXED NUMBERS

Joe says that the length of this bat is $\frac{7}{4}$ units. Maria says its length is $1\frac{3}{4}$ units.

$$\frac{7}{4} = 1\frac{3}{4}$$

fraction   mixed number

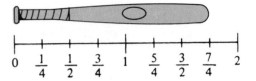

$\frac{7}{4}$ means $7 \div 4$
$$\begin{array}{r} 1\frac{3}{4} \\ 4\overline{)7} \\ 4 \\ \hline 3 \end{array}$$

Study and Learn

A. Write mixed numbers. Simplify if possible.

1. $\frac{3}{2}$ 2. $\frac{6}{4}$ 3. $\frac{7}{3}$ 4. $\frac{9}{6}$ 5. $\frac{13}{5}$

B. Write fractions.

6. $3\frac{1}{6} = 3 + \frac{1}{6}$
$\quad = \frac{18}{6} + \frac{1}{6}$
$\quad = \underline{\quad ? \quad}$

7. $5\frac{2}{3} = 5 + \frac{2}{3}$
$\quad = \frac{15}{3} + \frac{2}{3}$
$\quad = \underline{\quad ? \quad}$

8. $7\frac{2}{5}$ 9. $1\frac{5}{6}$ 10. $3\frac{3}{4}$ 11. $6\frac{1}{2}$

C. Here's a quick way to write fractions for mixed numbers.

Examples $6\frac{2}{3} = \frac{?}{3}$ $5\frac{1}{2} = \frac{?}{2}$

$6\frac{2}{3} = \frac{3 \times 6 + 2}{3}$ $5\frac{1}{2} = \frac{2 \times 5 + 1}{2}$

$\quad = \frac{18 + 2}{3}$ $\quad = \frac{10 + 1}{2}$

$\quad = \frac{20}{3}$ $\quad = \frac{11}{2}$

Write fractions.

12. $3\frac{1}{4}$ 13. $5\frac{7}{8}$ 14. $4\frac{1}{6}$ 15. $2\frac{1}{2}$

Practice

Write mixed numbers. Simplify.

1. $\frac{5}{2}$ **2.** $\frac{5}{4}$ **3.** $\frac{5}{3}$ **4.** $\frac{8}{6}$ **5.** $\frac{12}{5}$

6. $\frac{8}{3}$ **7.** $\frac{9}{4}$ **8.** $\frac{10}{4}$ **9.** $\frac{14}{6}$ **10.** $\frac{18}{8}$

11. $\frac{18}{10}$ **12.** $\frac{9}{7}$ **13.** $\frac{13}{3}$ **14.** $\frac{24}{5}$ **15.** $\frac{26}{6}$

16. $\frac{16}{3}$ **17.** $\frac{50}{20}$ **18.** $\frac{56}{24}$ **19.** $\frac{34}{4}$ **20.** $\frac{30}{9}$

Write fractions.

21. $1\frac{1}{8}$ **22.** $1\frac{2}{5}$ **23.** $1\frac{3}{7}$ **24.** $1\frac{5}{6}$ **25.** $1\frac{6}{10}$

26. $2\frac{1}{8}$ **27.** $2\frac{3}{8}$ **28.** $2\frac{1}{3}$ **29.** $2\frac{3}{4}$ **30.** $2\frac{1}{5}$

31. $2\frac{3}{7}$ **32.** $2\frac{1}{5}$ **33.** $2\frac{4}{5}$ **34.** $2\frac{5}{6}$ **35.** $2\frac{6}{7}$

36. $3\frac{3}{4}$ **37.** $3\frac{2}{3}$ **38.** $3\frac{4}{5}$ **39.** $3\frac{2}{10}$ **40.** $3\frac{5}{9}$

41. $4\frac{3}{4}$ **42.** $5\frac{7}{8}$ **43.** $6\frac{5}{6}$ **44.** $7\frac{3}{4}$ **45.** $8\frac{1}{2}$

Solve Problems

46. A bench is $3\frac{5}{8}$ ft long. A second bench is $\frac{28}{8}$ ft long. Which bench is longer?

47. Lucy practiced batting for $3\frac{3}{4}$ hours. Jim practiced $\frac{13}{4}$ hours. Who practiced longer?

ADDING FRACTIONS

Anthony spent $\frac{3}{10}$ hour changing the tires on his bicycle. Then he spent $\frac{3}{10}$ hour pumping air into the tires. How much time did he spend on fixing his bike in all?

▶ To add fractions with the *same* denominators:
Add the numerators.
Keep the common denominator. $\quad \frac{3}{10} + \frac{3}{10} = \frac{6}{10}$, or $\frac{3}{5}$

So, Anthony spent $\frac{3}{5}$ hour fixing his bike.

Study and Learn

A. Add and simplify.

1. $\frac{3}{8}$
   $+\frac{2}{8}$

2. $\frac{3}{4}$
   $+\frac{1}{4}$

3. $\frac{1}{9}$
   $+\frac{2}{9}$

4. $\frac{9}{10}$
   $+\frac{4}{10}$

5. $\frac{5}{6}$
   $+\frac{3}{6}$

6. $\frac{3}{8} + \frac{1}{8}$

7. $\frac{3}{10} + \frac{4}{10} + \frac{2}{10}$

8. $\frac{3}{4} + \frac{3}{4} + \frac{3}{4}$

B. Add $\frac{1}{2} + \frac{2}{5}$. Complete.

Think: You want to rename each fraction so that both have the same denominator.

9. The LCM of 2 and 5 is ___?___ .

10. $\frac{1}{2} = \frac{?}{10}$

11. $\frac{2}{5} = \frac{?}{10}$

12. $\frac{5}{10} + \frac{4}{10} = $ ___?___

13. So, $\frac{1}{2} + \frac{2}{5} = $ ___?___

C. Add and simplify.

14. $\frac{1}{8}$
    $+\frac{3}{4}$

15. $\frac{1}{6}$
    $+\frac{3}{8}$

16. $\frac{1}{3}$
    $+\frac{5}{6}$

17. $\frac{3}{4}$
    $+\frac{7}{10}$

18. $\frac{1}{2}$
    $\frac{1}{3}$
    $+\frac{1}{4}$

168

Practice

Add and simplify.

1. $\dfrac{4}{9}$ $+\dfrac{1}{9}$

2. $\dfrac{3}{8}$ $+\dfrac{4}{8}$

3. $\dfrac{5}{10}$ $+\dfrac{2}{10}$

4. $\dfrac{3}{8}$ $+\dfrac{3}{8}$

5. $\dfrac{1}{9}$ $+\dfrac{2}{9}$

6. $\dfrac{7}{10}$ $+\dfrac{4}{10}$

7. $\dfrac{3}{4}$ $+\dfrac{3}{4}$

8. $\dfrac{1}{24}$ $\dfrac{7}{24}$ $+\dfrac{5}{24}$

9. $\dfrac{3}{8}$ $\dfrac{3}{8}$ $+\dfrac{3}{8}$

10. $\dfrac{3}{10}$ $\dfrac{5}{10}$ $+\dfrac{8}{10}$

11. $\dfrac{3}{8} + \dfrac{6}{8}$

12. $\dfrac{3}{4} + \dfrac{2}{4} + \dfrac{3}{4}$

13. $\dfrac{5}{6} + \dfrac{1}{6} + \dfrac{4}{6}$

14. $\dfrac{1}{6}$ $+\dfrac{3}{8}$

15. $\dfrac{2}{5}$ $+\dfrac{1}{4}$

16. $\dfrac{1}{3}$ $+\dfrac{1}{2}$

17. $\dfrac{3}{4}$ $+\dfrac{1}{8}$

18. $\dfrac{3}{5}$ $+\dfrac{2}{10}$

19. $\dfrac{3}{4}$ $+\dfrac{1}{6}$

20. $\dfrac{7}{8}$ $+\dfrac{3}{4}$

21. $\dfrac{7}{10}$ $+\dfrac{1}{2}$

22. $\dfrac{1}{2}$ $+\dfrac{2}{3}$

23. $\dfrac{3}{4}$ $+\dfrac{4}{5}$

24. $\dfrac{3}{10}$ $\dfrac{1}{4}$ $+\dfrac{1}{2}$

25. $\dfrac{3}{4}$ $\dfrac{1}{2}$ $+\dfrac{1}{3}$

26. $\dfrac{1}{4}$ $\dfrac{3}{8}$ $+\dfrac{1}{2}$

27. $\dfrac{1}{2}$ $\dfrac{2}{3}$ $+\dfrac{5}{6}$

28. $\dfrac{5}{9}$ $\dfrac{1}{2}$ $+\dfrac{1}{6}$

29. $\dfrac{5}{6} + \dfrac{1}{12}$

30. $\dfrac{2}{3} + \dfrac{3}{5} + \dfrac{1}{6}$

31. $\dfrac{3}{10} + \dfrac{1}{2} + \dfrac{3}{4}$

Solve Problems

32. Audrey rode her bicycle for $\dfrac{3}{4}$ hour to the library and then for $\dfrac{3}{4}$ hour back home. How long did she ride in all?

33. Luis rode his bicycle $\dfrac{1}{2}$ mi to school, $\dfrac{3}{4}$ mi to a ballfield and $\dfrac{7}{10}$ mi home. How far did he ride in all?

ADDING MIXED NUMBERS

Ms. Andrews owns a stock which advanced $1\frac{3}{8}$ points one week and advanced another $2\frac{1}{8}$ points the next week. What was the total advance for the 2 weeks?

To add mixed numbers:
  Add the fractions.
  Add the whole numbers.

$$\begin{array}{r} 1\frac{3}{8} \\ + 2\frac{1}{8} \\ \hline 3\frac{4}{8} = 3\frac{1}{2} \end{array}$$

Answer: $3\frac{1}{2}$ points

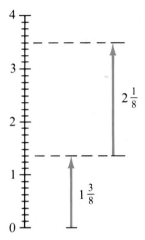

Study and Learn

✓ **A.** Add and simplify.

1. $\begin{array}{r} 3\frac{3}{8} \\ + 2\frac{4}{8} \\ \hline \end{array}$
2. $\begin{array}{r} 1\frac{1}{4} \\ + 2\frac{1}{2} \\ \hline \end{array}$
3. $\begin{array}{r} 3\frac{4}{8} \\ + 2\frac{1}{4} \\ \hline \end{array}$
4. $\begin{array}{r} 4\frac{1}{6} \\ + 2\frac{3}{8} \\ \hline \end{array}$
5. $\begin{array}{r} 5\frac{2}{4} \\ + 2\frac{1}{6} \\ \hline \end{array}$

✓ **B.** Sometimes the sum of the fractions is more than 1.

Examples

$\begin{array}{r} 2\frac{3}{8} \\ + 4\frac{5}{8} \\ \hline 6\frac{8}{8} = 7 \left[\frac{8}{8} = 1\right] \end{array}$ THINK:

$\begin{array}{r} 3\frac{5}{6} \\ + 2\frac{3}{6} \\ \hline 5\frac{8}{6} = 6\frac{2}{6} \left[\frac{8}{6} = 1\frac{2}{6}\right] \\ = 6\frac{1}{3} \end{array}$ THINK:

Add and simplify.

6. $\begin{array}{r} 1\frac{3}{4} \\ + 2\frac{2}{4} \\ \hline \end{array}$
7. $\begin{array}{r} 2\frac{3}{4} \\ + 3\frac{3}{4} \\ \hline \end{array}$
8. $\begin{array}{r} 4\frac{1}{2} \\ + 3\frac{5}{8} \\ \hline \end{array}$
9. $\begin{array}{r} 3\frac{1}{2} \\ 4\frac{1}{3} \\ + 5\frac{3}{4} \\ \hline \end{array}$
10. $\begin{array}{r} 1\frac{3}{4} \\ 3\frac{5}{8} \\ + 6\frac{1}{2} \\ \hline \end{array}$

C. Add and simplify.

11. $3\frac{3}{5} + 2\frac{3}{5}$

✓ 12. $4\frac{1}{2} + 3\frac{6}{8}$

✓ 13. $2\frac{1}{2} + 3\frac{3}{4} + 4$

170

Practice

Add and simplify.

1. $2\frac{1}{4}$
   $+ 3\frac{2}{4}$

2. $3\frac{1}{8}$
   $+ 2\frac{1}{8}$

3. $5\frac{1}{3}$
   $+ 2\frac{2}{3}$

4. $6\frac{1}{10}$
   $+ 3\frac{3}{10}$

5. $1\frac{2}{5}$
   $+ 1\frac{2}{5}$

6. $4\frac{1}{2}$
   $+ 3\frac{1}{4}$

7. $3\frac{3}{8}$
   $+ 2\frac{1}{2}$

8. $6\frac{3}{8}$
   $+ 2\frac{1}{4}$

9. $2\frac{2}{5}$
   $+ 3\frac{1}{10}$

10. $1\frac{5}{6}$
   $+ 2\frac{1}{3}$

11. $4\frac{3}{4}$
   $+ 2\frac{1}{4}$

12. $6\frac{2}{3}$
   $+ 2\frac{5}{6}$

13. $3\frac{1}{2}$
   $+ 2\frac{4}{8}$

14. $7\frac{3}{8}$
   $+ 4\frac{5}{6}$

15. $6\frac{2}{3}$
   $+ 7\frac{2}{5}$

16. $3\frac{5}{6}$
   $+ 2\frac{3}{4}$

17. $7\frac{1}{2}$
   $+ 6\frac{7}{10}$

18. $6\frac{3}{5}$
   $+ 4\frac{3}{4}$

19. $12\frac{3}{5}$
   $+ 7\frac{11}{15}$

20. $4\frac{3}{4}$
   $+ 2\frac{5}{6}$

21. $2\frac{3}{8}$
   $3\frac{2}{8}$
   $+ 4\frac{5}{8}$

22. $7\frac{1}{2}$
   $3\frac{3}{4}$
   $+ 2\frac{3}{4}$

23. $6\frac{1}{2}$
   $3\frac{1}{3}$
   $+ 4\frac{5}{6}$

24. $3\frac{3}{8}$
   $3\frac{3}{4}$
   $+ 2\frac{1}{2}$

25. $4\frac{3}{5}$
   $2\frac{3}{10}$
   $+ 1\frac{1}{2}$

26. $3\frac{3}{4} + 1\frac{1}{2}$

27. $7\frac{1}{2} + 3\frac{1}{2} + 1\frac{1}{4}$

28. $6\frac{3}{8} + 2\frac{1}{2} + 3\frac{3}{4}$

Something Extra
Non-Routine Problems

Look at this sequence: $\frac{1}{2}, \frac{1}{4}, \frac{1}{8}, \frac{1}{16}, \ldots$

The number after $\frac{1}{16}$ is $\frac{1}{2} \times \frac{1}{16}$, or $\frac{1}{32}$.

1. What is the number after $\frac{1}{32}$?

2. Add $\frac{1}{2} + \frac{1}{4}$.

3. Add $\frac{1}{2} + \frac{1}{4} + \frac{1}{8}$.

4. Add $\frac{1}{2} + \frac{1}{4} + \frac{1}{8} + \frac{1}{16}$.

5. If you add more and more of the numbers of this sequence, the sum gets closer and closer to what number?

Midchapter Review

Find the GCF. *(162)*

1. 10, 8   **2.** 15, 25   **3.** 18, 27   **4.** 12, 16, 28

Find the LCM. *(158)*

5. 2, 6   **6.** 4, 7   **7.** 9, 12   **8.** 3, 7, 2

Find equivalent fractions with the given denominators. *(160)*

9. $\frac{2}{3} = \frac{x}{9}$ **10.** $\frac{1}{2} = \frac{x}{12}$ **11.** $\frac{3}{4} = \frac{x}{16}$ **12.** $\frac{3}{5} = \frac{x}{25}$

Simplify. *(162)*

13. $\frac{3}{9}$ **14.** $\frac{6}{24}$ **15.** $\frac{6}{15}$ **16.** $\frac{12}{30}$

Compare. Use $>$, $<$, or $=$. *(164)*

17. $\frac{3}{8} \equiv \frac{1}{8}$ **18.** $\frac{3}{4} \equiv \frac{6}{8}$ **19.** $\frac{1}{2} \equiv \frac{2}{5}$ **20.** $\frac{5}{6} \equiv \frac{3}{4}$

Add and simplify. *(168, 170)*

21. $\begin{array}{r} \frac{3}{9} \\ + \frac{1}{9} \\ \hline \end{array}$
22. $\begin{array}{r} \frac{3}{8} \\ + \frac{3}{8} \\ \hline \end{array}$
23. $\begin{array}{r} \frac{3}{4} \\ + \frac{1}{6} \\ \hline \end{array}$
24. $\begin{array}{r} \frac{2}{3} \\ + \frac{4}{5} \\ \hline \end{array}$
25. $\begin{array}{r} \frac{1}{2} \\ \frac{3}{4} \\ + \frac{5}{8} \\ \hline \end{array}$

26. $\begin{array}{r} 3\frac{1}{4} \\ + 4\frac{1}{4} \\ \hline \end{array}$
27. $\begin{array}{r} 3\frac{3}{8} \\ + 6\frac{5}{8} \\ \hline \end{array}$
28. $\begin{array}{r} 7\frac{5}{6} \\ + 2\frac{3}{8} \\ \hline \end{array}$
29. $\begin{array}{r} 5\frac{2}{5} \\ + 3\frac{2}{3} \\ \hline \end{array}$
30. $\begin{array}{r} 3\frac{1}{2} \\ 2\frac{5}{6} \\ + 4\frac{2}{3} \\ \hline \end{array}$

Something Extra

Calculator Activity

Find the following products: $1 \times 142{,}857$; $2 \times 142{,}857$; $3 \times 142{,}857$; $4 \times 142{,}857$; $5 \times 142{,}857$; and $6 \times 142{,}857$ using a calculator. What do you notice about the products?

172

SUBTRACTING FRACTIONS

Jan plays the piano $\frac{3}{4}$ hour each day. She has played for $\frac{1}{4}$ hour. How much longer will she play?

$$
\begin{array}{r}
\frac{3}{4} \\
-\frac{1}{4} \\
\hline
\frac{2}{4}, \text{ or } \frac{1}{2}
\end{array}
$$

> The denominators are the same. Subtract the numerators.

Study and Learn

A. Subtract and simplify.

1. $\begin{array}{r}\frac{5}{6}\\-\frac{1}{6}\\\hline\end{array}$

2. $\begin{array}{r}\frac{9}{10}\\-\frac{3}{10}\\\hline\end{array}$

3. $\begin{array}{r}\frac{5}{8}\\-\frac{1}{8}\\\hline\end{array}$

4. $\frac{3}{5}-\frac{1}{5}$

5. $\frac{7}{9}-\frac{1}{9}$

B. When the denominators are not the same, rename each fraction so that the denominators are the same. Subtract and simplify.

6. $\begin{array}{r}\frac{3}{5}\\-\frac{1}{4}\\\hline\end{array}$

7. $\begin{array}{r}\frac{9}{10}\\-\frac{2}{5}\\\hline\end{array}$

8. $\begin{array}{r}\frac{3}{8}\\-\frac{1}{6}\\\hline\end{array}$

9. $\frac{2}{3}-\frac{1}{5}$

10. $\frac{3}{4}-\frac{1}{6}$

Practice

Subtract and simplify.

1. $\begin{array}{r}\frac{9}{12}\\-\frac{5}{12}\\\hline\end{array}$

2. $\begin{array}{r}\frac{4}{5}\\-\frac{2}{5}\\\hline\end{array}$

3. $\begin{array}{r}\frac{7}{8}\\-\frac{1}{8}\\\hline\end{array}$

4. $\frac{4}{6}-\frac{1}{6}$

5. $\frac{3}{4}-\frac{1}{6}$

6. $\begin{array}{r}\frac{1}{2}\\-\frac{1}{4}\\\hline\end{array}$

7. $\begin{array}{r}\frac{3}{4}\\-\frac{5}{8}\\\hline\end{array}$

8. $\begin{array}{r}\frac{7}{8}\\-\frac{3}{12}\\\hline\end{array}$

9. $\frac{5}{6}-\frac{1}{12}$

10. $\frac{5}{6}-\frac{3}{10}$

Solve Problems

11. Steve ran for $\frac{7}{8}$ hour and walked for $\frac{1}{2}$ hour to practice for a race. How much longer did he run than walk?

SUBTRACTING MIXED NUMBERS

Debbie had $4\frac{1}{2}$ boxes of books. She unpacked $2\frac{1}{4}$ of the boxes. How many boxes are still to be unpacked?

$$4\frac{1}{2} = 4\frac{2}{4}$$
$$- 2\frac{1}{4} = 2\frac{1}{4}$$ same denominator
$$\overline{\phantom{-2\frac{1}{4} = } 2\frac{1}{4}}$$

So, $2\frac{1}{4}$ boxes are still to be unpacked.

Study and Learn

A. Subtract and simplify.

1. $2\frac{3}{4}$
$- 1\frac{1}{4}$
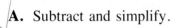

2. $6\frac{1}{2}$
$- 3\frac{1}{3}$

3. $7\frac{3}{8}$
$- 4\frac{1}{6}$

4. $6\frac{7}{12} - 3\frac{1}{4}$

5. $4\frac{3}{4} - 2\frac{1}{6}$

B. Whole numbers and fractions can be subtracted.

Examples

$3\frac{1}{6}$
$- 2$
$\overline{1\frac{1}{6}}$

$6 = 5\frac{2}{2}$ ← THINK $6 = 5 + 1 = 5 + \frac{2}{2}$
$- 2\frac{1}{2} = 2\frac{1}{2}$
$\overline{3\frac{1}{2}}$

Subtract.

6. 6
$- \frac{5}{8}$

7. 9
$- 3\frac{5}{6}$

8. $4\frac{1}{2}$
$- 2$

9. $2 - \frac{5}{8}$

10. $8 - 3\frac{4}{5}$

C. Sometimes you must rename the larger mixed number before you can subtract.

Example

$5\frac{1}{3} = 4\frac{4}{3}$
$- 1\frac{2}{3} = 1\frac{2}{3}$
$\overline{3\frac{2}{3}}$

THINK
$5 = 4\frac{3}{3}$
$5 + \frac{1}{3} = 4\frac{3}{3} + \frac{1}{3}$
$\phantom{5 + \frac{1}{3}} = 4\frac{4}{3}$

Subtract.

11. $7\frac{3}{5}$
$- 2\frac{4}{5}$

12. $7\frac{1}{2}$
$- 3\frac{7}{8}$

13. $4\frac{1}{3}$
$- 2\frac{3}{5}$

14. $8\frac{1}{6} - 3\frac{3}{8}$

15. $9\frac{2}{3} - 6\frac{3}{4}$

Practice

Subtract. Simplify.

1. $3\frac{3}{8}$
   $-2\frac{1}{8}$

2. $4\frac{4}{9}$
   $-2\frac{2}{9}$

3. $6\frac{5}{6}$
   $-2\frac{1}{6}$

4. $2\frac{3}{5} - 1\frac{2}{5}$

5. $8\frac{7}{10} - 3\frac{3}{10}$

6. $3\frac{3}{4}$
   $-1\frac{2}{3}$

7. $6\frac{7}{8}$
   $-3\frac{3}{4}$

8. $8\frac{5}{6}$
   $-2\frac{1}{4}$

9. $7\frac{2}{3} - 3\frac{3}{5}$

10. $6\frac{5}{6} - 3\frac{3}{8}$

11. 7
    $-\frac{5}{6}$

12. 9
    $-7\frac{3}{4}$

13. $5\frac{3}{4}$
    -3

14. $6 - \frac{7}{8}$

15. $9 - 3\frac{5}{6}$

16. $6\frac{3}{8}$
    $-2\frac{7}{8}$

17. $5\frac{1}{6}$
    $-2\frac{5}{6}$

18. $8\frac{7}{9}$
    $-3\frac{8}{9}$

19. $3\frac{1}{4} - 1\frac{3}{4}$

20. $8\frac{3}{5} - 2\frac{4}{5}$

21. $5\frac{1}{4}$
    $-2\frac{7}{8}$

22. $6\frac{1}{2}$
    $-4\frac{7}{20}$

23. $7\frac{1}{3}$
    $-3\frac{5}{6}$

24. $5\frac{5}{12} - 3\frac{5}{6}$

25. $8\frac{2}{5} - 6\frac{2}{3}$

26. $6\frac{1}{5}$
    $-2\frac{3}{4}$

27. $8\frac{1}{4}$
    $-2\frac{1}{3}$

28. $9\frac{1}{6}$
    $-6\frac{3}{8}$

★ 29. $8\frac{1}{6} - 3\frac{3}{4} + 2\frac{1}{2}$

★ 30. $9\frac{1}{12} - 5\frac{3}{8} - 1\frac{3}{4}$

Solve Problems

31. Frank had $2\frac{1}{2}$ cartons of orange juice. He used $\frac{3}{4}$ of a carton for breakfast. How much orange juice is left?

32. Blanche cut a $4\frac{1}{2}$ in. piece of metal for a machine from a piece $7\frac{3}{10}$ in. long. How long is the remaining piece?

● 33. Luisa ate $2\frac{1}{2}$ peaches and Eduardo ate $1\frac{1}{4}$ peaches. How many peaches in all were eaten?

● 34. Michelle bought 3 gallons of paint. She already had $\frac{3}{4}$ of a gallon. How much paint does she have now?

175

Problem-Solving Skills

Extra Information

Sometimes a problem can have too much information.

A supermarket ordered 17 dozen packages of light bulbs. The light bulbs are sold at 2 packages for $1.69. How much will 6 packages of light bulbs cost?

READ   The problem asks:
How much will 6 packages cost?

PLAN   The fact you need to know is:
2 packages cost $1.69

SOLVE $6 \div 2 = 3$, so 6 packages cost 1.69×3, or $5.07

You didn't need to know that the store had ordered 17 dozen packages.

Study and Learn

A. There are 24 cans of corn in a case. The cans of corn are selling at 2 for $0.79. What is the cost of 10 cans?

   1. What are you asked to find?

   2. What facts do you need to solve the problem?

   3. What facts don't you need?

   4. Solve.

B. At a sale a large bottle of apple juice sold for $0.99. The next smaller size cost $0.72. What is the cost of a dozen bottles of the large size?

   5. What are you asked to find?

   6. What facts do you need to solve the problem?

   7. What facts don't you need?

   8. Solve.

Practice

Identify only that information which is needed to solve each problem. Solve.

1. There are 48 cans of soup in a case. The cans of soup are selling at 3 for $0.89. What is the cost of a dozen cans of soup?

2. A package of a dozen muffins sells for $0.59. A package of 16 muffins sells for $0.75. What is the cost of 6 packages of a dozen muffins?

3. One brand of coffee is selling for $3.49 a can. A more expensive brand of coffee is selling for $4.19 a can. How much does a half dozen cans of the more expensive brand cost?

4. A 16-slice package of cheese is on sale for $0.99. A carton contains 48 of the 16-slice packages. How many slices of cheese are in a dozen packages?

5. Lu Chou bought peanuts at $0.69 a jar and cups at 2 packages for a dollar. The total purchases came to $3.38. How much change did she receive from a $10 bill?

6. Mr. Daniels bought 3 jars of mayonnaise at $0.89 each and 2 bottles of catsup at $0.86 each. How much change did he receive from a $20 bill?

7. At a sale a large bottle of shampoo was selling for $1.88. The next smaller size cost $1.29. Find the cost of 3 of the large size bottles of shampoo.

8. A can of tuna fish sells for $0.59 with a discount coupon. Without the coupon, the can of tuna costs $0.99. What is the cost of 8 cans of tuna without any coupons?

9. A brand of cat food is selling at 4 cans for $0.89. A brand of dog food is selling at 3 cans for $0.89. What does a half dozen cans of the dog food cost?

10. David bought milk for $0.51, cream cheese for $1.05, and butter for $2.00. What was the total cost of the milk and cream cheese?

11. A camera which lists for $179.95 is on sale for $139.50. A zoom lens for the camera costs $69.50. How much is saved by buying the camera on sale?

12. A portable electric typewriter has an original price of $439. It is on sale for $269. The manufacturer is giving a $20 rebate. What is the cost of the typewriter (without tax)?

MULTIPLYING FRACTIONS

Kaoni plays the violin $\frac{2}{3}$ hour a day. He has completed $\frac{1}{2}$ of his playing. What part of an hour has he played?

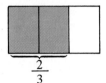 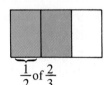

$\frac{1}{2}$ of $\frac{2}{3}$

$\frac{1}{2} \times \frac{2}{3} = \frac{1 \times 2}{2 \times 3}$ ← Multiply numerators.
← Multiply denominators.

$= \frac{2}{6}$, or $\frac{1}{3}$ Simplify.

Study and Learn

A. Multiply and simplify.

1. $\frac{1}{3} \times \frac{2}{5}$ **2.** $\frac{3}{4} \times \frac{1}{10}$ **3.** $\frac{2}{3} \times \frac{1}{8}$ **4.** $\frac{2}{5} \times \frac{1}{2} \times \frac{1}{4}$

B. Multiply and simplify.

Examples
$$\frac{2}{3} \times 9 = \frac{2}{3} \times \frac{9}{1} \qquad 3 \times \frac{2}{5} = \frac{3}{1} \times \frac{2}{5}$$
$$= \frac{2 \times 9}{3 \times 1} \qquad\qquad = \frac{3 \times 2}{1 \times 5}$$
$$= \frac{18}{3}, \text{ or } 6 \qquad\quad = \frac{6}{5}, \text{ or } 1\frac{1}{5}$$

5. $\frac{3}{4} \times 8$ **6.** $\frac{4}{5} \times 10$ **7.** $4 \times \frac{1}{4}$ **8.** $\frac{3}{10} \times 15$

Here's a shortcut when multiplying fractions.

Long way   *Short way*

$\frac{2}{3} \times \frac{1}{4} = \frac{2}{12}$, or $\frac{1}{6}$ $\overset{1}{\cancel{2}} \times \frac{1}{\underset{2}{\cancel{4}}} = \frac{1}{6}$

C. Complete.

9. $\frac{5}{\underset{3}{\cancel{9}}} \times \overset{1}{\cancel{\frac{2}{8}}} = \underline{\quad?\quad}$ **10.** $\frac{\overset{3}{\cancel{9}}}{10} \times \frac{7}{\underset{4}{\cancel{12}}} = \underline{\quad?\quad}$ **11.** $\frac{\overset{1}{\cancel{2}}}{\underset{1}{\cancel{3}}} \times \frac{\overset{3}{\cancel{9}}}{\underset{5}{\cancel{10}}} = \underline{\quad?\quad}$

D. Multiply and simplify.

12. $\frac{3}{4} \times \frac{5}{6}$ **13.** $\frac{3}{4} \times \frac{8}{9}$ **14.** $\frac{1}{2} \times \frac{8}{9} \times \frac{3}{5}$ **15.** $\frac{2}{3} \times 6$

178

Practice

Multiply and simplify.

1. $\frac{1}{3} \times \frac{1}{2}$ **2.** $\frac{1}{4} \times \frac{1}{8}$ **3.** $\frac{1}{3} \times \frac{1}{8}$ **4.** $\frac{1}{4} \times \frac{1}{6}$

5. $\frac{1}{2} \times \frac{3}{4}$ **6.** $\frac{1}{2} \times \frac{3}{5}$ **7.** $\frac{2}{3} \times \frac{5}{7}$ **8.** $\frac{3}{4} \times \frac{9}{10}$

9. $\frac{5}{6} \times \frac{7}{8}$ **10.** $\frac{7}{9} \times \frac{3}{5}$ **11.** $\frac{5}{6} \times \frac{11}{12}$ **12.** $\frac{5}{8} \times \frac{5}{9}$

13. $\frac{2}{3} \times 4$ **14.** $\frac{3}{4} \times 5$ **15.** $\frac{5}{6} \times 8$ **16.** $\frac{1}{2} \times 5$

17. $\frac{5}{7} \times 14$ **18.** $\frac{2}{3} \times 9$ **19.** $5 \times \frac{1}{3}$ **20.** $6 \times \frac{2}{3}$

21. $7 \times \frac{1}{2}$ **22.** $8 \times \frac{3}{4}$ **23.** $9 \times \frac{2}{3}$ **24.** $12 \times \frac{1}{4}$

25. $\frac{2}{3} \times \frac{6}{7}$ **26.** $\frac{3}{9} \times \frac{1}{2}$ **27.** $\frac{2}{5} \times \frac{1}{6}$ **28.** $\frac{4}{5} \times \frac{3}{8}$

29. $\frac{3}{5} \times \frac{2}{9}$ **30.** $\frac{3}{4} \times \frac{8}{11}$ **31.** $\frac{5}{6} \times \frac{7}{10}$ **32.** $\frac{2}{3} \times \frac{1}{4}$

33. $\frac{1}{2} \times \frac{4}{5}$ **34.** $\frac{3}{4} \times \frac{8}{9}$ **35.** $\frac{2}{3} \times \frac{9}{20}$ **36.** $\frac{4}{5} \times \frac{15}{16}$

37. $\frac{1}{5} \times \frac{2}{3} \times \frac{1}{4}$ **38.** $\frac{3}{4} \times \frac{1}{6} \times \frac{2}{5}$ **39.** $\frac{1}{2} \times \frac{9}{10} \times \frac{2}{3}$

★**40.** $\left(\frac{1}{2} + \frac{1}{3}\right) \times \left(\frac{1}{4} + \frac{1}{2}\right)$ ★**41.** $\left(\frac{9}{10} - \frac{3}{5}\right) \times \left(1 - \frac{4}{9}\right)$

Solve Problems

42. Bill plays the piano $\frac{3}{4}$ hour a day. Cathy, his sister, plays $\frac{1}{2}$ as long. How long does Cathy play each day?

43. Jill signed up for 48 piano lessons. She took $\frac{2}{3}$ of them by September. How many piano lessons did she take by September?

44. In water, sound travels approximately $\frac{9}{10}$ of a mile per second. How far will it travel in $\frac{1}{2}$ second?

45. A package of typing paper costs $1.60. During a sale, Leroy paid $\frac{3}{4}$ of the cost. How much did he pay?

MULTIPLYING MIXED NUMBERS

Joe needs $3\frac{1}{2}$ cans of paint to paint each room in his house. How many cans of paint will he need for 8 rooms?

$$8 \times 3\frac{1}{2} = \overset{4}{\cancel{\frac{8}{1}}} \times \frac{7}{\underset{1}{\cancel{2}}}$$
$$= 28$$

So, 28 cans are needed for 8 rooms.

Study and Learn

A. To multiply mixed numbers, first rename them as fractions. Then multiply.

Examples
$$\frac{3}{4} \times 1\frac{1}{2} = \frac{3}{4} \times \frac{3}{2}$$
$$= \frac{9}{8}, \text{ or } 1\frac{1}{8}$$

$$1\frac{1}{2} \times 2\frac{1}{3} = \overset{1}{\cancel{\frac{3}{2}}} \times \frac{7}{\underset{1}{\cancel{3}}}$$
$$= \frac{7}{2}, \text{ or } 3\frac{1}{2}$$

1. $6 \times 4\frac{2}{3}$ **2.** $3\frac{2}{5} \times 8$ **3.** $7 \times 3\frac{3}{4}$ **4.** $\frac{2}{3} \times 2\frac{1}{2}$

5. $\frac{3}{4} \times 2\frac{1}{6}$ **6.** $2\frac{2}{3} \times \frac{5}{8} \times 6$ **7.** $3\frac{2}{3} \times 1\frac{1}{4}$ **8.** $2\frac{2}{5} \times 1\frac{5}{6} \times 1\frac{1}{2}$

Practice

Multiply.

1. $9 \times 1\frac{1}{3}$ **2.** $8 \times 3\frac{3}{4}$ **3.** $12 \times 6\frac{1}{2}$ **4.** $10 \times 5\frac{3}{10}$

5. $5 \times 1\frac{1}{4}$ **6.** $4 \times 2\frac{2}{3}$ **7.** $5 \times 3\frac{1}{2}$ **8.** $6 \times 5\frac{3}{5} \times 1\frac{2}{3}$

9. $2\frac{1}{2} \times 4$ **10.** $3\frac{1}{3} \times 4$ **11.** $1\frac{2}{5} \times 10$ **12.** $2\frac{1}{4} \times 8$

13. $2\frac{1}{2} \times 3$ **14.** $1\frac{1}{3} \times 4$ **15.** $3\frac{1}{4} \times 5$ **16.** $6\frac{2}{3} \times 7 \times 1\frac{1}{5}$

17. $\frac{3}{4} \times 1\frac{1}{2}$ **18.** $\frac{4}{5} \times 2\frac{1}{4}$ **19.** $\frac{7}{8} \times 3\frac{2}{3}$ **20.** $\frac{5}{6} \times 3\frac{1}{8}$

21. $2\frac{1}{2} \times 1\frac{1}{4}$ **22.** $3\frac{1}{3} \times 1\frac{3}{4}$ **23.** $4\frac{2}{3} \times 2\frac{3}{5}$ **24.** $2\frac{1}{6} \times 1\frac{7}{8} \times 5\frac{1}{3}$

RECIPROCALS

A. Multiply. What is true of each product?

    1. $\frac{2}{3} \times \frac{3}{2}$ **2.** $\frac{5}{4} \times \frac{4}{5}$ **3.** $\frac{1}{2} \times 2$ **4.** $3\frac{1}{2} \times \frac{2}{7}$

    ▶ If the product of two numbers is 1, one number is the **reciprocal** of the other.

B. Find the reciprocals.

    5. $\frac{2}{3}$ **6.** $\frac{5}{7}$ **7.** 6   **8.** $3\frac{1}{4}$

C. Fractions like $\dfrac{\frac{2}{7}}{\frac{3}{4}}$ are called **complex fractions.** Simplify $\dfrac{\frac{2}{7}}{\frac{3}{4}}$.

    9. What is the reciprocal of the denominator?

    10. Multiply the numerator and the denominator by $\frac{4}{3}$.

    11. Simplify numerator and denominator. $\dfrac{\frac{2}{7} \times \frac{4}{3}}{\frac{3}{4} \times \frac{4}{3}} = \underline{\quad ? \quad}$

D. Simplify.

    12. $\dfrac{\frac{1}{3}}{\frac{1}{2}}$ **13.** $\dfrac{\frac{2}{3}}{\frac{4}{3}}$ **14.** $\dfrac{\frac{1}{4}}{\frac{3}{2}}$ **15.** $\dfrac{\frac{1}{10}}{\frac{3}{5}}$

Practice

Find the reciprocals.

    1. $\frac{1}{2}$ **2.** $\frac{1}{3}$ **3.** $\frac{3}{4}$ **4.** 3   **5.** $1\frac{1}{2}$

Simplify.

    6. $\dfrac{\frac{3}{5}}{\frac{5}{9}}$ **7.** $\dfrac{\frac{2}{3}}{\frac{6}{9}}$ **8.** $\dfrac{\frac{3}{10}}{\frac{2}{5}}$ **9.** $\dfrac{\frac{5}{6}}{\frac{1}{2}}$ **10.** $\dfrac{\frac{1}{3}}{\frac{3}{5}}$

DIVIDING FRACTIONS AND MIXED NUMBERS

Multiplication can be used to divide fractions.

How many $\frac{1}{3}$'s in 1?

$1 \div \frac{1}{3}$

$1 \times \frac{3}{1} = \frac{3}{1} = 3$

↑
reciprocal

How many $\frac{1}{4}$'s in $\frac{1}{2}$?

$\frac{1}{2} \div \frac{1}{4}$

$\frac{1}{2} \times \frac{4}{1} = \frac{4}{2} = 2$

↑
reciprocal

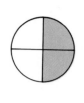

▶ To divide any number by a fraction, multiply by the reciprocal of the divisor.

Study and Learn

A. Here's how to divide $\frac{2}{3} \div \frac{1}{6}$.

   1. What is the reciprocal of $\frac{1}{6}$?

   2. Rewrite the problem as multiplication.
     $\frac{2}{3} \div \frac{1}{6} = \frac{2}{3} \times \underline{\quad ? \quad}$

   3. Multiply $\frac{2}{3} \times \frac{6}{1}$.

   4. What is $\frac{2}{3} \div \frac{1}{6}$?

B. Divide.

   5. $3 \div \frac{1}{2}$ **6.** $\frac{9}{10} \div 6$ **7.** $\frac{3}{4} \div \frac{1}{8}$ **8.** $\frac{3}{5} \div \frac{2}{3}$

C. To divide mixed numbers, change to fractions.

   Examples $2\frac{1}{2} \div \frac{3}{4} = \frac{5}{2} \div \frac{3}{4}$ $3\frac{1}{2} \div 1\frac{1}{5} = \frac{7}{2} \div \frac{6}{5}$

                               $= \frac{5}{2} \times \frac{4}{3}$ $= \frac{7}{2} \times \frac{5}{6}$

                               $= \frac{10}{3}$, or $3\frac{1}{3}$ $= \frac{35}{12}$, or $2\frac{11}{12}$

   Divide.

   9. $5\frac{3}{4} \div \frac{2}{3}$ **10.** $1\frac{2}{3} \div 6$ **11.** $2\frac{3}{4} \div 1\frac{1}{2}$ **12.** $5\frac{5}{6} \div 2\frac{2}{3}$

Practice

Divide.

1. $1 \div \frac{1}{2}$

2. $1 \div \frac{1}{5}$

3. $3 \div \frac{1}{3}$

4. $3 \div \frac{1}{10}$

5. $\frac{1}{4} \div \frac{1}{8}$

6. $\frac{1}{3} \div \frac{1}{6}$

7. $\frac{3}{4} \div \frac{1}{2}$

8. $\frac{5}{6} \div \frac{1}{3}$

9. $\frac{2}{3} \div \frac{1}{2}$

10. $\frac{3}{4} \div \frac{7}{8}$

11. $\frac{4}{5} \div \frac{7}{8}$

12. $\frac{5}{6} \div \frac{2}{3}$

13. $\frac{2}{3} \div \frac{3}{4}$

14. $\frac{9}{10} \div \frac{3}{5}$

15. $\frac{3}{8} \div \frac{3}{4}$

16. $\frac{7}{12} \div \frac{14}{15}$

17. $2\frac{1}{2} \div \frac{3}{4}$

18. $3\frac{1}{5} \div \frac{1}{2}$

19. $2\frac{3}{4} \div \frac{7}{8}$

20. $3\frac{5}{6} \div \frac{2}{3}$

21. $2\frac{3}{4} \div \frac{2}{5}$

22. $3\frac{5}{7} \div \frac{1}{2}$

23. $2\frac{5}{6} \div \frac{3}{4}$

24. $5\frac{1}{2} \div \frac{3}{5}$

25. $\frac{3}{4} \div 2$

26. $\frac{5}{6} \div 3$

27. $4 \div \frac{8}{9}$

28. $10 \div \frac{5}{6}$

29. $1\frac{4}{5} \div 3$

30. $2\frac{6}{7} \div 2$

31. $12 \div 1\frac{1}{7}$

32. $6 \div 2\frac{1}{4}$

33. $\frac{2}{3} \div 1\frac{1}{2}$

34. $\frac{5}{8} \div 1\frac{1}{4}$

35. $\frac{9}{10} \div 1\frac{3}{5}$

36. $\frac{8}{12} \div 1\frac{3}{4}$

37. $2\frac{1}{2} \div 2\frac{1}{2}$

38. $3\frac{3}{4} \div 1\frac{1}{2}$

39. $2\frac{1}{4} \div 3\frac{1}{3}$

40. $1\frac{3}{5} \div 1\frac{3}{5}$

41. $3\frac{1}{2} \div 4\frac{1}{2}$

42. $\frac{5}{6} \div 1\frac{1}{9}$

43. $9 \div \frac{3}{7}$

44. $\frac{7}{8} \div \frac{3}{4}$

★ **45.** $\left(\frac{9}{10} \times \frac{5}{8}\right) \div \frac{3}{8}$

★ **46.** $\left(3\frac{2}{3} \div 5\frac{1}{2}\right) \div \left(4\frac{1}{2} \div \frac{3}{4}\right)$

★ **47.** $\left(\frac{3}{5} + \frac{1}{3}\right) \div \left(\frac{3}{4} - \frac{7}{10}\right)$

Solve Problems

48. Carl has 24 apples. He uses $2\frac{2}{3}$ apples to make a dessert. How many desserts can he make?

49. Gloria bought $40\frac{1}{2}$ ft of wire mesh to fence in her vegetable gardens. She cut the wire mesh into $4\frac{1}{2}$ ft pieces. How many pieces does she have?

FRACTIONS AND DECIMALS

Sue Ann is working in shop class. She is told to bore a hole $\frac{1}{2}$ in. wide. Is this the same as 0.5 in. wide?

To change a fraction to a decimal, divide.

$$\frac{1}{2} \text{ means } 1 \div 2 \text{ or } 2\overline{)1} \longrightarrow 2\overline{)1.0} \quad \begin{array}{r} 0.5 \\ \hline \end{array}$$

So, $\frac{1}{2} = 0.5$

Study and Learn

A. Write decimals.

  1. $\frac{1}{5}$ **2.** $\frac{2}{5}$ **3.** $\frac{3}{2}$ **4.** $\frac{6}{5}$ **5.** $\frac{7}{2}$

B. Sometimes you need to divide to hundredths.

  Example $\frac{3}{4}$ means $3 \div 4$ or

$$\begin{array}{r} 0.75 \\ 4\overline{)3.00} \\ \underline{2\,8} \\ 20 \\ \underline{20} \\ 0 \end{array}$$

Write decimals.

  6. $\frac{1}{4}$ **7.** $\frac{1}{25}$ **8.** $\frac{2}{50}$ **9.** $\frac{4}{25}$ **10.** $\frac{5}{4}$

C. Sometimes you need to divide to thousandths.

  Example $\frac{1}{8} \longrightarrow$

$$\begin{array}{r} 0.125 \\ 8\overline{)1.000} \\ \underline{8} \\ 20 \\ \underline{16} \\ 40 \\ \underline{40} \\ 0 \end{array}$$

Write decimals.

  11. $\frac{3}{8}$ **12.** $\frac{5}{8}$ **13.** $\frac{1}{40}$ **14.** $\frac{3}{40}$ **15.** $\frac{1}{125}$

184

D. Write decimals.

Examples $1\frac{1}{2} = 1.5$ $2\frac{3}{4} = 2.75$

16. $2\frac{1}{2}$ **17.** $3\frac{3}{4}$ **18.** $4\frac{1}{5}$ **19.** $4\frac{1}{25}$ **20.** $3\frac{1}{8}$

Practice

Write decimals.

1. $\frac{5}{2}$ **2.** $\frac{7}{5}$ **3.** $\frac{7}{2}$ **4.** $\frac{7}{10}$ **5.** $\frac{7}{28}$

6. $\frac{3}{25}$ **7.** $\frac{3}{125}$ **8.** $\frac{7}{4}$ **9.** $\frac{9}{4}$ **10.** $\frac{7}{8}$

11. $\frac{5}{40}$ **12.** $\frac{27}{40}$ **13.** $2\frac{1}{5}$ **14.** $3\frac{1}{2}$ **15.** $4\frac{3}{10}$

16. $3\frac{1}{50}$ **17.** $1\frac{7}{25}$ **18.** $2\frac{3}{4}$ ★ **19.** $\frac{1}{80}$ ★ **20.** $2\frac{23}{80}$

Solve Problems

21. A drill size is $\frac{3}{4}$ in. Will the hole made by this drill be larger or smaller than 0.80 in.?

22. A drill size is $\frac{1}{8}$ in. A hole is made with this drill. Will it be larger or smaller than 0.1 in.?

Skills Review

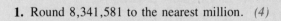

1. Round 8,341,581 to the nearest million. *(4)*

Solve. *(60, 66, 70)*

2. $5x = 30$ **3.** $x + 9 = 24$

4. $7x + 2 = 23$ **5.** $\frac{x}{5} + 2 = 6$

Compare. Use >, <, or = . *(82)*

6. $0.4 \equiv 0.9$ **7.** $0.52 \equiv 0.520$ **8.** $0.29 \equiv 0.3$ **9.** $0.1 \equiv 0.08$

Write in scientific notation. *(92)*

10. 900   **11.** 12,000   **12.** 80,000   **13.** 4,620

Problem-Solving Applications

Career

1. Information to a computer may be put on magnetic tape. The computer reads the information at a rate of 600,000 characters per second. How many characters can it read in a minute? [HINT: Multiply by 60.]

2. Early computers took 0.1 second to multiply two 10-digit numbers. A modern computer takes about 1 nanosecond (0.000000001) to find the same product. How many times faster is the modern computer?

3. Ms. Block worked 240 hours on a program. She is paid $25 per hour. How much did she earn?

4. A machine used to read punched cards can read 895 cards in 1 minute. How many cards can it read in an hour?

5. Computers are programmed to read the magnetic ink symbols found at the bottom of a check. A machine can sort 2,100 checks in one minute. How many checks can the machine sort in an 8-hour day?

6. Mr. Anderson worked $7\frac{3}{4}$ hours one day. He earned $124. How much was he paid per hour?

Chapter Review

Find equivalent fractions with the given denominators. *(160)*

1. $\frac{4}{5} = \frac{x}{15}$ **2.** $\frac{9}{10} = \frac{x}{20}$

Simplify. *(162)*

3. $\frac{4}{6}$ **4.** $\frac{15}{20}$ **5.** $\frac{18}{24}$

Compare. Use $>$, $<$, or $=$. *(164)*

6. $\frac{1}{2} \equiv \frac{2}{3}$ **7.** $\frac{3}{4} \equiv \frac{6}{8}$ **8.** $\frac{2}{3} \equiv \frac{3}{5}$

Add and simplify. *(168, 170)*

9. $\begin{array}{r} \frac{3}{8} \\ + \frac{4}{8} \\ \hline \end{array}$ **10.** $\begin{array}{r} 11\frac{5}{8} \\ + \ 5\frac{5}{6} \\ \hline \end{array}$ **11.** $\frac{2}{3} + \frac{1}{6} + \frac{5}{12}$

Subtract and simplify. *(173, 174)*

12. $\begin{array}{r} \frac{7}{16} \\ - \frac{3}{16} \\ \hline \end{array}$ **13.** $\begin{array}{r} 4\frac{3}{4} \\ - 2\frac{2}{3} \\ \hline \end{array}$ **14.** $3\frac{1}{2} - 1\frac{3}{4}$

Multiply and simplify. *(178, 180)*

15. $\frac{2}{5} \times \frac{1}{4}$ **16.** $\frac{3}{4} \times 12$ **17.** $\frac{1}{2} \times 3\frac{3}{4}$

Divide. *(182)*

18. $\frac{3}{4} \div \frac{3}{5}$ **19.** $5 \div 1\frac{2}{3}$ **20.** $3\frac{2}{3} \div 1\frac{1}{2}$

Write decimals. *(184)*

21. $\frac{9}{2}$ **22.** $\frac{7}{25}$ **23.** $6\frac{5}{8}$

Solve. *(176, 186)*

24. Lois bought juice at 6 cans for $1.59 and picnic plates at $0.89 a package. The total came to $6.74. How much change did she receive from a $20 bill?

25. A systems analyst earns a salary of $30,000 a year and $6,600 from a second job at a college. What is her average monthly earnings?

Chapter Test

Find equivalent fractions with the given denominators. *(160)*

1. $\frac{5}{8} = \frac{x}{16}$

2. $\frac{7}{12} = \frac{x}{24}$

Simplify. *(162)*

3. $\frac{6}{8}$

4. $\frac{12}{15}$

5. $\frac{10}{12}$

Compare. Use >, <, or =. *(164)*

6. $\frac{2}{3} \equiv \frac{8}{12}$

7. $\frac{3}{8} \equiv \frac{5}{6}$

8. $\frac{3}{5} \equiv \frac{4}{7}$

Add and simplify. *(168, 170)*

9. $\begin{array}{r} \frac{3}{4} \\ + \frac{1}{5} \\ \hline \end{array}$

10. $\begin{array}{r} 2\frac{1}{6} \\ + 3\frac{3}{4} \\ \hline \end{array}$

11. $\frac{5}{8} + \frac{3}{4} + \frac{1}{2}$

Subtract and simplify. *(173, 174)*

12. $\begin{array}{r} \frac{5}{8} \\ - \frac{1}{8} \\ \hline \end{array}$

13. $\begin{array}{r} 7\frac{1}{3} \\ - 2\frac{3}{4} \\ \hline \end{array}$

14. $7 - 5\frac{2}{3}$

Multiply and simplify. *(178, 180)*

15. $\frac{2}{3} \times \frac{3}{4}$

16. $\frac{4}{5} \times 20$

17. $6\frac{1}{2} \times 4$

Divide. *(182)*

18. $\frac{2}{3} \div \frac{1}{2}$

19. $6 \div 2\frac{1}{3}$

20. $4\frac{1}{2} \div 2\frac{1}{3}$

Write decimals. *(184)*

21. $\frac{11}{4}$

22. $\frac{9}{125}$

23. $2\frac{11}{50}$

Solve. *(176, 186)*

24. The supermarket has 4 dozen cases of canned milk, 3 dozen cases of powdered milk, and 2 dozen cases of cocoa. How many cases of cocoa does it have?

25. A computer programmer earned $24,400 last year. This year, he received a raise of $\frac{1}{4}$ last year's salary. What is his salary this year?

Skills Check

1. Multiply.

$$\begin{array}{r} 0.347 \\ \times\ 0.36 \\ \hline \end{array}$$

A 0.12392 B 0.12492

C 1.2392 D 124.92

---

2. Multiply.

$$\begin{array}{r} \$423.62 \\ \times\ 24 \\ \hline \end{array}$$

E $10,166.88 F $8,166.78

G $6,166.78 H none of
the above

---

3. Divide.

$$6\overline{)4.764}$$

A 0.974 B 0.864

C 0.794 D 0.791

---

4. Divide.

$$0.07\overline{)3.843}$$

E 0.538 F 5.49

G 54.9 H 53,800

---

5. Divide.

$$7\overline{)\$65.73}$$

A $4.93 B $6.77

C $8.63 D $9.39

---

6. Add.

$$\begin{array}{r} \frac{3}{10} \\ +\ \frac{5}{6} \\ \hline \end{array}$$

E $3\frac{4}{15}$ F $2\frac{8}{15}$

G $2\frac{2}{15}$ H $1\frac{2}{15}$

---

7. Add.

$$\begin{array}{r} 3\frac{4}{9} \\ +\ 2\frac{5}{6} \\ \hline \end{array}$$

A $5\frac{3}{5}$ B $5\frac{5}{9}$

C $6\frac{5}{18}$ D $6\frac{5}{9}$

---

8. Subtract.

$$\begin{array}{r} \frac{19}{25} \\ -\ \frac{31}{50} \\ \hline \end{array}$$

E $3\frac{1}{3}$ F $2\frac{1}{6}$

G $\frac{7}{50}$ H $\frac{1}{50}$

---

9. Subtract.

$$\begin{array}{r} 4\frac{1}{2} \\ -\ 1\frac{3}{4} \\ \hline \end{array}$$

A $3\frac{1}{4}$ B $2\frac{3}{4}$

C $2\frac{1}{2}$ D $2\frac{1}{4}$

---

10. Subtract. $7 - 3\frac{2}{3}$

E $3\frac{1}{3}$ F $3\frac{5}{6}$

G $4\frac{1}{3}$ H $10\frac{2}{3}$

---

11. Multiply. $\frac{3}{4} \times \frac{8}{15}$

A $1\frac{2}{5}$ B $\frac{4}{5}$

C $\frac{2}{5}$ D $\frac{1}{5}$

---

12. Multiply. $6 \times \frac{4}{9}$

E $1\frac{1}{3}$ F $1\frac{2}{3}$

G $2\frac{1}{3}$ H $2\frac{2}{3}$

---

13. Multiply. $2\frac{1}{3} \times \frac{3}{4}$

A $7\frac{3}{4}$ B $2\frac{1}{4}$

C $1\frac{3}{4}$ D $1\frac{5}{12}$

---

14. Divide. $\frac{3}{5} \div \frac{9}{10}$

E $2\frac{1}{3}$ F $1\frac{2}{3}$

G $\frac{2}{3}$ H none of
the above

---

15. Divide. $6 \div 1\frac{1}{2}$

A 4 B $4\frac{1}{2}$

C 9 D 12

---

16. Solve. 42% of 700

E 306 F 296

G 294 H 194

189

8 GEOMETRY AND CONSTRUCTIONS

The space shuttle takes off from the Kennedy Space Center, Florida.

ANGLES

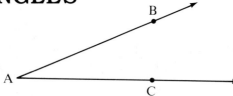

Ray AB ($\overrightarrow{AB}$) and ray AC ($\overrightarrow{AC}$) have a common endpoint. The angle formed is named $\angle BAC$, or $\angle CAB$, or $\angle A$. Point A is the vertex. $\overrightarrow{AB}$ and $\overrightarrow{AC}$ are the sides.

A protractor is used to measure angles. $m\angle COB = 50°$

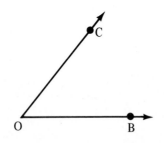 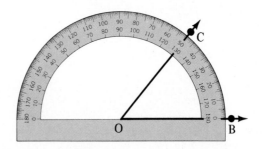

Study and Learn

A. Look at the angle at the right.

   1. Name the vertex.

   2. Name the sides.

   3. Name the angle in 3 ways.

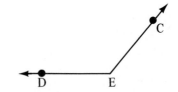

B. Draw angles with these measures.

   4. 30°   **5.** 90°   **6.** 140°   **7.** 65°   **8.** 170°

C. Give the measures of the angles. $m\angle AOB$ is read measure of angle AOB.

   9. $m\angle AOB =$ ___?___   **10.** $m\angle AOC =$ ___?___

   11. $m\angle AOD =$ ___?___   **12.** $m\angle AOE =$ ___?___

   13. $m\angle GOF =$ ___?___   **14.** $m\angle GOE =$ ___?___

   15. $m\angle GOC =$ ___?___   **16.** $m\angle DOG =$ ___?___

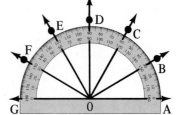

▶ Two angles with the same measure are congruent.

$$\angle AOB \cong \angle GOF$$

192

D. Measure these angles. Which are congruent?

17.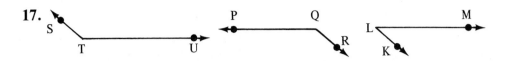

Practice

Name each angle in 3 ways.

1.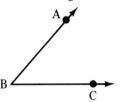

2.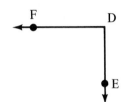

3.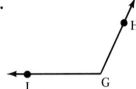

Draw angles with these measures.

4. 40°   **5.** 80°   **6.** 150°   **7.** 62°   **8.** 20°

Measure these angles.

9.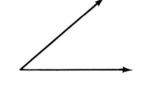

10.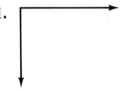

11.

12.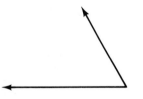

Which angles are congruent?

13.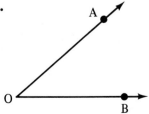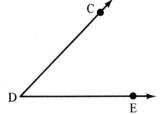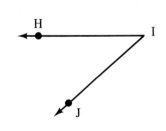

ANGLES AND TRIANGLES

Angles and triangles can be classified by the measures of their angles.

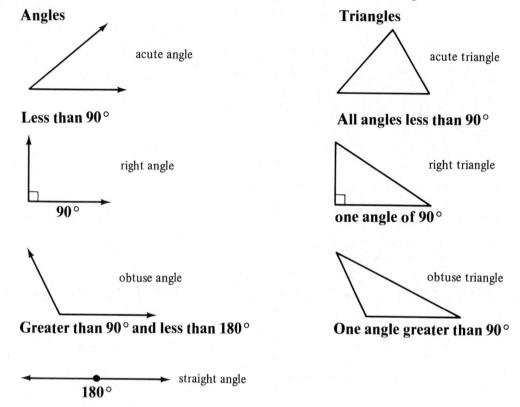

Angles

acute angle

Less than 90°

right angle

90°

obtuse angle

Greater than 90° and less than 180°

straight angle
180°

Triangles

acute triangle

All angles less than 90°

right triangle

one angle of 90°

obtuse triangle

One angle greater than 90°

Study and Learn

A. Classify these angles.

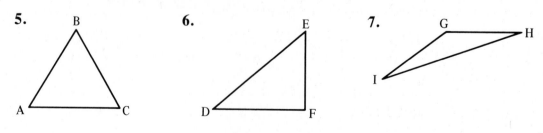

1.   **2.**   **3.**   **4.**

B. Classify these triangles by the measures of their angles.

5. B A C   **6.** E D F   **7.** G H I

194

Triangles can also be classified by the measures of their sides.

scalene triangle

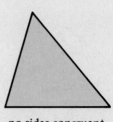

no sides congruent

isosceles triangle

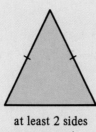

at least 2 sides
congruent

equilateral triangle

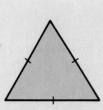

all sides congruent

C. Classify these triangles by the measures of their sides.

8. sides: 8 cm, 8 cm, 6 cm

9. sides: 2 in., 3 in., 4 in.

10. sides: 15 mm, 15 mm, 15 mm

11. sides: 3 cm, 4 cm, 5 cm

Practice

Classify these angles.

1.

2.

3.

4.

5. 153°

6. 12°

7. 90°

8. 180°

Classify these triangles by the measures of their angles.

9.

10.

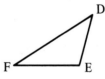

11.

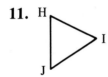

12. 40°, 40°, 100°

13. 30°, 60°, 90°

14. 24°, 85°, 71°

Classify these triangles by the measures of their sides.

15.

8 cm   6 cm

10 cm

16.

8 in.

6 in.

8 in.

17.

20 mm   20 mm

20 mm

18. 10 cm, 10 cm, 10 cm

19. 4 ft, 6 ft, 8 ft

20. 3 m, 4 m, 3 m

195

ANGLE CONSTRUCTION

Materials: compass, straightedge

A. Here's how to construct an angle congruent to ∠ACB.

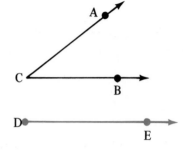

1. Draw $\overrightarrow{DE}$.

2. Put your compass point on C. Draw an arc. Mark points S and T.

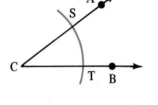

3. Keep your compass open to the same distance. Now put the compass point on D and draw an arc. Mark point U.

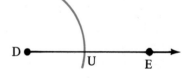

4. Put your compass point on T and open it enough to reach S.

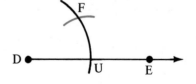

5. Using the distance from T to S, put your compass point on U and draw an arc. Mark point F.

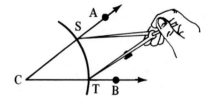

6. Draw $\overrightarrow{DF}$.
   ∠FDE ≅ ∠ACB

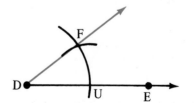

B. Here's how to bisect ∠ *XYZ*.

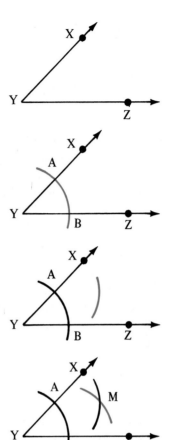

7. Put your compass point on *Y*.
Draw an arc. Mark points *A*
and *B*.

8. With your compass point on
A, draw a small arc in the
interior of the angle.

9. With your compass point on *B*
and keeping the same distance
on the compass, draw another
arc that intersects the first
arc. Mark point *M*.

10. Draw $\overrightarrow{YM}$.
$\overrightarrow{YM}$ bisects ∠ *XYZ*.

Practice

1. Draw ∠ *MNO*. Construct an angle congruent to it.

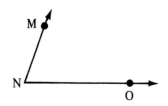

2. Draw an acute angle. Bisect it.

POLYGONS

Polygons are made up of line segments. They are classified by the number of line segments.

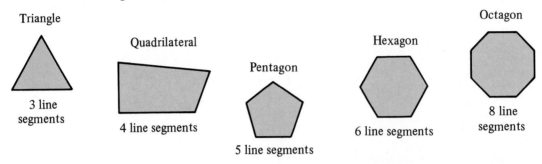

Triangle

3 line segments

Quadrilateral

4 line segments

Pentagon

5 line segments

Hexagon

6 line segments

Octagon

8 line segments

Study and Learn

A. Name the polygons.

1.   2.   3.

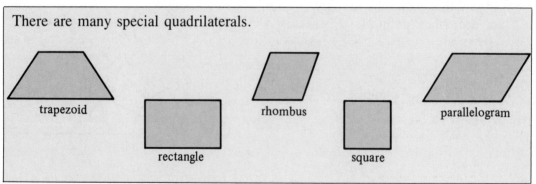

There are many special quadrilaterals.

trapezoid

rectangle

rhombus

square

parallelogram

B. Look at the quadrilaterals above.

   4. How many pairs of sides are parallel in a trapezoid?

   5. In a parallelogram, how many pairs of sides are parallel?

   6. In a parallelogram, what is true about the opposite sides and opposite angles?

   7. In what way is a rhombus a special parallelogram?

   8. In what way is a rectangle a special parallelogram?

   9. In what way is a square a special rectangle?

C. Find x.

10.
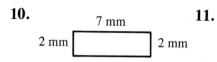
7 mm
2 mm   2 mm
x

11.
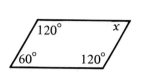
120° x
60°   120°

12.

5 cm
5 cm x
5 cm

Practice

Name the polygons.

1.
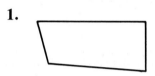

2.

3.

4.

5.

6.

Name the quadrilaterals.

7.
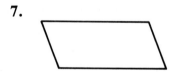

8.

9.

10.

11.
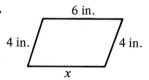

12.
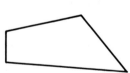

Find x.

13.
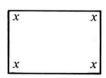
5 cm
3 cm x
5 cm

14.
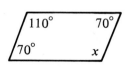
6 in.
4 in.   4 in.
x

15.
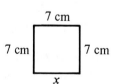
7 cm
7 cm   7 cm
x

16.

x x
x x

17.
110°   70°
70° x

18.
x x
x x

ANGLES OF A POLYGON

Discovery of geometric relationships may be made by experimentation.

Perform this experiment.

Draw a large triangle.

Cut out the triangle.

Cut off the 3 angles and place them side by side.

Notice that a straight angle is formed. The measure of a straight angle is 180°.

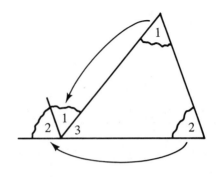

▶ The sum of the measures of the angles of a triangle is 180°.

Study and Learn

A. In △ABC, m∠A = 63° and m∠B = 59°. What is m∠C?

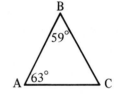

   1. What is the sum of m∠A + m∠B + m∠C?

   2. What is the sum of m∠A + m∠B?

   3. What is m∠C?

B. Quadrilateral $ABCD$ has 4 vertices.

   4. Name the 4 vertices of $ABCD$.

   5. $\overline{AC}$ is a diagonal. Name the other diagonal.

C. Quadrilateral $ABCD$ is divided into 2 triangles.

   6. What is the sum of the measures of the angles of each triangle?

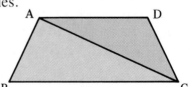

   7. What is the sum of the measures of the angles of the quadrilateral?

200

D. A **regular polygon** has the same size angles and sides. Find the measures of each angle of these regular polygons.

8.
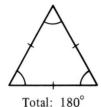
Total: 180°

9.
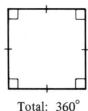
Total: 360°

10.

Total: 540°

Practice

The measures of 2 angles of a triangle are given. Find the measure of the third angle.

1. 50°, 50°

2. 10°, 140°

3. 90°, 50°

4. 60°, 70°

5. 24°, 37°

6. 18°, 90°

7. 73°, 46°

8. 131°, 17°

9. 47°, 101°

Draw diagonals. Find the sum of the measures of the angles of each polygon.

10.
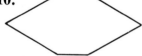

11.
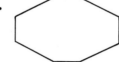

Solve Problems

12. One acute angle of a right triangle is 30°. What is the measure of the other acute angle?

13. An equilateral triangle is a regular polygon. What is the measure of each angle of an equilateral triangle?

14. One angle of a triangle measures 88°. The other 2 angles each have the same measure. What is the measure of each of the other 2 angles?

15. Each of 2 angles of a triangle measures 77°. What is the measure of the third angle of the triangle?

★**16.** What is the measure of each angle of a regular hexagon?

★**17.** What is the sum of the measures of the angles of a polygon with 10 sides?

PERIMETER OF A POLYGON

An iron fence will be placed around
a pool as shown. How much fence
will be needed?

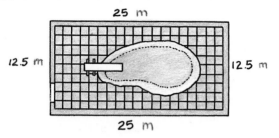

perimeter = 25 + 12.5 + 25 + 12.5
         = 75

The perimeter is 75 m.

▶ The distance around a polygon is called the **perimeter** of the
polygon. The perimeter is found by adding the measures of
the sides.

Study and Learn

A. Find the perimeters.

1.
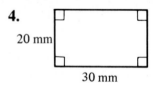

2.
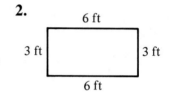

3.
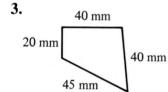

B. There are formulas to find the perimeters of some polygons.

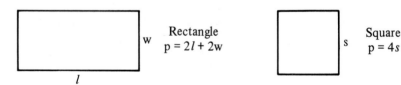

Rectangle
p = 2l + 2w

Square
p = 4s

Find the perimeters.

4.
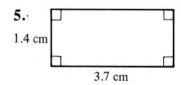
20 mm
30 mm

5.
1.4 cm
3.7 cm

6.
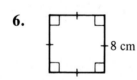
8 cm

C. Here is an equilateral triangle.

7. Write a formula for finding the perimeter
of an equilateral triangle.

8. What is the perimeter of an equilateral
triangle with one side 24 cm long?

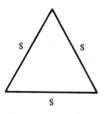

Practice

Find the perimeters.

1.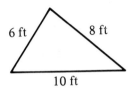

2.

3.

4.

5.

6.

7.

8.

9.

10.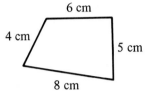

11.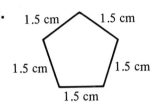

12.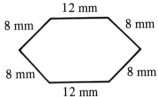

Computer

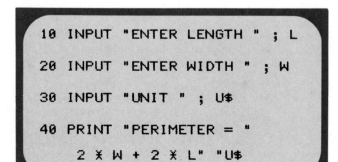

```
10 INPUT "ENTER LENGTH " ; L

20 INPUT "ENTER WIDTH " ; W

30 INPUT "UNIT " ; U$

40 PRINT "PERIMETER = "
   2 * W + 2 * L" "U$
```

This program computes the perimeter of a rectangle.

The length, width, and unit of measurement are stored. Then the answer is shown. You can use this program to check **5–9.**

Skills Review

Compare. Use >, <, or = . *(82)*

1. $0.4 \equiv 0.8$

2. $0.73 \equiv 0.730$

3. $0.29 \equiv 0.3$

4. $0.01 \equiv 0.009$

5. Round 3.247 to the nearest tenth. *(84)*

Add. Simplify when possible. *(168, 170)*

6. $\frac{3}{4}$
   $+ \frac{2}{3}$

7. $3\frac{5}{6}$
   $+ 2\frac{3}{8}$

8. $4\frac{1}{2}$
   $+ 3\frac{3}{4}$

9. $\frac{5}{8} + 1\frac{1}{2} + 2\frac{5}{6}$

Subtract. Simplify when possible. *(173, 174)*

10. $\frac{4}{5}$
   $- \frac{1}{4}$

11. $5\frac{2}{3}$
   $- 2\frac{3}{8}$

12. $3\frac{1}{2}$
   $- 1\frac{3}{4}$

13. $8 - 3\frac{5}{6}$

Multiply. *(36, 112, 178)*

14. 624
   $\times 35$

15. 427
   $\times 30$

16. $8,912$
   $\times 605$

17. 286
   $\times 542$

18. 0.6×0.06

19. 0.6×0.9

20. 0.23×4.5

21. $\frac{1}{2} \times \frac{3}{4}$

22. $\frac{2}{3} \times \frac{9}{2}$

23. $\frac{4}{5} \times 2\frac{1}{7}$

24. $5\frac{1}{2} \times 1\frac{1}{3}$

Divide. *(40, 118, 182)*

25. $7\overline{)364}$

26. $14\overline{)224}$

27. $33\overline{)10,032}$

28. $215\overline{)8,600}$

29. $0.3\overline{)495}$

30. $43\overline{)0.0129}$

31. $7.1\overline{)24.85}$

32. $0.34\overline{)32.81}$

33. $\frac{1}{2} \div \frac{4}{5}$

34. $\frac{3}{5} \div \frac{1}{5}$

35. $\frac{6}{7} \div \frac{3}{14}$

36. $2\frac{1}{4} \div 1\frac{1}{5}$

Divide. Round the quotient to the nearest tenth. *(120)*

37. $6\overline{)1.4}$

38. $0.4\overline{)3.112}$

39. $1.6\overline{)57}$

40. $0.09\overline{)0.631}$

Divide. Round the quotient to the nearest hundredth. *(120)*

41. $9\overline{)6.521}$

42. $0.7\overline{)16.056}$

43. $2.9\overline{)15.38}$

44. $0.52\overline{)0.976}$

Find equivalent fractions with the given denominators. *(160)*

45. $\frac{1}{2} = \frac{x}{6}$

46. $\frac{5}{6} = \frac{x}{24}$

47. $\frac{3}{10} = \frac{x}{30}$

48. $\frac{7}{12} = \frac{x}{48}$

Midchapter Review

Measure these angles. *(192)*

1.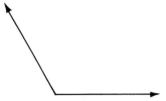

2.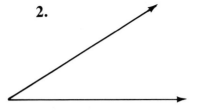

3.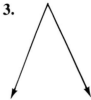

Classify these triangles by the measures of their angles. *(194)*

4.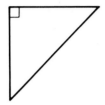

5.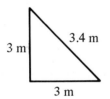

6.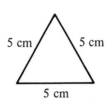

Classify these triangles by the measures of their sides. *(195)*

7.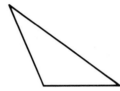

8.

9.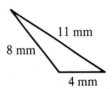

Find the missing measures. *(198, 200)*

10.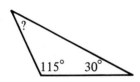

11.

12.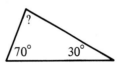

Find the perimeters. *(202)*

13.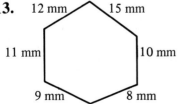

14.

15.

Problem-Solving Skills
Missing Information

The length of a rectangle is 40 mm. What is the perimeter of the rectangle?

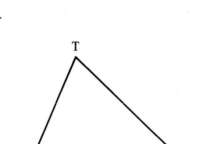

40 mm

PLAN perimeter = $2l + 2w$

You are told the length, but not the width. You don't have enough information to solve this problem.

Study and Learn

A. $\triangle RST$ is an acute triangle. m$\angle R = 43°$. What is the measure of $\angle S$?

1. What information do you need to solve this problem?

2. What information is given?

3. Do you have enough information to solve the problem?

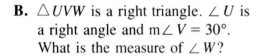

B. $\triangle UVW$ is a right triangle. $\angle U$ is a right angle and m$\angle V = 30°$. What is the measure of $\angle W$?

4. What information do you need to solve this problem?

5. What information is given?

6. Do you have enough information to solve the problem?

7. Solve.

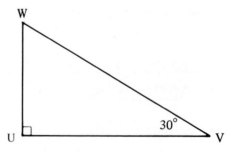

C. The width of a rectangle is 34 cm. What is the perimeter of the rectangle?

8. Do you have enough information to solve this problem?

9. What information is missing?

10. Supply the missing information and solve.

Practice

Decide if there is enough information to solve the problem.
If so, solve.

1. The length of a rectangle is 30 mm. What is the perimeter of the rectangle?

2. In scalene triangle *RST*, *RS* = 1.5 cm. What is the perimeter of the triangle?

3. In right triangle *DEF*, ∠*D* is a right angle. m∠*E* = 22°. What is the measure of ∠*F*?

4. In △*ABC*, m∠*A* = 53°. What is the measure of ∠*B*?

5. A side of a regular pentagon is 10 cm long. What is the perimeter of the pentagon?

6. △*ABC* is a right triangle. ∠*C* is a right angle. What is the sum of the measures of ∠*A* and ∠*B*?

Tell what additional information is needed to solve the problem.
Supply the information and solve.

7. The width of a rectangle is 24 cm. What is the perimeter of the rectangle?

8. In △*ABC*, *AB* = 13 mm and *AC* = 15 mm. What kind of triangle is △*ABC*?

9. The perimeter of a rectangle is 120 ft. What is the width?

10. In right triangle *MNO*, ∠*O* is a right angle. What is the measure of ∠*M*?

11. A side of a hexagon is 12 cm long. What is the perimeter of the hexagon?

12. The perimeter of a triangle is 42 cm long. How long is each side?

CIRCLES AND CIRCUMFERENCE

Every point on this circle is 3 cm away from the center, O. $\overline{OA}$ is a radius. $\overline{BC}$ is a diameter. The length of the radius is 3 cm. The length of the diameter is 6 cm.

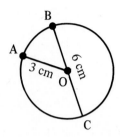

▶ The length of a diameter of a circle is twice the length of a radius of that circle.

Study and Learn

A. Find the length of the diameter for the given radii.

  1. 62 mm   **2.** 0.5 m   **3.** 2.1 cm   **4.** 9.6 in.

B. Find the length of the radius for the given diameters.

  5. 32 m   **6.** 6 cm   **7.** 15 mm   **8.** 3.8 in.

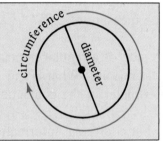

The distance around a circle is called the **circumference.** The ratio of the circumference to the diameter is the same for all circles. $\dfrac{\text{circumference}}{\text{diameter}} = \pi$

$$\pi \doteq 3.14$$

▶ The formula for finding the circumference is circumference = $\pi \cdot$ diameter, or $C = \pi \cdot d$. The formula $C = 2 \cdot \pi \cdot r$ is also used for finding the circumference of a circle.

C. Find the circumference. Give the answer in terms of π.

  Examples $d = 14$ mm $r = 6$ cm

                       $C = \pi \cdot d$ $C = 2 \cdot \pi \cdot r$

                          $= \pi \cdot 14$, or 14 π mm $= 2 \cdot \pi \cdot 6$, or 12 π cm

  9. $d = 10$ m   **10.** $d = 3.5$ cm   **11.** $r = 4$ mm   **12.** $r = 5.4$ m

D. Find the circumference. Use $\pi \doteq 3.14$.

Examples
$$d = 8 \text{ cm} \qquad\qquad r = 5 \text{ mm}$$
$$C = \pi \cdot d \qquad\qquad C = 2 \cdot \pi \cdot r$$
$$ = 3.14 \cdot 8 \qquad\qquad = 2 \cdot 3.14 \cdot 5$$
$$ = 25.12 \text{ cm} \qquad\qquad = 31.4 \text{ mm}$$

13. $d = 11$ cm **14.** $d = 21$ mm **15.** $r = 9$ mm **16.** $r = 1.5$ in.

Practice

Find the length of the diameter for the given radii.

1. 7 cm **2.** 0.4 m **3.** 2.3 cm **4.** 1.3 ft

Find the length of the radius for the given diameters.

5. 24 mm **6.** 2.6 cm **7.** 8.6 m **8.** 27 in.

Find the circumferences. Give the answer in terms of π.

9. $d = 18$ cm **10.** $d = 9$ mm **11.** $d = 24$ m **12.** $d = 3.4$ in.

13. $r = 9$ m **14.** $r = 14$ cm **15.** $r = 31$ mm **16.** $r = 4.5$ ft

Find the circumferences. Use $\pi \doteq 3.14$.

17. $d = 3$ m **18.** $d = 50$ cm **19.** $d = 17$ cm **20.** $d = 1{,}000$ in.

21. $r = 10$ km **22.** $r = 25$ cm **23.** $r = 30$ mm **24.** $r = 3.5$ ft

Solve Problems

25. A wheel has a diameter which is 30.8 cm long. How long is a radius of the wheel?

26. A wheel has a diameter which is 50 cm long. How far will the wheel travel when it makes 1 complete turn?

★ **27.** The circumference of a circle is 25.12 cm long. How long is the diameter of the circle?

★ **28.** The circumference of a circle is 47.1 m long. How long is the radius of the circle?

PAIRS OF ANGLES

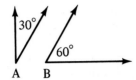

30°

60°

A B

two complementary angles
sum = 90°
∠A is the complement of ∠B.

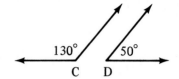

130° 50°

C D

two supplementary angles
sum = 180°
∠C is the supplement of ∠D.

Study and Learn

A. Complementary or supplementary?

 1. 20°, 70°   **2.** 146°, 34°   **3.** 120°, 60°   **4.** 88°, 2°

B. Find the complements.

 5. 30°   **6.** 24°   **7.** 67°   **8.** 54°   **9.** 89°

C. Find the supplements.

 10. 30°   **11.** 120°   **12.** 54°   **13.** 117°   **14.** 179°

D. △ABC is a right triangle. ∠A is a right angle.

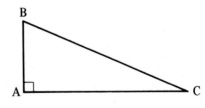

 15. What is the measure of ∠A?

 16. What is the sum of the measures of the angles in △ABC?

 17. What is the sum of the measures of ∠B and ∠C?

▶ The two acute angles of a right triangle are complementary.
▶ Consecutive angles of a parallelogram are supplementary.

E. Find the missing measures.

18.

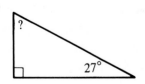

?

27°

19.

?

42°

20.

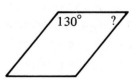

130° ?

210

Practice

Complementary or supplementary?

1. 24°, 66°   **2.** 80°, 10°   **3.** 80°, 100°   **4.** 52°, 38°

5. 74°, 16°   **6.** 90°, 90°   **7.** 40°, 140°   **8.** 91°, 89°

9. 37°, 53°   **10.** 89°, 1°   **11.** 179°, 1°   **12.** 17°, 163°

Find the complements.

13. 14°   **14.** 62°   **15.** 71°   **16.** 17°   **17.** 49°

18. 64°   **19.** 24°   **20.** 53°   **21.** 81°   ★ **22.** 90°

Find the supplements.

23. 80°   **24.** 75°   **25.** 141°   **26.** 74°   **27.** 114°

28. 178°   **29.** 40°   **30.** 125°   **31.** 119°   **32.** 0°

Find the missing measures.

33. 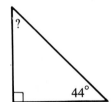   **34.** 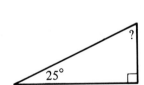   **35.**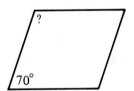

Something Extra

Non-Routine Problem

The diagram shows that all triangles and all quadrilaterals are polygons. It also shows that some polygons are neither triangles nor quadrilaterals.

Which of the following are true?

a. Some polygons are triangles.
b. Some quadrilaterals are triangles.
c. Some triangles are quadrilaterals.
d. Some polygons are quadrilaterals.

PARALLEL LINES

transversal

$m \parallel n$

is parallel to

Study and Learn

A. In the figure above, $m \parallel n$.

    1. $\angle x$ and $\angle y$ are called **alternate interior angles.** Name another pair of alternate interior angles.

    2. $\angle b$ and $\angle y$ are called **corresponding angles.** Name 3 more pairs of corresponding angles.

B. Draw a pair of parallel lines and a transversal.

    3. Measure a pair of alternate interior angles. Measure another pair of alternate interior angles.

    4. What is true of each pair of alternate interior angles?

    5. Measure a pair of corresponding angles. Measure another pair of corresponding angles.

    6. What is true of each pair of corresponding angles?

    ▶ If 2 parallel lines are cut by a transversal, the alternate interior angles are congruent.

    ▶ If 2 parallel lines are cut by a transversal, the corresponding angles are congruent.

C. Look at the figure above. Complete.

    7. If $m\angle x = 50°$, $m\angle y =$ ___?___   **8.** If $m\angle a = 130°$, $m\angle w =$ ___?___

D. When 2 lines intersect, vertical angles are formed.
$\angle a$ and $\angle b$ are vertical angles.
$\angle c$ and $\angle d$ are vertical angles.
Vertical angles are congruent.

Complete.

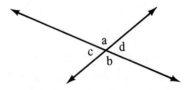

    9. If $m\angle a = 120°$, $m\angle b =$ ___?___   **10.** If $m\angle c = 60°$, $m\angle d =$ ___?___

Practice

$l \parallel m$. t is a transversal.

1. Name the pairs of alternate interior angles.

2. Name the pairs of corresponding angles.

3. Name the pairs of vertical angles.

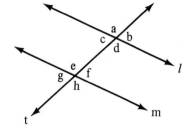

$x \parallel y$. t is a transversal. True or false?

4. $\angle f \cong \angle c$ 5. $\angle f \cong \angle b$

6. $\angle e \cong \angle d$ 7. $\angle e \cong \angle g$

8. $\angle a \cong \angle g$ 9. $\angle c \cong \angle h$

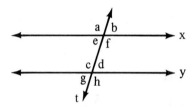

$p \parallel q$. Find m$\angle x$. Do not use a protractor.

10. 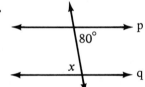   11. 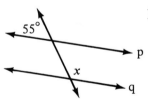   12.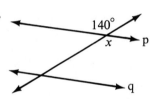

13. 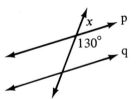   14. 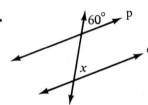   15.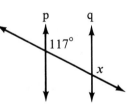

$s \parallel t$. m$\angle a = 112°$. Find the measures of the following angles.

16. $\angle b$ 17. $\angle c$

18. $\angle d$ 19. $\angle e$

20. $\angle f$ 21. $\angle g$

22. $\angle h$

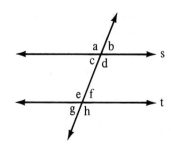

213

PERPENDICULAR LINES

Perpendicular lines

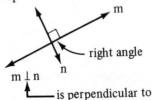

m ⊥ n

└── is perpendicular to

Perpendicular segments

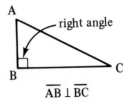

right angle

$\overline{AB} \perp \overline{BC}$

Study and Learn

A. $\overleftrightarrow{UV} \perp \overleftrightarrow{WX}$.

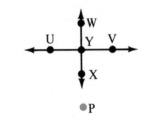

1. What is the measure of ∠ *WYU*?

2. What is the measure of ∠ *XYV*?

B. Construct the one line through point *P* perpendicular to line *m*.

3. Draw line *m* and point *P* as shown.

4. Place the point of your compass at *P* and make arcs which cut line *m*.

5. Keep the opening of the compass the same. Place the point of the compass at *A* and draw an arc below the line. Do the same at *B*.

6. Draw $\overleftrightarrow{PQ}$. $\overleftrightarrow{PQ} \perp m$.

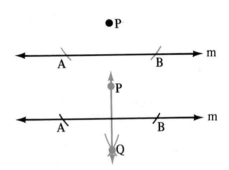

C. Construct a line through point *S* perpendicular to line *n*.

7. Draw line *n* and point *S* as shown.

8. Place the point of your compass at *S* and make arcs which cut line *n*.

9. Keep the opening of the compass at *A* and draw arcs above and below the line. Do the same at *B*.

10. Draw $\overleftrightarrow{TU}$. $\overleftrightarrow{TU} \perp n$.

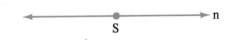

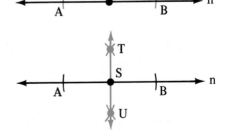

D. Construct a square.

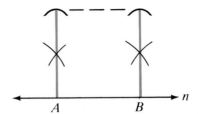

    11. Draw line *n* and $\overline{AB}$ as shown.
    12. Construct perpendiculars at
         A and *B*.
    13. Complete the square.

Practice

Which are perpendicular?

1.

2.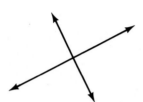

3.

How many right angles are in each figure?

4.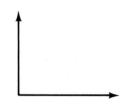

5.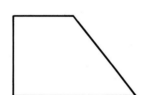

6.

    7. Draw line *n* and point *P* as shown. From point *P* construct a
    line perpendicular to line *n*.

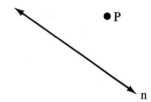

    8. Construct a right triangle *ABC* with right angle at *C*.
    (HINT: Draw $\overline{CB}$ and construct perpendicular at *C*.)

  ★ **9.** Draw two different segments on your paper. Construct a
      rectangle, using the segments as the length and width.

CONSTRUCTING PARALLEL LINES

There are 2 ways to construct parallel lines.
One way uses corresponding angles.

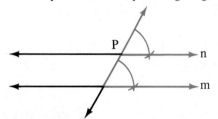

One way uses alternate interior angles.

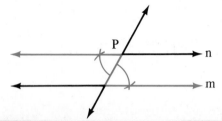

Study and Learn

A. Look at the figure at the right.

  1. How many possible lines through point P can be parallel to line m?

B. Construct the one line through P parallel to line m. Use the idea that corresponding angles are congruent.

  2. Draw line m and point P as shown.

  3. Through point P draw a transversal which intersects line m at Q.

  4. Construct an angle congruent to ∠PQR at point P.

  5. Draw line n. Line n through point P is parallel to line m.

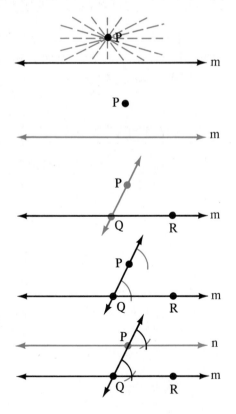

Practice

1. Draw a line *n* and point *P* as shown. Construct a line through point *P* parallel to line *n* using corresponding angles.

● P

⟵————————————————⟶ n

2. Draw a line *n* and point *P* as shown. Construct a line through point *P* parallel to line *n* using alternate interior angles.

● P

⟵————————————————⟶ n

★ **3.** Draw a line *n* and point *P* not on line *n*. Construct a line through *P* parallel to line *n*. Then construct a line through *P* perpendicular to line *n*.

Problem-Solving Applications

Using a Chart

60-Month Car Loan Rates	
Amount Loaned	Monthly Payment
$3,500	$79.25
$4,400	$99.63
$6,600	$149.45

1. Mr. Ames bought a $5,000 car. He put down $1,500. He took a $3,500 car loan for a 60-month period. What are his payments per month? [HINT: Look at the chart above.]

2. What will be Mr. Ames's total loan payments in 60 months?

3. What was the total amount that Mr. Ames paid for the car?

4. How much more did Mr. Ames pay for the car than the original cost of the car? (Note: This is called the finance charge.)

5. Mrs. Rodriguez bought a car for $7,500. How much cash must she put down if she wishes to take a $4,400 loan for 60 months?

6. How much must Mrs. Rodriguez put down if she wishes to take a $6,600 loan?

7. On a $4,400 loan, how much less cash will Mrs. Rodriguez pay each month than for a $6,600 loan?

8. What are the finance charges on the $4,400 loan for 60 months?

9. What are the finance charges on the $6,600 loan for 60 months?

10. How much will Mrs. Rodriguez save on finance charges with the $4,400 loan over the $6,600 loan?

11. Mr. Sloan bought a $13,000 car. He put down $6,400 and took a $6,600 car loan for a 60-month period. How much more than $13,000 will he pay for the car because of the loan?

Chapter Review

Measure these angles. *(192)*

1.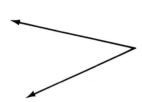

2.

3.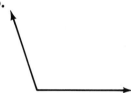

Classify these triangles by the measures of their angles. *(194)*

4.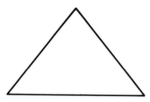

5.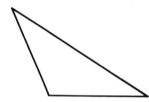

Classify these triangles by the measures of their sides. *(195)*

6.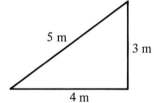

5 m
3 m
4 m

7.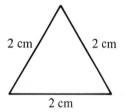

2 cm 2 cm
2 cm

Identify these polygons. *(198)*

8.

9.

10.

Find the missing measures. *(198, 200)*

11.

?
25°

12.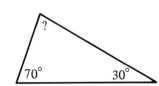

?
70° 30°

13.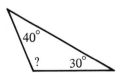

40°
? 30°

Chapter Review continues

Find the perimeters. *(202)*

14.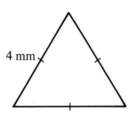

15.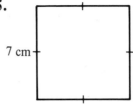

16.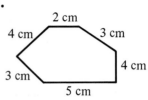

Find the circumference. Use $\pi \doteq 3.14$. *(208)*

17. $d = 3$ m

18. $r = 1$ cm

19. $r = 7$ mm

Find the complements. *(210)*

20. 45°   **21.** 50°   **22.** 3°   **23.** 23°

Find the supplements. *(210)*

24. 45°   **25.** 50°   **26.** 103°   **27.** 127°

$m \parallel n$ *(212)*

28. What is the measure of $\angle a$?

29. What is the measure of $\angle b$?

30. What is the measure of $\angle c$?

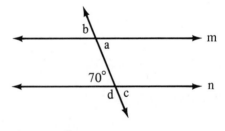

31. Draw a line l. Draw a point S on the line. Construct a
(214) perpendicular to line l through point S.

Solve, if possible. *(206, 218)*

32. The perimeter of a rectangle is
64 cm. The length is 20 cm.
What is the width?

33. Ms. Citera pays $137.50 each
month for her car loan. How
much does she pay in 1 year?

Chapter Test

Measure these angles. *(192)*

1.

2.

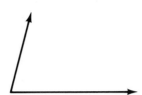

3.

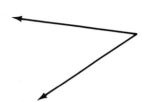

Classify these triangles by the measures of their angles. *(194)*

4.

5.

Classify these triangles by the measures of their sides. *(195)*

6.

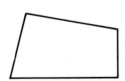

3 m   3 m

4 m

7.

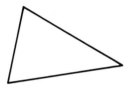

130 mm   200 mm

220 mm

Identify these polygons. *(198)*

8.

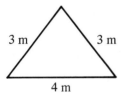

9.

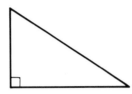

10.

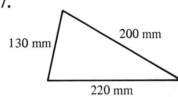

Find the missing measures. *(198, 200)*

11.

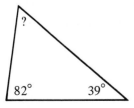

82°   39°

12.

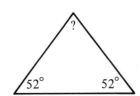

52°   52°

13.

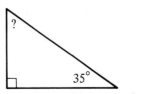

35°

Chapter Test continues

Chapter Test   continued

Find the perimeters. *(202)*

14.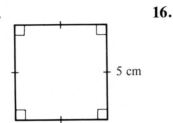
5 cm
2 cm
2 cm
4 cm
3 cm

15.
5 cm

16.
2 m
7 m

Find the circumference. Use $\pi \doteq 3.14$. *(208)*

17. $d = 11$ cm

18. $r = 4$ mm

19. $r = 6$ cm

Find the complements. *(210)*

20. $26°$

21. $35°$

22. $74°$

23. $41°$

Find the supplements. *(210)*

24. $60°$

25. $80°$

26. $103°$

27. $135°$

$m \parallel n$ *(212)*

28. What is the measure of $\angle a$?

29. What is the measure of $\angle b$?

30. What is the measure of $\angle c$?

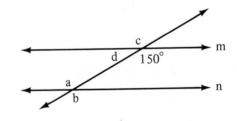

c
d   150°   m
a
b   n

31. Draw a line m. Draw a point P above the line. Construct a
(214) perpendicular to line m through point P.

Solve, if possible. *(206, 218)*

32. In right triangle PQR, $\angle Q$ is a
right angle. What is the measure
of $\angle R$?

33. Mr. Henn bought a car for
$7,800. He put down $2,900.
How much of a loan did he take?

Skills Check

1. Bill has saved $56.95 for a bicycle that costs $88.50. How much more does he need to buy the bicycle?

    A $31.45   B $31.50

    C $31.55   D $32.55

2. Carla had $643.27 in her checking account. She made out a check for $37.29. What is the balance in her checking account?

    E $604.98   F $605.98

    G $606.98   H $680.56

3. It was ⁻23°F in Nome, Alaska when it was 23°F in New York City. How much colder was it in Nome?

    A 0°   B 23°

    C 46°   D 50°

4. October 2 is a Monday. What day of the week is October 17th?

    E Monday   F Tuesday

    G Wednesday   H Thursday

5. What is the perimeter of the rectangle?

    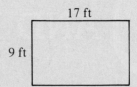

    A 153 ft   B 52 ft

    C 34 ft   D 26 ft

6. Bob jogged 850 m, 400 m, and 300 m one afternoon. How many kilometers did he jog altogether?

    E 1,550 km   F 155 km

    G 15.5 km   H 1.55 km

7. What seems to be the trend in the sale of magazines?

    A increasing   B decreasing

    C constant   D up and down

8. Which of the following units would you use to measure the amount of water in a raindrop?

    E mg   F mL

    G L   H kg

PROBLEM SOLVING
AND PROPORTIONS

The first commercial motorcycle could attain a speed of about 25 mph.

WRITING AND EVALUATING EXPRESSIONS

In solving problems, it is often necessary to translate word expressions to algebraic expressions.

Examine the chart below. *n* represents a number.

Word expressions	Algebraic expressions
2 more than a number	$n + 2$
a number increased by 2	$n + 2$
the sum of a number and 2	$n + 2$
3 less than a number	$n - 3$
twice a number	$2n$
5 more than twice a number	$2n + 5$
a number divided by 2	$\frac{n}{2}$

Study and Learn

A. Write algebraic expressions. Use *n* to represent a number.

1. 4 less than a number

2. 3 more than a number

3. the quotient of a number and 3

4. 3 more than twice a number

5. $\frac{1}{2}$ of a number

6. the square of a number

B. Compute.

$$2 \times (3 + 4)$$
$$= 2 \times 7$$
$$= 14$$

$$3 \times 6 + 4 \times 7$$
$$= 18 + 28$$
$$= 46$$

▶ Rules for order of operations: (1) Operate within parentheses.
(2) Multiply and divide from left to right.
(3) Add and subtract from left to right.

7. $3 \times (2 + 4)$ **8.** $8 \times (7 - 1)$ **9.** $10 + 7 \cdot 2$ **10.** $14 \div 7 - 2$

C. Evaluate.

11. $\frac{n}{2} \cdot (a + l)$ if $n = 8$, $a = 1$, and $l = 15$

12. $4a - b$ if $a = 6$, $b = 7$

Practice

Write algebraic expressions. Use n to represent a number.

1. 5 more than a number

2. a number increased by 12

3. the sum of a number and 9

4. 6 less than a number

5. a number decreased by 6

6. the quotient of a number and 9

7. 6 more than twice a number

8. 4 less than three times a number

9. $\frac{1}{3}$ of a number

10. $\frac{2}{5}$ of a number

11. the square of a number increased by 1

12. 14 decreased by a number

13. two times a number, decreased by 7

14. 6 less than $\frac{1}{2}$ of a number

15. $\frac{1}{2}$ of a number, increased by 6

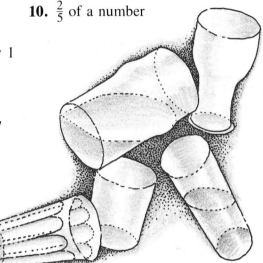

Compute.

16. $7 \times (9 - 3)$

17. $5 \times (6 + 7)$

18. $(16 - 4) - (18 - 16)$

19. $5 \cdot 9 + 6$

20. $18 - 12 \div 6$

21. $9 \cdot (2 \cdot 5 - 3)$

Evaluate.

22. $\frac{n}{2} \cdot (a + l)$ if $n = 6$, $a = 4$, and $l = 9$

23. $6s + t$ if $s = 3$ and $t = 4$

24. $50 - 25p$ if $p = 2$

★ **25.** $\frac{w + 6}{z}$ if $w = 4$ and $z = 2$

26. In a restaurant, the number of glasses can be found by computing $19 \cdot 6 + 4$. How many glasses are there?

WRITING EQUATIONS

Word sentences can be translated into equations.

Twice a number, increased by 7 is 23.

$$2 \qquad x \qquad + \qquad 7 = 23 \qquad 2x + 7 = 23$$

Seven less than a number is 12.

$$x \qquad - \qquad 7 \qquad = 12 \qquad x - 7 = 12$$

Study and Learn

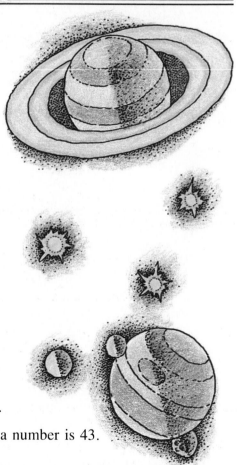

A. Write equations. Use n for a number.

  1. Six more than a number is 47.

  2. Eight less than a number is 63.

  3. A number divided by eight is 18.

  4. Three times a number is 129.

B. Write an equation for this sentence.
Four less than twice a number is 18.

  5. Use n for a number. What is twice a number?

  6. Write an algebraic expression for 4 less than twice a number.

  7. Write an equation for the sentence.

C. Write equations. Use x for a number.

  8. Twice a number, decreased by 9 is 47.

  9. Seven more than the product of 3 and a number is 43.

  10. Five less than, 7 times a number is 23.

Practice

Write equations. Use n for a number.

1. A number increased by 8 is 32.
2. A number decreased by 7 is 49.
3. Twice a number is 34.
4. Three times a number is 36.
5. A number divided by 4 is 18.
6. Six more than a number is 31.
7. Six less than a number is 24.
8. The quotient of a number and 7 is 5.
9. The sum of a number and 9 is 23.
10. The difference of a number and 8 is 32.
11. Four times a number, increased by 7 is 43.
12. Three times a number, decreased by 3 is 24.
13. Eight more than, 3 times a number is 44.
14. Six less than, 4 times a number is 42.
15. Twice a number, decreased by 6 is 34.
16. Twice a number, increased by 9 is 33.
17. Six more than, 4 times a number is 38.
18. Five less than, twice a number is 18.
19. The sum of 3 times a number and 8 is 20.
20. Five less than, the product of 6 and a number is 37.

Problem-Solving Skills

Writing Equations from Word Problems

Mrs. Jackson bought 3 containers of milk and a package of bread for $2.18. The bread cost $0.65. What was the cost of each container of milk?

PLAN   Use x for the cost of each container of milk.

$3x$ is an expression to represent the cost of 3 containers of milk.

THINK the cost of 3 containers of milk + cost of bread = total cost

$$3x \qquad\qquad + \qquad 65 \qquad = \qquad 218$$

The equation is $3x + 65 = 218$.

SOLVE $3x + 65 = 218$

$$3x + 65 - 65 = 218 - 65$$
$$3x = 153$$
$$x = 51$$

Each container of milk costs $0.51.

Study and Learn

A. Abe works in a supermarket. He receives $4 an hour. Last week he earned $92. How many hours did he work? Write an equation and solve.

    1. What does the problem ask?

    2. Let x represent the number of hours Abe worked. Write an expression to represent his pay for the week.

    3. Write an equation and solve.

B. Ellen bought 3 shirts, each at the same price. She paid a $5 deposit. This left a balance of $43. What was the cost of each shirt? Write an equation and solve.

    4. What does the problem ask?

    5. Let n represent the cost of each shirt. Write an expression to represent the cost of 3 shirts.

    6. Write an expression to represent the cost of 3 shirts less a $5 deposit.

    7. Write an equation and solve.

Practice

Write equations. Solve.

1. Josie bought 3 pens and a pencil for $2.08. The price of each pen was the same. The pencil cost $0.10. What was the cost of each pen?

2. Kathleen won a school election by receiving 17 more votes than Ken. Kathleen received 214 votes. How many votes did Ken receive?

3. Sam delivers papers and magazines. He has 43 customers for papers and 7 customers for magazines. Last week he received $14.25. He received $10.75 for delivering papers. How much does he receive from each of the 7 magazine customers?

4. Andy's father rented a car for a family trip. The car rented for $84 a week plus the cost of gas. Andy's father found it cost him $126 for the week. If he used 40 gal of gasoline, what was the cost of each gallon?

5. Ann's father is 50 years old. He is 8 years more than 3 times Ann's age. How old is Ann?

6. Marcia's bowling score was 28 pins more than Marty's score. Marcia's bowling score was 151. What was Marty's bowling score?

7. Gene bought a pair of shoes and 3 pairs of socks for $30.34. The shoes were $25 and the price of each pair of socks was the same. What was the cost of each pair of socks?

8. A wire was 20 m long. It was cut into 3 pieces of the same length and a piece 2 m long. How long was each of the 3 pieces of the same length?

Problem-Solving Skills

Making Up Problems from Equations

One equation may solve many problems. Here are 2 problems that can be solved by $2x + 4 = 10$.

Amy scored a total of 10 points in basketball. She scored 4 points for foul shots. How many baskets did she make? (Each basket is 2 points.) Let x be the number of baskets.

2 points for each basket	+	points for fouls	=	point total
$2x$	+	4	=	10

Jeff bought 2 records. The price of each is the same. He also bought a $4 tape. The total cost was $10. How much did each record cost? Use x for the cost of one record.

cost of 2 records	+	cost of tape	=	total cost
$2x$	+	4	=	10

Practice

Make up two problems for each equation.

1. $x + 24 = 37$

2. $x + 34 = 102$

3. $x - 5 = 18$

4. $x - 15 = 76$

5. $2x + 4 = 36$

6. $3x + 8 = 71$

7. $3x - 9 = 39$

8. $4x - 12 = 60$

9. $\frac{x}{3} + 2 = 5$

10. $\frac{x}{2} - 12 = 14$

Midchapter Review

Compute. *(226)*

1. $4 \times (8 - 3)$ **2.** $24 \div (1 + 5)$ **3.** $18 - 6 \cdot 2$

Evaluate. *(226)*

4. $4a + b$ if $a = 6$ and $b = 7$ **5.** $n - n + k$ if $n = 14$ and $k = 3$

Write equations. Solve. *(228, 230)*

6. A number increased by 9 is 53.

7. Five less than a number is 37.

8. Three times a number, increased by 9 is 18.

9. Twice a number, decreased by 5 is 29.

10. Sonia wants to buy a radio for $78. She earns $7 a week baby-sitting, and has saved $43. How many more weeks must she work to buy the radio?

11. Henry bought 3 packs of notebook paper and a notebook for $4.75. The notebook cost $1.00. What is the price of each pack of notebook paper?

Something Extra
Non-Routine Problem

Mike bought 3 shirts and 2 pairs of jeans. The cost of the 3 shirts was the same as the cost of the 2 pairs of jeans.

Which of the following can be determined from the above information?

a. The cost of 1 pair of jeans is the same as the cost of 1 shirt.

b. A pair of jeans costs less than a shirt.

c. A shirt costs $\frac{1}{3}$ of the cost of a pair of jeans.

d. A pair of jeans costs $1\frac{1}{2}$ times the cost of a shirt.

REASONING

Given: All eighth-grade students were present today. Henry is an eighth-grade student.

Conclusion: Henry was present today.

A diagram can help you decide if the conclusion is necessarily true.

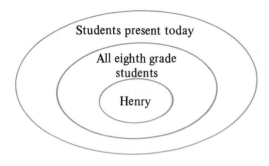

The conclusion is necessarily true.

Study and Learn

A. Given: All baseball players eat "High Power." Mary eats "High Power."

Conclusion: Mary is a baseball player.

1. Is it possible that Mary is a baseball player?

2. Is it possible that Mary is not a baseball player?

3. Is the conclusion necessarily true?

B. Given: All movie stars use "Clean" toothpaste. John uses "Clean."

Conclusion: John is a movie star.

4. Is the conclusion necessarily true?

234

Practice

Decide whether the conclusions are necessarily true.

1. Given: All intelligent students can pass mathematics.
Maria is an intelligent student.
Conclusion: Maria can pass mathematics.

2. Given: All intelligent students can pass mathematics.
Walter can pass mathematics.
Conclusion: Walter is an intelligent student.

3. Given: All athletes eat "Strongies" for breakfast.
Jane is an athlete.
Conclusion: Jane eats "Strongies" for breakfast.

4. Given: All athletes eat "Strongies" for breakfast.
Tony eats "Strongies" for breakfast.
Conclusion: Tony is an athlete.

5. Given: All students on the math team are eighth-grade students.
Adele is on the math team.
Conclusion: Adele is an eighth-grade student.

6. Given: All students on the math team are eighth-grade students.
Adele is an eighth-grade student.
Conclusion: Adele is on the math team.

7. Given: All movie stars eat lunch at "Oscars."
Sonia eats lunch at "Oscars."
Conclusion: Sonia is a movie star.

8. Given: All movie stars eat soup for lunch.
Sue is a movie star.
Conclusion: Sue eats soup for lunch.

9. Given: All diameters are chords.
$\overline{AB}$ is a chord.
Conclusion: $\overline{AB}$ is a diameter.

Problem-Solving Applications
The Distance Formula

A jet plane flew at the rate of 950 km/h. How far did it fly in 9 hours?

Distance formula: distance = rate · time
$$d = r \cdot t$$

To solve the problem:

Step 1 Write the formula. $d = r \cdot t$

Step 2 Substitute the given information. $d = 950 \cdot 9$

Step 3 Solve the equation. $d = 8{,}550$ km

So the plane had flown 8,550 km.

Study and Learn

A. Solve. Use the distance formula.

Example   Lucian left home at 8:00 am and arrived at the office at 10:00 am. He lives 64 miles from the office. What was his average rate of speed?

Step 1   Write the distance formula. $d = r \cdot t$

Step 2   Substitute the given information. $64 = r \cdot 2$

Step 3   Solve the equation. $\frac{64}{2} = \frac{2r}{2}$

                                             $32 = r$

So, Lucian's average rate of speed was 32 mph.

Mary drove 180 mi in 4 hours. At what rate did she drive?

1. Write the distance formula.

2. Substitute the given information.

3. Solve $180 = r \cdot 4$.

B. Dave drove 260 km at 65 km/h. How long did the trip take?

4. Write the distance formula.

5. Substitute the given information

6. Solve $260 = 65 \cdot t$.

Practice

Solve. Use the distance formula.

1. A commuter drove from her home to her place of work in $\frac{1}{2}$ hour. She drove at the rate of 76 km/h. What is the distance from her home to her office?

2. The highway distance from Chicago, Illinois, to Phoenix, Arizona, is 1,755 mi. Ms. Gomez made the trip in 30 hours over a 3-day period. What was her average speed for the trip?

3. May and Sam took a bicycle trip. They rode at the rate of 15 km/h. They rode from 8:00 am to 4:00 pm that afternoon. How far did they ride?

4. A ship left port at 9:00 pm and at 2:00 am the next morning was 80 mi out at sea. At what average rate of speed was the ship traveling?

5. A bus driver averaged 80 km/h during a 6-hour driving day. How far did he drive?

6. Ms. Joly rode a bicycle 96 km at an average rate of 16 km/h. How long did the trip take?

7. Mr. Perez walked from 8:30 am to noon along a hiking trail. He walked at an average rate of 3 mph. How far did he walk?

8. On a $4\frac{1}{2}$-hour air trip, the pilot announced that their average speed was 500 mph. How far was the trip?

9. Pete plans to take a car trip of 900 km. He plans to average 75 km/h. How long will the trip take?

10. The average speed during a trip was 54.7 mph. The time for the trip was 3 hours. How long was the trip?

11. Maria saw lightning and then heard thunder 5 seconds later. How far from the lightning was she if sound travels about 1,100 feet per second?

12. A bus left Town A at 8:30 am and arrived in Town B, 175 miles away, at noon. What was the average speed of the bus?

★ 13. A man drove 200 miles in 4 hours. He drove the return trip of 200 miles in 5 hours. What was his average speed for the round trip?

237

Skills Review

Add. (102, 170)

1. 36.1
    84.3
    61.4
    + 81.4

2. 64.183
    2.409
    31.604
    + 0.786

3. 24.38162
    0.48239
    9.14613
    + 0.32007

4. 1.4 + 6.39 + 0.7

5. 0.7 + 0.06 + 0.931

6. $3\frac{3}{8}$
   $+ 2\frac{1}{4}$

7. $7\frac{1}{2}$
   $+ 2\frac{3}{4}$

8. $4\frac{7}{8}$
   $+ 3\frac{1}{2}$

9. $7\frac{3}{5}$
   $+ 2\frac{3}{4}$

10. $25\frac{3}{4}$
    $+ 12\frac{1}{3}$

Subtract. (104, 174)

11. 64.1
    − 38.9

12. 815.6
    − 314.8

13. 0.9004
    − 0.8966

14. 6.3967
    − 2.6808

15. 4.3578
    − 1.0681

16. 0.8 − 0.5

17. 0.017 − 0.009

18. 5 − 0.87

19. 9 − 0.656

20. $5\frac{5}{8}$
    $- 2\frac{1}{8}$

21. $12\frac{3}{4}$
    $- 9\frac{3}{8}$

22. 5
    $- 2\frac{1}{2}$

23. 9
    $- 8\frac{7}{9}$

24. $8\frac{1}{3}$
    $- 4\frac{5}{6}$

Multiply. (110, 178)

25. 29.6
    × 7

26. 86.5
    × 5

27. 0.127
    × 3

28. 0.424
    × 8

29. 136.8
    × 9

30. 0.34
    × 2.1

31. 0.29
    × 0.8

32. 23.86
    × 4.21

33. 2.809
    × 0.82

34. 0.064
    × 0.37

35. $\frac{2}{3} \times \frac{5}{6}$

36. $\frac{2}{3} \times 12$

37. $3\frac{1}{2} \times 1\frac{2}{3}$

38. $4\frac{1}{8} \times 6\frac{1}{2}$

Divide. (116, 182)

39. $4\overline{)17.6}$

40. $9\overline{)0.081}$

41. $18\overline{)94.14}$

42. $26\overline{)241.8}$

43. $87\overline{)562.89}$

44. $0.3\overline{)0.24}$

45. $1.7\overline{)255}$

46. $0.08\overline{)0.048}$

47. $0.006\overline{)30}$

48. $6.9\overline{)310.5}$

49. $\frac{1}{2} \div \frac{1}{3}$

50. $\frac{7}{8} \div \frac{2}{5}$

51. $6 \div \frac{1}{4}$

52. $3\frac{1}{2} \div 2$

RATIO

6 girls and 8 boys joined the school running team. The ratio of
the number of girls to the number of boys on the team is 6 to 8.

The ratio 6 to 8 may be written as $6:8$ or $\frac{6}{8}$.

Study and Learn

A. Write each ratio in 2 ways.
The track team won 9 races and lost 3.

  1. What is the ratio of the number of races won to the number
  of races lost?

  2. What is the ratio of the number of races won to the total
  number of races?

B. Two ratios are equal if they can be written as equivalent fractions.

6 is to 8   3 is to 4

$\lfloor$equal ratios$\rfloor$ since $\frac{6}{8} = \frac{3}{4}$

Which ratios are equal to 10 is to 15?

  3. 6 is to 9   **4.** 5 is to 10   **5.** 20 is to 30

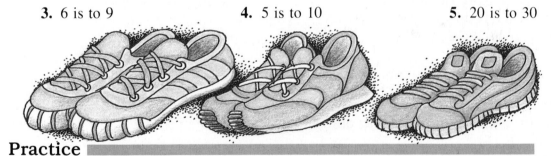

Practice

Leonard is on the basketball team. He played in 6 games and did
not play in 4.

  1. What is the ratio of the number
  of games played to the total
  number of games?

  2. What is the ratio of the number of
  games not played to the number
  of games played?

Which ratios are equal to 7 is to 14?

  3. 6 is to 42   **4.** 12 is to 24   **5.** 8 is to 15

PROPORTION

Steve's team won 7 games and lost 3 games.
Emma's team won 14 games and lost 6 games.
The ratios of games won to games lost are equal.

$\frac{7}{3} = \frac{14}{6}$ $7:3 = 14:6$ ⟵ read 7 is to 3 as 14 is to 6

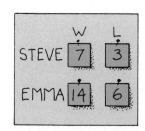

Study and Learn

A. A proportion is a statement of equal ratios.

$\frac{7}{3} = \frac{14}{6}$ is a true proportion.

Which are true proportions?

1. $6:3 = 5:2$ **2.** $8:24 = 6:18$ **3.** $\frac{3}{10} = \frac{9}{30}$

B. A proportion has 2 means and 2 extremes.

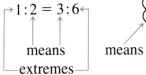

Identify the means.   Identify the extremes.

4. $\frac{1}{2} = \frac{2}{4}$ **5.** $1:3 = 2:6$ **6.** $\frac{8}{4} = \frac{4}{2}$ **7.** $5:4 = 50:40$

C. In a true proportion, the product of the means equals the product of the extremes.

8. What is the product of the means in $\frac{8}{4} = \frac{4}{2}$?

9. What is the product of the extremes in $\frac{8}{4} = \frac{4}{2}$?

10. Are the products equal?

D. Solve the proportions.

Examples $2:3 = 10:x$ | $\frac{1}{2} = \frac{x}{10}$

Step 1 Find products. $30 = 2x$ | $10 = 2x$
Step 2 Solve. $15 = x$ | $5 = x$

11. $1:2 = 10:x$ **12.** $8:6 = x:3$ **13.** $\frac{x}{12} = \frac{3}{4}$

240

Practice

Which are true proportions?

1. $1:3 = 4:12$ **2.** $2:3 = 4:5$ **3.** $\frac{4}{6} = \frac{2}{3}$ **4.** $\frac{8}{3} = \frac{24}{9}$

5. $3:4 = 9:15$ **6.** $8:2 = 4:2$ **7.** $\frac{6}{8} = \frac{9}{12}$ **8.** $\frac{12}{8} = \frac{9}{6}$

Solve the proportions.

9. $1:2 = 2:x$ **10.** $2:3 = x:6$ **11.** $2:x = 4:10$ **12.** $x:3 = 6:9$

13. $5:4 = x:8$ **14.** $8:2 = 12:x$ **15.** $4:x = 3:6$ **16.** $x:9 = 4:6$

17. $6:5 = 12:x$ **18.** $8:x = 4:5$ **19.** $x:8 = 3:12$ **20.** $15:10 = 3:x$

21. $\frac{6}{8} = \frac{3}{x}$ **22.** $\frac{2}{3} = \frac{x}{6}$ **23.** $\frac{8}{x} = \frac{4}{5}$ **24.** $\frac{x}{8} = \frac{3}{2}$

25. $\frac{x}{6} = \frac{12}{9}$ **26.** $\frac{3}{x} = \frac{4}{8}$ **27.** $\frac{5}{4} = \frac{x}{8}$ **28.** $\frac{2}{6} = \frac{3}{x}$

29. $\frac{5}{x} = \frac{2}{4}$ **30.** $\frac{9}{6} = \frac{12}{x}$ **31.** $\frac{x}{28} = \frac{2}{7}$ ★ **32.** $\frac{x}{12} = \frac{3}{x}$

Something Extra

Calculator Activity

1. Find these products on a calculator.

$15{,}873 \times 7 = \underline{\ ?\ }$
$15{,}873 \times 14 = \underline{\ ?\ }$
$15{,}873 \times 21 = \underline{\ ?\ }$
$15{,}873 \times 28 = \underline{\ ?\ }$
$15{,}873 \times 35 = \underline{\ ?\ }$
$15{,}873 \times 42 = \underline{\ ?\ }$
$15{,}873 \times 49 = \underline{\ ?\ }$
$15{,}873 \times 56 = \underline{\ ?\ }$

2. What do you notice?

3. What would be the next problem in the sequence? What is the product?

Problem-Solving Applications

The Lever Formula

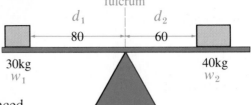

Two different masses can be balanced
on a seesaw by adjusting their
distances from the fulcrum.

The greater mass should be closer to
the fulcrum.

The seesaw will be in balance when the masses and distances are in
this proportion:

$$\frac{w_1}{w_2} = \frac{d_2}{d_1} \qquad \frac{30}{40} = \frac{60}{80}$$

Since the product of the extremes equals the product of the means,
the lever formula can be written this way:

$$w_1 \cdot d_1 = w_2 \cdot d_2 \qquad 30 \cdot 80 = 40 \cdot 60$$
$$2{,}400 = 2{,}400$$

Study and Learn

A. The scale is balanced. What
is the mass of the rocks?

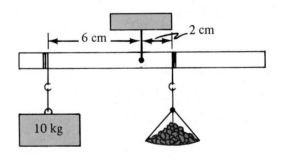

   1. Write the lever formula
   in product form.

   2. Substitute the given
   information in the formula.

   3. Solve. $10 \cdot 6 = x \cdot 2$ or
   $60 = 2x$

B. A 16-kg mass is 5 cm from the fulcrum. What mass 4 cm from
the fulcrum will balance it?

   4. Write the lever formula in product form.

   5. Substitute the given information in the formula.

   6. Solve $16 \cdot 5 = x \cdot 4$.

Practice

The levers are balanced. Find x.

1.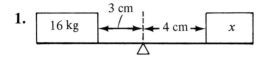

2.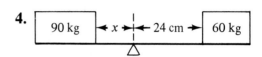

3.

12 lb
| 6 in. | x | 24 lb |

4.

| 90 kg | x | 24 cm | 60 kg |

Solve. Use the lever formula.

5. A 36-g mass is 4 cm from the fulcrum. What mass 6 cm from the fulcrum will balance it?

6. An 81-lb weight is 2 yd from the fulcrum. What weight 3 yd from the fulcrum will balance it?

7. Mike has a mass of 32 kg. He is sitting 300 cm from the fulcrum. José has a mass of 24 kg. The seesaw is balanced. How far is José sitting from the fulcrum?

8. An 18-g mass 4 cm from the fulcrum balances a 12-g mass. How far from the fulcrum is the 12-g mass?

9. A 9-g mass 12 cm from the fulcrum balances a 6-g mass. How far from the fulcrum is the 6-g mass?

10. A 300-lb rock is 4 ft from the fulcrum. What weight 12 ft from the fulcrum will balance it?

Something Extra
Non-Routine Problem

Complete the sequences.

(A)$\frac{1}{2}$, (C)$\frac{1}{4}$, (E)$\frac{1}{6}$, _?_ , _?_

(A)$1\frac{1}{2}$, (B)$2\frac{1}{2}$, (C)$3\frac{1}{2}$, _?_ , _?_

(Z)$\frac{1}{5}$, (Y)$\frac{1}{10}$, (X)$\frac{1}{20}$, (W)$\frac{1}{40}$, _?_ , _?_

Problem-Solving Applications

Using Proportions

Class 8-7 held an election for class president. Ann received 3 votes for each 2 votes that Jim received. Ann received 15 votes. How many votes did Jim receive?

To solve the problem, set up a proportion.

Ann $\longrightarrow$ $\dfrac{3}{2} = \dfrac{15}{x}$
Jim $\longrightarrow$

$30 = 3x$ $\longleftarrow$ product of means = product of extremes
$10 = x$

So Jim received 10 votes.

Practice

1. Class 8-3 held a bazaar. They sold records at 3 for $1.25. What was the cost of a dozen records?

2. The class bazaar was advertised on rectangular signs. The ratio of the width to the length of the sign was 3 to 5. The length of the sign was 35 cm. What was the width?

3. In home economics class, Class 8-3 made class pennants. They sold at the bazaar at 2 for 55¢. What was the cost of 6 pennants?

4. Class 8-4 won 2 out of every 5 races at the annual school picnic. They won 6 races. How many races were held?

5. For the annual school picnic, Class 8-6 brought 8 sandwiches for every 5 students. 25 students came to the picnic from Class 8-6. How many sandwiches were brought?

6. Class 8-5 sells greeting cards. For each $7 sale, $5 goes to charity. How much goes to charity if a sale is $21?

7. In Class 8-1, 5 out of every 7 students arrive by school bus. How many students arrive by bus if the class has 28 students?

8. The ratio of the number of tapes sold to the number of records sold was 4:3. If 24 tapes were sold, how many records were sold?

Chapter Review

Compute. *(226)*

1. $7 \times (9 - 3)$

2. $5 \times 2 + 8$

Evaluate. *(226)*

3. $7n + 3$ if $n = 4$

4. $3 \cdot (a \cdot b)$ if $a = 8$ and $b = 7$

Write an equation. *(228)*

5. Four times a number, decreased by 7 is 21.

Write an equation. Solve. *(230)*

6. Steve bought 3 records and a tape for $17.10. The tape cost $5.25. Each record was the same price. What was the price of each record?

Solve. Use the distance formula. *(236)*

7. Beth drove from 9:00 am to 3:00 pm. Her average speed was 64 km/h. How far did she drive?

Solve the proportion. *(240)*

8. $\frac{5}{8} = \frac{10}{x}$

Solve. Use the lever formula. *(242)*

9. A 24-g mass is 4 cm from the fulcrum. What mass 6 cm from the fulcrum will balance it?

Solve. *(244)*

10. A clothing warehouse carries dresses and suits in a ratio of 3 dresses to 2 suits. If they have 210 dresses, how many suits do they have?

Chapter Test

Compute. *(226)*

  1. $5 \times (11 - 6)$

                                2. $18 \div 3 + 3$

Evaluate. *(226)*

  3. $5x - y$ if $x = 8$ and $y = 9$

                                4. $\frac{p}{2} \cdot (q - 3)$ if $p = 8$ and $q = 4$

Write an equation. *(228)*

  5. Six less than 7 times a number is 29.

Write an equation. Solve. *(230)*

  6. Martha bought 4 cans of soup and a loaf of bread for $1.71. The loaf of bread cost $0.75. What was the price of each can of soup?

Solve. Use the distance formula. *(236)*

  7. Ms. Miles drove from her home to work and back again. The total distance driven was 78 mi. She drove at an average speed of 52 mph. How long did she drive?

Solve the proportion. *(240)*

  8. $3:2 = x:6$

Solve. Use the lever formula. *(242)*

  9. An 18-g mass is 2 cm from the fulcrum. How far from the fulcrum must a 4-g mass be placed to balance it?

Solve. *(244)*

  10. Eva needs 3 parts of red paint to 4 parts of blue paint to get a certain shade of purple. How much blue paint does she need to add to 24 L of red paint?

Skills Check

1. What type of angle is ∠*ABC*?

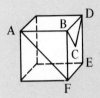

A	B	C	D
right	obtuse	acute	straight

2. Which line segment represents the edge of a cube?

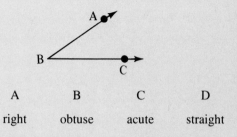

E	F	G	H
$\overline{AF}$	$\overline{AB}$	$\overline{BC}$	none of the above

3. What type of figure is shown?

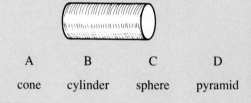

A	B	C	D
cone	cylinder	sphere	pyramid

4. Which is an isosceles triangle?

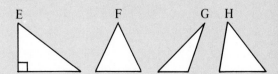

5. The radius of a circle is 24 mm. What is the diameter?

A	B	C	D
12 mm	24 mm	36 mm	48 mm

6. Which line segment is the altitude of △*GHI*?

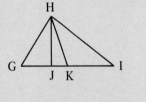

E	F	G	H
$\overline{HJ}$	$\overline{HK}$	$\overline{GI}$	$\overline{HI}$

7. Which angle is congruent to ∠1?

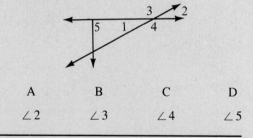

A	B	C	D
∠2	∠3	∠4	∠5

8. Which of the figures is not symmetric?

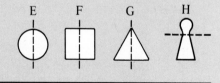

9. Which of the angles is 150°?

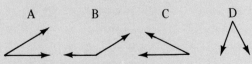

The modern oil industry dates from 1859.

FRACTIONS AND PERCENTS

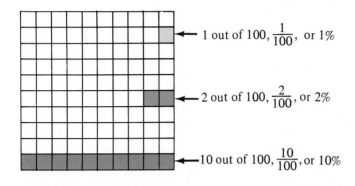

1 out of 100, $\frac{1}{100}$, or 1%

2 out of 100, $\frac{2}{100}$, or 2%

10 out of 100, $\frac{10}{100}$, or 10%

$\frac{18}{100} = 18\%$

$50\% = \frac{50}{100}$, or $\frac{1}{2}$

Study and Learn

A. Write the simplest fractions or whole numbers.

    1. 10%   **2.** 25%   **3.** 40%   **4.** 100%   **5.** 200%

B. Write the simplest fractions.

    Example $33\frac{1}{3}\% = 33\frac{1}{3} \cdot 1\%$

                      $= 33\frac{1}{3} \cdot \frac{1}{100}$

                      $= \frac{\overset{1}{\cancel{100}}}{3} \cdot \frac{1}{\underset{1}{\cancel{100}}} = \frac{1}{3}$ So, $33\frac{1}{3}\% = \frac{1}{3}$

    6. $12\frac{1}{2}\%$ **7.** $16\frac{2}{3}\%$ **8.** $4\frac{1}{2}\%$ **9.** $\frac{1}{4}\%$ **10.** $\frac{1}{2}\%$

C.

> Look at these methods for writing fractions as percents.
>
> **Method 1**
>
> $\frac{3}{5} = \frac{3 \cdot 20}{5 \cdot 20}$
>
> $= \frac{60}{100}$
>
> $= 60\%$
>
> **Method 2**
>
> $\frac{3}{5}$ or $5\overline{)3}$ $5\overline{)3.00}$ ← hundredths, quotient 0.60
>
> $\frac{3}{5} = 0.60 = \frac{60}{100} = 60\%$

    11. Use both methods to change $\frac{3}{4}$ to a percent.

D. Write percents.

    12. $\frac{1}{2}$ **13.** $\frac{1}{8}$ **14.** $\frac{1}{200}$ **15.** $\frac{300}{100}$ **16.** $\frac{7}{4}$

Practice

Write the simplest fractions or whole numbers.

1. 75% **2.** 24% **3.** 55% **4.** 92% **5.** 72%

6. 16% **7.** 65% **8.** 81% **9.** 300% **10.** 400%

11. $17\frac{1}{2}\%$ **12.** $8\frac{1}{3}\%$ **13.** $6\frac{1}{4}\%$ **14.** $\frac{2}{3}\%$ **15.** $\frac{1}{10}\%$

16. $\frac{1}{5}\%$ **17.** 250% **18.** 12% **19.** $\frac{1}{3}\%$ **20.** $14\frac{1}{2}\%$

Write percents.

21. $\frac{1}{4}$ **22.** $\frac{2}{5}$ **23.** $\frac{3}{5}$ **24.** $\frac{7}{10}$ **25.** $\frac{3}{25}$ **26.** $\frac{11}{50}$

27. $\frac{3}{8}$ **28.** $\frac{1}{12}$ **29.** $\frac{1}{3}$ **30.** $\frac{3}{200}$ **31.** $\frac{1}{400}$ **32.** $\frac{3}{400}$

33. $\frac{200}{100}$ **34.** $\frac{500}{100}$ **35.** $\frac{6}{5}$ **36.** $\frac{5}{4}$ ★**37.** $\frac{8}{1,000}$ ★**38.** $\frac{3}{10,000}$

Here are tables of commonly used percents. Complete.

	Percent	Fraction
39.	25%	
40.	50%	
41.	75%	
42.	$33\frac{1}{3}\%$	
43.	$66\frac{2}{3}\%$	
44.	10%	
45.	20%	

	Percent	Fraction
46.	30%	
47.	40%	
48.	60%	
49.	70%	
50.	80%	
51.	90%	
52.	100%	

	Percent	Fraction
53.	$12\frac{1}{2}\%$	
54.	$37\frac{1}{2}\%$	
55.	$62\frac{1}{2}\%$	
56.	$16\frac{2}{3}\%$	
57.	$83\frac{1}{3}\%$	
58.	$87\frac{1}{2}\%$	
★**59.**	$\frac{1}{8}\%$	

Solve.

60. A record was marked $33\frac{1}{3}\%$ off. What fractional part of the cost is saved?

DECIMALS AND PERCENTS

Percent means hundredth. $1\% = 0.01$
Here is a method for changing percents to decimals:

$$140\% = 140 \cdot 1\% \qquad\qquad 0.3\% = 0.3 \cdot 1\%$$
$$= 140 \cdot 0.01 \qquad\qquad = 0.3 \cdot 0.01$$
$$= 1.40 \qquad\qquad\qquad = 0.003$$

Study and Learn

A. Change to decimals.

   1. 7%   **2.** 81%   **3.** 130%   **4.** 0.4%   **5.** 2.6%

B. Here is a quick way to change a percent to a decimal.

   Examples $140\% = 1.40$ **Step 1**   Move the decimal point 2 places to the left.

                       $0.3\% = 0.003$ **Step 2**   Drop the % sign.

                       $9\% = 0.09$

Change to decimals.

   6. 63%   **7.** 150%   **8.** 0.5%   **9.** 2.3%   **10.** 3.04%

C. Percents with fractions may be changed to decimals.

   Example $\frac{1}{2}\%$ **Step 1**   Change the fraction to a decimal.

$$\frac{1}{2} \text{ is } 2\overline{)1.0}^{\,0.5} \qquad \text{So, } \frac{1}{2}\% = 0.5\%$$

                       Step 2   Change the percent to a decimal.

$$0.5\% = 0.005 \qquad \text{So, } \frac{1}{2}\% = 0.005$$

Change to decimals.

   11. $\frac{1}{4}\%$ **12.** $\frac{1}{5}\%$ **13.** $\frac{3}{4}\%$ **14.** $2\frac{1}{2}\%$ **15.** $5\frac{1}{4}\%$

D. The quick way is reversed to change decimals to percents.

   Examples $0.06 \rightarrow 0.06 = 6\%$ **Step 1**   Move the decimal point 2 places to the right.

                  $0.3 \rightarrow 0.30 = 30\%$ **Step 2**   Write the % sign.

Change to percents.

   16. 0.78   **17.** 0.5   **18.** 1   **19.** 2.4   **20.** 0.075

E. Percents, fractions, and decimals are related. Complete the table.

	%	Fraction	Decimal
21.	20%		
22.		$\frac{1}{2}$	
23.			0.25

Practice

Change to decimals.

1. 2%   **2.** 6%   **3.** 9%   **4.** 10%   **5.** 23%

6. 34%   **7.** 42%   **8.** 60%   **9.** 80%   **10.** 100%

11. 120%   **12.** 140%   **13.** 210%   **14.** 250%   **15.** 300%

16. 0.2%   **17.** 0.6%   **18.** 0.9%   **19.** 3.1%   **20.** 8.6%

21. 9.5%   **22.** 0.03%   **23.** 0.06%   **24.** 2.04%   **25.** 5.25%

26. $\frac{1}{5}$%   **27.** $\frac{3}{10}$%   **28.** $2\frac{1}{4}$%   **29.** $5\frac{1}{2}$%   ★**30.** $6\frac{5}{8}$%

Change to percents.

31. 0.08   **32.** 0.64   **33.** 0.93   **34.** 0.6   **35.** 0.7

36. 0.9   **37.** 2   **38.** 2.8   **39.** 3.5   **40.** 4

41. 0.025   **42.** 0.050   **43.** 0.064   **44.** 0.001   **45.** 0.0075

46. 0.0061   ★**47.** $0.02\frac{1}{2}$ ★**48.** $0.05\frac{1}{4}$ ★**49.** $0.12\frac{1}{2}$ ★**50.** $0.66\frac{2}{3}$

Complete the table.

	%	Fraction	Decimal
51.	75%		
52.			0.40
★**53.**		$\frac{1}{6}$	

FINDING A PERCENT OF A NUMBER

Mrs. Perez bought a bicycle for $85.
She had to pay 7% sales tax. How
much sales tax did she pay?

THINK: 7% of $85 is what?

Equation: $7\% \cdot 85 = n$

$$0.07 \cdot 85 = n$$
$$5.95 = n$$

The sales tax was $5.95.

Study and Learn

A. Compute.

    1. 8% of 65   **2.** 12% of 150   **3.** 0.3% of 6,000

B. Find 140% of 65.

    4. THINK: 140% of 65 is what? Write an equation.

    5. Write a decimal for 140%.

    6. Solve the equation $1.40 \cdot 65 = n$.

    7. What is 140% of 65?

C. Compute.

    Example $\frac{1}{4}\%$ of 600 $\frac{1}{4}\% \cdot 600 = n$

$$0.0025 \cdot 600 = n$$
$$1.5 = n$$

    8. $\frac{1}{4}\%$ of 800   **9.** $\frac{1}{2}\%$ of 800   **10.** $3\frac{1}{2}\%$ of 450

D. Sometimes it is easier to change the percent to a fraction.

    Example   25% of 60 $25\% \cdot 60 = n$

$$\frac{1}{4} \cdot 60 = n$$
$$15 = n$$

Compute.

    11. 75% of 40   **12.** $12\frac{1}{2}\%$ of 72   **13.** $33\frac{1}{3}\%$ of 24

Practice

Compute.

1. 6% of 40
2. 9% of 100
3. 8% of 700
4. 10% of 45
5. 12% of 76
6. 56% of 850
7. 76% of 80
8. 87% of 240
9. 86% of 750
10. 100% of 80
11. 110% of 700
12. 120% of 40
13. 150% of 640
14. 225% of 80
15. 375% of 800
16. $\frac{1}{2}$% of 60
17. $\frac{1}{4}$% of 800
18. $\frac{3}{4}$% of 6,000
19. $2\frac{1}{2}$% of 80
20. $3\frac{1}{4}$% of 500
21. $5\frac{1}{4}$% of 2,500
22. 0.2% of 600
23. 0.5% of 950
24. 0.7% of $675
25. 0.06% of 842
26. 0.08% of 1,000
27. 0.07% of 8,000
28. 50% of 60
29. 5% of 200
30. 25% of 8
31. $33\frac{1}{3}$% of 9
32. $66\frac{2}{3}$% of 12
33. $12\frac{1}{2}$% of 16
34. $37\frac{1}{2}$% of 24
35. $62\frac{1}{2}$% of 80
36. $16\frac{2}{3}$% of 24
37. 25% of 120
38. 175% of 12
39. $\frac{1}{4}$% of 20
40. $2\frac{1}{2}$% of 90
41. 0.4% of 60
42. $87\frac{1}{2}$% of 100
★ 43. 20% of 0.5
★ 44. $33\frac{1}{2}$% of $16\frac{1}{2}$
★ 45. 400% of $\frac{1}{4}$

Solve Problems

46. Juan bought a television set for $685. The sales tax was 4% of the selling price. How much sales tax did he pay?

47. Maria bought a clock radio for $54. She paid a sales tax of 7% of the selling price. How much sales tax did she pay?

★ 48. Central Junior High won 90% of the 20 basketball games played last year. How many games did they lose last year?

★ 49. 25% of the 312 students in the eighth grade entered the science fair. How many of the eighth graders did not enter the science fair?

FINDING PERCENTS

At the Low Company, 16 of the 25 employees had perfect attendance records. What percent of the employees had perfect attendance records?

THINK: What percent of 25 is 16?

Equation: $n \cdot 25 = 16$

Solve: $n = \frac{16}{25} \longrightarrow 25\overline{)16.00}^{\,0.64}$

$n = 0.64$

$n = 64\%$

Answer: 64% of the employees had perfect attendance records.

Study and Learn

A. What percent of 30 is 12?

    1. Write an equation for the problem.

    2. Solve the equation $n \cdot 30 = 12$.

    3. Write a percent for 0.40.

B. Compute.

    4. What percent of 25 is 3?   **5.** 15 is what percent of 24?

C. Sometimes the percent is greater than 100%.

    Example   What percent of 4 is 8? $n \cdot 4 = 8$

$$n = \frac{8}{4}$$

$$n = 2$$

$$n = 200\%$$

    6. What percent of 3 is 9?   **7.** 21 is what percent of 4?

D. What percent of 10 is 0.05?

    8. Write an equation for the problem.

    9. Solve the equation $n \cdot 10 = 0.05$.

    10. Write a percent for 0.005.

E. Compute.

    11. What percent of 12 is 0.3?   **12.** 0.01 is what percent of 4?

Practice

Compute.

1. What percent of 2 is 1?

2. 4 is what percent of 8?

3. What percent of 10 is 3?

4. 2 is what percent of 5?

5. What percent of 25 is 4?

6. 6 is what percent of 8?

7. What percent of 15 is 9?

8. 1 is what percent of 8?

9. What percent of 3 is 1?

10. 9 is what percent of 16?

11. What percent of 4 is 12?

12. 24 is what percent of 12?

13. What percent of 9 is 18?

14. 10 is what percent of 2?

15. What percent of 2 is 15?

16. 9 is what percent of 4?

17. What percent of 5 is 7?

18. 9 is what percent of 5?

19. What percent of 100 is 0.05?

20. 0.3 is what percent of 10?

21. What percent of 15 is 0.3?

22. 0.4 is what percent of 5?

23. What percent of 100 is 0.08?

24. 0.07 is what percent of 14?

25. What percent of 5 is 8?

26. 0.8 is what percent of 16?

27. 24 is what percent of 42?

28. What percent of 7 is 14?

29. 0.02 is what percent of 12?

30. What percent of 7 is 3?

31. 9 is what percent of 16?

32. What percent of 20 is 0.5?

Solve Problems

33. At the company picnic, 40 of the 60 people came by car. What percent of the people came by car?

34. Team A made 7 baskets out of 25 tries. Team B made 10 baskets out of 30 tries. Who had a better scoring average?

FINDING THE NUMBER

Ralph bought a sweater on sale. He saved $6. During the sale everything was marked down 25%. What was the regular price of the sweater?

THINK: 25% of the regular price is saved.

Equation:
$$25\% \cdot n = 6$$
$$\tfrac{1}{4} \cdot n = 6$$
$$n = 24$$

Check: $\tfrac{1}{4} \cdot 24 = 6$

The regular price of the sweater was $24.

25% off

regular price of sweater

Study and Learn

A. Compute. 48 is 40% of what number?

  1. Write an equation.

  2. Write a decimal for 40%.

  3. Solve the equation $48 = 0.40 \cdot n$.

  4. Check the solution.

B. Compute.

  5. 75% of what number is 18?

  6. $33\tfrac{1}{3}\%$ of what number is 9?

  7. 6.3 is 15% of what number?

  8. 16.2 is 30% of what number?

C. Compute.

Examples

10 is 250% of what number?

$$250\% \cdot n = 10$$
$$2.5 \cdot n = 10$$
$$n = \tfrac{10}{2.5}$$
$$n = 4$$

$\tfrac{1}{2}\%$ of what number is 40?

$$\tfrac{1}{2}\% \cdot n = 40$$
$$0.005 \cdot n = 40$$
$$n = \tfrac{40}{0.005}$$
$$n = 8,000$$

  9. 12 is 300% of what number?

  10. 150% of what number is 6?

  11. $1\tfrac{1}{2}\%$ of what number is 12?

  12. 104 is 5.2% of what number?

Practice

Compute.

1. 25% of what number is 8?

2. 9 is 25% of what number?

3. 75% of what number is 9?

4. 30 is 50% of what number?

5. 10% of what number is 80?

6. 24 is 40% of what number?

7. $33\frac{1}{3}$% of what number is 3?

8. 80 is $66\frac{2}{3}$% of what number?

9. $12\frac{1}{2}$% of what number is 4?

10. 30 is $62\frac{1}{2}$% of what number?

11. 24% of what number is 15.6?

12. 36.1 is 38% of what number?

13. 100% of what number is 5?

14. 70 is 100% of what number?

15. 125% of what number is 6?

16. 18 is 150% of what number?

17. 220% of what number is 550?

18. 33 is 132% of what number?

19. $\frac{1}{4}$% of what number is 6?

20. 30 is $1\frac{1}{2}$% of what number?

21. 0.4% of what number is 3.8?

22. 12 is 2.4% of what number?

23. 0.1% of what number is 40?

24. 58.9 is 62% of what number?

25. 9 is $4\frac{1}{2}$% of what number?

26. 75% of what number is 9?

27. 8 is 250% of what number?

28. 2.5% of what number is 14?

Solve Problems

29. Ludmila bought a coat on sale. She saved $20. During the sale, everything was marked down $33\frac{1}{3}$%. What was the regular price of the coat?

30. Henry bought a jacket on sale. He saved $17. During the sale, everything was marked down 20%. What was the regular price of the jacket?

Skills Review

Add. *(102)*

1.   6.3
      4.0
      8.6
   + 7.9

2.   9.83
      7.48
      6.36
   + 5.79

3.   3.10746
      9.51387
      6.71283
   + 4.00196

4. $4 + 0.1 + 3.47 + 0.008$

5. $0.0164 + 0.24 + 0.9 + 5$

Subtract. *(104)*

6.   7.804
   − 3.460

7.   4.3160
   − 2.4094

8. $0.5 - 0.291$

9. $7 - 2.23$

Multiply. *(110, 112)*

10. 4.06
    × 7

11. 0.349
    × 18

12. 10×0.8

13. 100×0.47

14. 0.46
   × 0.19

15. 2.13
   × 4.01

16.   61.4
  × 0.923

17. 40.01
  × 1.90

Divide.   *(116, 118)*

18. $4\overline{)0.08}$

19. $6\overline{)0.126}$

20. $7\overline{)0.021}$

21. $9\overline{)0.0027}$

22. $0.3\overline{)2.4}$

23. $0.03\overline{)6}$

24. $0.4\overline{)2.88}$

25. $3.1\overline{)17.36}$

Divide. Give the quotient to the nearest tenth. *(120)*

26. $0.6\overline{)0.8}$

27. $0.04\overline{)0.029}$

28. $5\overline{)3.2}$

29. $4.2\overline{)8.71}$

Divide. Give the quotient to the nearest hundredth. *(120)*

30. $3\overline{)1.658}$

31. $0.9\overline{)0.3251}$

32. $1.1\overline{)8.346}$

33. $0.53\overline{)0.0972}$

Estimate to the nearest whole number. *(106, 121)*

34. 8.816
    4.500
   + 9.319

35. 7.4163
   − 2.9184

36. 4.913
   × 8.7

37. $4.3\overline{)7.651}$

Write decimals. *(184)*

38. $\frac{3}{5}$

39. $\frac{1}{4}$

40. $\frac{5}{8}$

41. $3\frac{1}{2}$

Midchapter Review

Write percents.

1. $\frac{7}{100}$
(250)

2. $\frac{86}{100}$
(250)

3. $\frac{3}{4}$
(250)

4. $\frac{7}{25}$
(250)

5. 0.07
(252)

6. 0.74
(252)

7. 3.4
(252)

8. $0.04\frac{1}{4}$
(252)

Write simplest fractions or whole numbers. *(250)*

9. 25%

10. 40%

11. 200%

12. $12\frac{1}{2}\%$

Write decimals. *(252)*

13. 100%

14. 250%

15. 5.3%

16. $\frac{1}{4}\%$

Compute.

17. 8% of 60
(254)

18. 120% of 80
(254)

19. What percent of 12 is 4?
(256)

20. 12 is what percent of 9?
(256)

21. 25% of what number is 12?
(258)

22. 39 is 0.3% of what number?

Something Extra

Non-Routine Problems

Choose the *opposite* of the first picture.

1.

a
b
c
d

2.

a
b
c
d

Problem-Solving Skills
Writing Mini-Problems

Jean bought a pair of gloves for $8.00. A day later she returned the gloves in exchange for a pair costing $9.25. How much more must she pay?

PLAN   Final pair cost $9.25
Original pair cost $8.00
How much more must she pay?

Write an equation.
$9.25 − $8.00 = x

SOLVE $9.25 − $8.00 = x
$1.25 = x

Jean must pay $1.25 more.

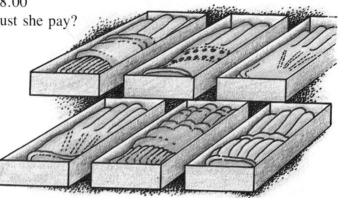

Study and Learn

A. Write a mini-problem. Solve.

1. Mr. Guiterez bought a hammer for $5.95. He found the handle was split. He returned the hammer and bought a screwdriver for $3.75. How much money was returned to him?

B. In multi-step problems, it helps to write more than 1 mini-problem.

Example   Bill bought a radio for $30. He exchanged it for a better radio costing $37.50. How much change did he receive from a $10 bill?

Mini-problem 1
Second radio cost $37.50
First radio cost $30
Additional cost? $7.50

Mini-problem 2
Gave $10
Additional cost $7.50
Change? $2.50

Write 2 mini-problems. Solve.

2. Mrs. Artuso bought a chair for $125. It was damaged so she exchanged it for a chair that cost $149.50. She gave the salesperson a $50 bill. How much change did she receive?

Practice

Write mini-problems. Solve.

1. On Saturday Mrs. Glinka bought a lantern for $16.95. On Monday she returned it and exchanged it for a lantern costing $22. How much more did she pay?

2. Pete bought a record for $5.95. He found that it was cracked. He returned it and was given a credit toward another album. He bought a record set for $12. How much more did he pay?

3. Mr. Benitez bought a jacket for $65. He returned it and bought a different jacket. 10% of his money was returned. How much money was returned to him?

4. The Stone family bought a 21″ television set for $495. They returned it for a 17″ set. 5% of their money was returned. How much money was returned to the Stone family?

5. Mrs. Peterson bought a suit for $39.50. She decided that she didn't like it and returned it. In exchange she bought a suit for $45.75. She gave the salesperson a $20 bill. How much change did she receive?

6. Mr. DeAngelo bought a box of 3 handkerchiefs for $2.49. He exchanged them for a box of 3 handkerchiefs for $4.75. He gave the salesperson a $10 bill. How much change did he receive?

7. Paula bought 3 pairs of knee socks at $1.25 each. She decided to exchange them for 3 pairs of pantyhose costing $1.75 each. She gave the salesperson a $5 bill. How much change did she receive?

DISCOUNTS AND SALES

At a sale, every item is sold at a discount of 20%.
What is the sale price of a belt marked $8?

Step 1 Find the discount.
discount = rate of discount · price
$$x = 20\% \cdot 8$$
$$x = 0.20 \cdot 8$$
$$x = 1.60$$
discount = $1.60

Step 2 Find the sale price.
$8.00 − $1.60 = $6.40
The sale price is $6.40.

Study and Learn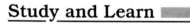

A. Find the discount and the sale price. The rate of discount is 25%.

    1. Pocketbook: $48   **2.** Skirt: $31   **3.** Shirt: $17.80

B. Here is a quick way to find the sale price.

    Example   The marked price of sneakers is $40 at a 30% sale.
               THINK: You save 30%, so you must pay 70% (100% − 30%).
                  sale price = rate · marked price
$$n = 70\% \cdot 40$$
$$n = 0.70 \cdot 40$$
$$n = 28$$
                 sale price = $28

Find the sale price. The rate of discount is 40%.

    4. Hat: $24   **5.** Shoes: $28   **6.** Pants: $32

C. A sweater was marked $16. It sells for $12 during a sale. Find
the rate of discount as a percent.

    7. Find the discount in dollars.

    8. Write an equation. [HINT: discount = rate of discount · marked price]

    9. Solve the equation.

    10. Write the rate of discount as a percent.

Practice

Find the discount and the sale price.
The rate of discount is 10%.

1. Jacket: $85

2. Coat: $95

3. Shirt: $18.50

Find the sale price. Use the quick way.

4. Shoes marked $28 at a 25%-off
sale

5. Tie marked $8 at a 20%-off sale

6. Sweater marked $18.10 at a 30%-off sale

Find the rate of discount as a percent.

7. Hat: was $15, now $10

8. Jacket: was $50, now $40

9. Skirt: was $32, now $24

Complete.

	Marked Price	*Rate of discount*	*Discount*	*Sales Price*
10.	$80	20%		
11.	$60		$12	
12.	$40			$30
13.	$28.50	40%		
14.	$150		$50	
★**15.**		25%		$72

Solve Problems

16. At a sale every item was sold at
a discount of 30%. What is the
sale price of a blouse marked
$12?

17. A pair of gloves was $10. It sold
for $8 during a sale. Find the
rate of discount as a percent.

BORROWING MONEY

Edna's father loaned her $200 to buy an electric guitar. Edna is to pay back the loan in 3 years at an interest rate of 6% per year. How much interest will she pay?

▶ interest = principal · rate (per year) · time (in years)

$i = p \cdot r \cdot t$
$= 200 \cdot 6\% \cdot 3$
$= 200 \cdot 0.06 \cdot 3$
$= 36.00$

So, Edna will pay $36 interest.

Study and Learn

A. Find the interest.

   1. $6,300 at 8% for 2 years

   2. $2,000 at 5.5% for 3 years

B. Find the total amount to be paid back.

   Example   Borrowed $800 at 8% per year for 3 months

   Step 1 $i = p \cdot r \cdot t$
   $= 800 \cdot 0.08 \cdot 0.25$ [HINT: 3 months = 0.25 year]
   $= 16.00$
   $= \$16$

   Step 2   Total amount $= p + i$
   $= \$800 + \$16$
   $= \$816$

   3. $3,500 at 9% for 1 year   **4.** $4,550 at 8.5% for 1 year

   5. $1,200 for 7.5% for 2 years   **6.** $500 at 7% for 6 months

   7. $6,500 at 12% for 3 months   **8.** $2,400 at $5\frac{1}{4}\%$ for 6 months

Practice

Find the interest.

1. $600 at 7% for 2 years
2. $4,800 at 8% for 6 months
3. $3,000 at 7.5% for 4 years
4. $8,000 at $7\frac{1}{4}$% for 3 months

Find the total amount to be paid back.

	Principal	Rate per year	Time
5.	$500	6%	1 year
6.	$800	5%	2 years
7.	$1,000	12%	3 years
8.	$1,200	6%	6 months
9.	$1,600	8%	3 months
10.	$2,100	12%	4 months
11.	$600	6.5%	2 years
12.	$1,000	17.25%	3 years
13.	$200	$5\frac{1}{2}$%	2 years
14.	$4,000	8%	6 months

Solve Problems

15. Bill borrowed $90 to buy a tape recorder. He agreed to pay back the loan at 7% interest in 2 years. How much interest must he pay?

16. Joanne's mother loaned her $8,000 for college. Joanne paid back her mother at 5% interest after 4 years. What was the total amount she repaid?

17. Ms. Smith borrowed $4,500 for a new car. The rate of interest was $1\frac{1}{2}$% per month. How much would she pay if the money was borrowed for 6 months?

★ 18. Mr. Boone charged $200 at a local department store. The rate of interest was $1\frac{1}{2}$% per month. How much interest would he pay if the money was borrowed for 2 months?

COMPOUND INTEREST

Mr. Chin deposited $530 in a bank savings account that pays 6% interest per year compounded annually. How much did Mr. Chin have in his account at the end of 2 years?

First year	**Second year**
$i = p \cdot r \cdot t$	$i = p \cdot r \cdot t$
$= 530 \cdot 0.06 \cdot 1$	$= 561.80 \cdot 0.06 \cdot 1$
$= 31.80$	$= 33.708 \rightarrow 33.71$
new amount $= 530 + 31.80$	new amount $= 561.80 + 33.71$
$= \$561.80$	$= \$595.51$

At the end of 2 years, Mr. Chin had $595.51 in his account.

Study and Learn

A. $1,000 is deposited in an account that pays 7% interest per year compounded annually.

    1. How much interest is earned in 1 year?

    2. What amount is in the account at the end of 1 year?

    3. The $1,070 is left in the account for another year. How much interest is earned the second year?

    4. What amount is in the account at the end of 2 years?

    5. What amount is in the account at the end of 3 years?

B. Interest may be compounded quarterly. This means that the interest is compounded every 3 months, or 4 times a year. Interest may also be compounded semiannually. This means that the interest is compounded every 6 months, or 2 times a year. What amount is in the account at the end of 1 year if the interest rate is 6% per year?

    Example $1,000, compounded semiannually

Six months	**One year**
$i = p \cdot r \cdot t$	$i = p \cdot r \cdot t$
$= 1,000 \cdot 0.06 \cdot 0.5$	$= 1,030 \cdot 0.06 \cdot 0.5$
$= 30$	$= 30.90$
new amount $= 1,000 + 30$	new amount $= 1,030 + 30.90$
$= \$1,030$	$= \$1,060.90$

    6. $800, compounded semiannually   **7.** $250, compounded quarterly

Practice

Complete.

Principal	Interest rate per year	Length of term	Compounding	Compounded amount
1. $1,000	5%	3 yr	annual	
2. $ 800	8%	1 yr	semiannual	
3. $ 400	7%	1 yr	quarterly	
★ **4.** $1,000	8%	$1\frac{1}{2}$ yr	semiannual	
★ **5.** $3,000	$12\frac{1}{4}\%$	9 mo	quarterly	

Computer

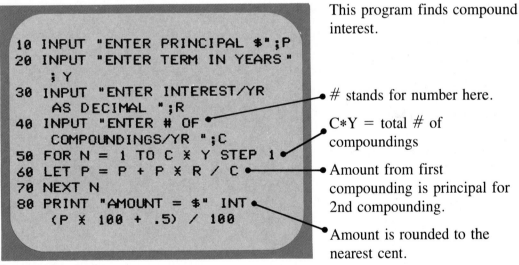

```
10 INPUT "ENTER PRINCIPAL $";P
20 INPUT "ENTER TERM IN YEARS "
   ;Y
30 INPUT "ENTER INTEREST/YR
   AS DECIMAL ";R
40 INPUT "ENTER # OF
   COMPOUNDINGS/YR ";C
50 FOR N = 1 TO C X Y STEP 1
60 LET P = P + P X R / C
70 NEXT N
80 PRINT "AMOUNT = $" INT
   (P X 100 + .5) / 100
```

This program finds compound interest.

stands for number here.

C*Y = total # of compoundings

Amount from first compounding is principal for 2nd compounding.

Amount is rounded to the nearest cent.

In exercise **1** above,

1. What would you enter in line 10?   line 20?   line 30?   line 40?

2. How many times would the computer loop?

3. What is in P after the first loop?   Second loop?   Third loop?

★ **4.** Use the program to check your answers to the Practice.

PERCENT CHANGE

On Monday Elena did 80 jumping jacks. On Tuesday she did 100 jumping jacks. What is the percent increase in the number of jumping jacks?

Step 1 The increase is $100 - 80 = 20$.
Step 2 Find the percent increase.
   amount of change = rate · original amount
   $$20 = r \cdot 80$$
   $$\frac{20}{80} = r$$
   $$\frac{1}{4} = r$$
   $$25\% = r$$ The percent increase is 25%.

Study and Learn

A. The basketball team scored 32 points in their first game. They scored 36 points their second game. What is the percent increase?

1. What was the increase in the number of points scored?

2. Write an equation to solve the problem.

3. Solve the equation.

4. What is the percent increase?

B. Find the percent decrease.

   Example Last year: won 6 games
   This year: won 4 games
   Step 1 The decrease is 2.
   Step 2 amount of change = rate · original amount $\quad 2 = r \cdot 6$
   $$\frac{2}{6} = r$$
   $$\frac{1}{3} = r$$
   $$33\frac{1}{3}\% = r$$

   The percent decrease is $33\frac{1}{3}\%$.

5. Price last week: $25
   Price this week: $20

6. Sold: 16 radios last week
   Sold: 12 radios this week

Practice

Find the percent increase.

1. Last year's cost: $6
   This year's cost: $8

2. 1st math test: 80
   2nd math test: 90

3. Last week's cereal: $0.80
   This week's cereal: $1.00

4. Last week: 64 students raced
   This week: 80 students raced

5. 9:00 am: 20°C
   Noon: 24°C

6. 1st game: 40,000
   2nd game: 50,000

Find the percent decrease.

7. Price last week: $60
   Price this week: $40

8. Population 10 years ago: 8,000
   Population now: 6,000

9. Race time last year: 25 sec
   Race time this year: 20 sec

10. Sold 640 records last month
   Sold 600 records this month

11. 720 customers last year
   600 customers this year

12. 44 workers on vacation last week
   40 workers on vacation this week

Complete.

BASEBALL LEAGUE POINTS EARNED

	Team	Last week	This week	Increase		Decrease	
				Number	%	Number	%
13.	Hornets	9	6				
14.	Wasps	4	6				
15.	Bees	3	5				
16.	Flies	1	2				
17.	Robins	4	2				
18.	Bats	5	3				
19.	Crows	10	8				
20.	Eagles	2	9				

Problem-Solving Applications

Commission

1. Ilga sold $4,000 worth of dental supplies to dentists. Her rate of commission is 9%. How much commission did she receive?
   [HINT: commission = rate of commission · sales]

2. Mr. Valenti earns commission at the rate of 40% of his sales. Last week he sold an insurance policy costing $500. What was his commission?

3. Peggy earns a commission of 12% of her total sales of office equipment. Last week she sold $84 worth of office equipment. What was her commission?

4. Mr. Gomez earns 6% commission for selling houses. He sold a house for $64,500. What was his commission?

5. Mr. Spalding receives $100 a week salary plus 4% of all sales over $5,000. Last week he sold $9,000 worth of groceries to supermarkets. How much did he earn last week?

6. Mrs. Janeway receives $50 a week salary plus 7% of total sales above $3,000. Last week her total sales amounted to $4,500. How much did she earn last week?

7. Ms. Soberjeski received $200 in commission. Her rate of commission was 25%. What were her total sales?

8. Pedro received $95 commission last week. His rate of commission is 10%. What were his total sales?

Chapter Review

Write percents.

1. $\frac{10}{100}$ *(250)*

2. 0.24 *(252)*

3. $\frac{1}{4}$ *(250)*

4. 0.047 *(252)*

Write simplest fractions or whole numbers. *(250)*

5. 7%

6. 75%

7. 300%

8. $37\frac{1}{2}$%

Write decimals. *(252)*

9. 9%

10. 350%

11. 6.7%

12. $\frac{1}{2}$%

13. 0.2%

Compute.

14. 7% of 80 *(254)*

15. $\frac{1}{2}$% of 60 *(254)*

16. What percent of 16 is 4? *(256)*

17. 0.5 is what percent of 50? *(256)*

18. 75% of what number is 90? *(258)*

19. 6% of what number is 4.8? *(258)*

Solve.

20. At a sale a jacket marked $75 was sold at a discount of 30%. What was the sale price of the jacket? *(264)*

21. A shirt marked $15 was sold for $12 during a sale. What was the rate of discount during the sale? *(264)*

22. Fred borrowed $800 at 8% annual interest for 2 years. How much interest did he pay for the use of the money? *(266)*

23. The price for a can of soup was 24¢ last year. This year the same can costs 30¢. What is the percent increase in the cost of the can of soup? *(270)*

Write a mini-problem. Solve. *(262)*

24. Jennifer bought a sweater for $12.50 last Friday. On Monday she returned it for a sweater costing $19.75. How much more did she pay?

Solve. *(272)*

25. Ms. Klein sold $6,500 worth of watchbands. Her rate of commission is 8%. How much commission did she receive?

Chapter Test

Write percents.

1. $\frac{5}{100}$ **2.** 0.42   **3.** $\frac{2}{5}$ **4.** 0.074
(250)   *(252)*   *(250)*   *(252)*

Write simplest fractions or whole numbers. *(250)*

5. 3%   **6.** 25%   **7.** 400%   **8.** $66\frac{2}{3}\%$

Write decimals *(252)*

9. 8%   **10.** 250%   **11.** 7.6%   **12.** $\frac{1}{4}\%$ **13.** 0.7%

Compute.

14. 13% of 70
(254)

15. 0.9% of 300
(254)

16. What percent of 24 is 18?
(256)

17. 0.2 is what percent of 50?
(256)

18. 25% of what number is 60?
(258)

19. 8% of what number is 7.2?
(258)

Solve.

20. At a sale a dress marked $60 was
(264) sold at a discount of 25%. What
was the sale price of the dress?

21. A tie marked $12 was sold for
(264) $10 during a sale. What was the
rate of discount during the sale?

22. Gwen borrowed $750 at 6% annual
(266) interest for 2 years. How much
interest did she pay for the use
of the money?

23. A supermarket sold 300 boxes of
(270) tissues last week. This week the
supermarket sold 250 boxes of
tissues. What is the percent
decrease in sale of tissues?

Write a mini-problem. Solve. *(262)*

24. Mr. Sanko bought a set of drinking glasses for $12.95. Some of
the glasses were cracked, so he returned them for a set costing
$15.25. How much more did he pay?

Solve. *(272)*

25. Miss Lowenthal received $360 in commission. Her rate of
commission was 8%. What were her total sales?

Skills Check

1. Soup is selling at 2 cans for 45¢. About how much will 6 cans of soup cost?

   A $3.00   B $2.00

   C $1.50   D $1.00

2. One year tests showed 23.9, 28.9, 22.3, and 19.7 units of radioactive materials in milk. What was the average number of radioactive units in the milk?

   E 23.4   F 23.6

   G 23.7   H 24.8

3. Five eggs contain 400 calories. Sue ate 2 eggs for breakfast. How many calories are in 2 eggs?

   A 100 calories   B 150 calories

   C 160 calories   D 200 calories

4. What is the area of a square with one side 12 cm?

   E 144 cm²   F 96 cm²

   G 48 cm²   H 24 cm²

5. What is the circumference of a circle with a radius of 14 cm? Use 3.14 for π. Round to the nearest whole number.

   A 616 cm   B 88 cm

   C 41 cm   D 22 cm

6. To tell the distance between two cities, which unit is used?

   E cm   F m

   G mm   H km

7. About how much money was spent on clothing?

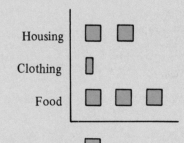

   Key: ☐ = $50

   A $100   B $75

   C $50   D $25

8. Mr. Jimenez bought 1.5 L of bleach and 750 mL of fabric softener. How much did he buy in all?

   E 2.25 mL   F 751.5 mL

   G 2.25 L   H 751.5 L

A typical hang glider has wings about 12 m long and 1.2 m wide.

OPPOSITES AND ABSOLUTE VALUE

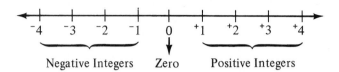

The integers are the numbers . . . , ⁻5, ⁻4, ⁻3, ⁻2, ⁻1, 0, ⁺1, ⁺2, ⁺3, ⁺4, ⁺5, . . .   On a number line, the point matched with 0 is called the *origin*.

Study and Learn

A. When comparing numbers, the greater of two integers is paired with a point farther to the right on a number line.

Examples   ⁻7 > ⁻10   ⁺4 > ⁺2   ⁻3 < ⁺1

Compare. Use > or <.

1. ⁺3 ☰ ⁺1   **2.** ⁺2 ☰ ⁺5   **3.** 0 ☰ ⁻3   **4.** ⁻4 ☰ ⁻1

▶ ⁻1 and ⁺1 are called **opposites.** ⁺4 and ⁻4 are opposites. Their points are on opposite sides of the origin and each point is the same distance from the origin. The opposite of 0 is 0.

B. Give the opposites.

5. ⁺6   **6.** ⁻7   **7.** ⁺15   **8.** ⁻11   **9.** ⁻43   **10.** ⁺73

▶ The **absolute value** of an integer is the number or its opposite, whichever is positive. The absolute value of 0 is 0.

C. |⁻4| is read absolute value of negative 4. It is equal to positive 4.

11. What is |⁻5|?   **12.** What is |⁺5|?

D. Find the absolute value.

13. |⁻9| **14.** |⁺7| **15.** |⁻21| **16.** |⁺16| **17.** |⁺37| **18.** |⁻86|

E. Compare. Use >, <, or =.

19. |⁺5| ☰ |⁻6| **20.** |⁻15| ☰ |⁺11| **21.** |⁻22| ☰ |⁺22|

Practice

Compare. Use > or <.

1. $^+1 \equiv {}^+4$　　　2. $^+7 \equiv {}^+4$　　　3. $^-6 \equiv {}^-7$　　　4. $^+4 \equiv 0$

5. $^-1 \equiv 0$　　　6. $^-4 \equiv {}^-2$　　　7. $^+3 \equiv {}^-7$　　　8. $^-6 \equiv {}^-8$

9. $0 \equiv {}^+6$　　　10. $^+3 \equiv {}^-4$　　　11. $^-9 \equiv {}^-1$　　　12. $^+1 \equiv {}^-1$

Give the opposites.

13. $^+3$　　14. $^-9$　　15. $^+8$　　16. $^+7$　　17. $^-6$　　18. 0

19. $^-14$　　20. $^+27$　　21. $^+56$　　22. $^-83$　　23. $^-124$　　24. $^+253$

Find the absolute value.

25. $|^+8|$　　　26. $|^-3|$　　　27. $|^-5|$　　　28. $|^+2|$　　　29. $|^+9|$

30. $|0|$　　　31. $|^-15|$　　　32. $|^+11|$　　　33. $|^-26|$　　　34. $|^-52|$

35. $|^+86|$　　　36. $|^-75|$　　　37. $|^+37|$　　　38. $|^-248|$　　　39. $|^+374|$

Compare. Use >, <, or =.

40. $|^+4| \equiv |^-2|$　　41. $|^-8| \equiv |^+7|$　　42. $|^+9| \equiv |^+15|$　　43. $|^-4| \equiv |^-10|$

44. $|^-6| \equiv |^+9|$　　45. $|^+10| \equiv |^-18|$　46. $|^-12| \equiv |^+6|$　　47. $|^+13| \equiv |^-14|$

48. $|^+25| \equiv |^-25|$　49. $|^-13| \equiv |^-15|$　50. $|^+18| \equiv |^-16|$　51. $|^-9| \equiv |^+6|$

52. $|^-11| \equiv |^+8|$　　53. $|^+20| \equiv |^-16|$　54. $|^-12| \equiv |^+9|$　　55. $|^-7| \equiv |^-9|$

★ The equation $|n| = {}^+2$ has solutions, $^+2$ and $^-2$, since $|^+2| = {}^+2$ and $|^-2| = {}^+2$. Find the solutions.

56. $|n| = {}^+5$　　　57. $|n| = {}^+6$　　　58. $|n| = 0$　　　59. $|n| = {}^+18$

ADDING INTEGERS

To add $^+2 + {}^+3$:

Start at 0.

Move 2 units to the right.

From $^+2$, move 3 units to the right.

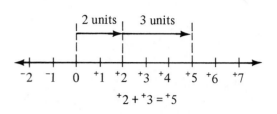

$$^+2 + {}^+3 = {}^+5$$

▶ The sum of two positive integers is a positive integer.

Study and Learn

A. Add. Use a number line.

1. $^+3 + {}^+1$ **2.** $^+2 + {}^+5$ **3.** $0 + {}^+3$ **4.** $^+4 + {}^+2$

B. Add.

5. $\begin{array}{r} {}^+36 \\ + {}^+51 \end{array}$ **6.** $\begin{array}{r} {}^+24 \\ + {}^+17 \end{array}$ **7.** $^+31 + {}^+41$ **8.** $^+16 + {}^+15$

Here is how to find the sum of 2 negative numbers.

Add $^-3 + {}^-4$.

Start at 0.

Move 3 units to the left.

From $^-3$, move 4 units to the left.

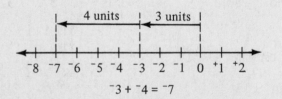

$$^-3 + {}^-4 = {}^-7$$

▶ The sum of two negative integers is a negative integer.

C. Add. Use a number line.

9. $^-2 + {}^-2$ **10.** $^-1 + {}^-5$ **11.** $0 + {}^-4$ **12.** $^-2 + {}^-4$

D. Add.

13. $\begin{array}{r} {}^-16 \\ + {}^-13 \end{array}$ **14.** $\begin{array}{r} {}^+35 \\ + {}^+47 \end{array}$ **15.** $^-22 + {}^-34$ **16.** $^+56 + {}^+17$

17. $^+7 + {}^+8 + {}^+9$ **18.** $^+11 + {}^+24 + {}^+9$ **19.** $^-12 + {}^-43 + {}^-24$

Practice

Add.

1. $^+2 + ^+1$ 2. $^+3 + ^+2$ 3. $^+6 + ^+2$ 4. $^+3 + ^+7$

5. $^+8 + ^+3$ 6. $^+4 + ^+8$ 7. $^+7 + ^+6$ 8. $^+8 + ^+9$

9. $^-2 + ^-1$ 10. $^-3 + ^-2$ 11. $^-1 + ^-4$ 12. $^-7 + ^-3$

13. $^-3 + ^-8$ 14. $^-4 + ^-7$ 15. $^-9 + ^-8$ 16. $^-5 + ^-5$

17. $^+10 + ^+12$ 18. $^+19 + ^+21$ 19. $^+21 + ^+31$ 20. $^+24 + ^+31$

21. $^+41 + ^+53$ 22. $^+32 + ^+49$ 23. $^+56 + ^+29$ 24. $^+20 + ^+86$

25. $^-22 + ^-14$ 26. $^-24 + ^-15$ 27. $^-15 + ^-29$ 28. $^-32 + ^-24$

29. $^-27 + ^-38$ 30. $^-62 + ^-14$ 31. $^-74 + ^-18$ 32. $^-65 + ^-19$

33. $^-4 + ^-9$ 34. $^-6 + ^-7$ 35. $^+15 + ^+6$ 36. $^+26 + ^+7$

37. $^-12 + ^-21$ 38. $^+35 + ^+43$ 39. $^-46 + ^-37$ 40. $^+38 + ^+77$

41. $^+3$ $+ ^+2$ 42. $^+3$ $+ ^+4$ 43. $^+6$ $+ ^+5$ 44. $^+8$ $+ ^+9$

45. $^+57$ $+ ^+29$ 46. $^+64$ $+ ^+37$ 47. $^+81$ $+ ^+29$ 48. $^+46$ $+ ^+28$

49. $^-34$ $+ ^-17$ 50. $^-64$ $+ ^-29$ 51. $^-64$ $+ ^-59$ 52. $^-67$ $+ ^-74$

53. $^+2 + ^+6 + ^+7$ 54. $^+29 + ^+17 + ^+8$

55. $^-3 + ^-4 + ^-5$ 56. $^-17 + ^-3 + ^-5$

Solve.

57. In Maine the average temperature was $^+5°C$ on Monday. The average temperature on Tuesday was 8° higher. What was the average temperature on Tuesday?

ADDING POSITIVE AND NEGATIVE INTEGERS

To add $^+3 + {}^-2$:

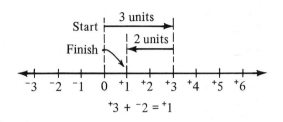

$$^+3 + {}^-2 = {}^+1$$

Start at 0.

Move 3 units to the right.

From $^+3$, move 2 units to the left.

Study and Learn

A. To add $^-5 + {}^+2$:

1. Draw a number line from $^-5$ to $^+5$.
2. Start at 0. Show $^-5$ by a move on the number line.
3. From $^-5$, move $^+2$ units or 2 units to the right.
4. The sum $^-5 + {}^+2 = \underline{\quad?\quad}$.

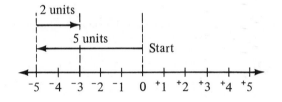

B. Add. Use a number line.

    5. $^+4 + {}^-3$ **6.** $^+4 + {}^-7$ **7.** $^-3 + {}^+5$

Notice a pattern for adding positive and negative integers.

$^+3 + {}^-2 = {}^+1$ The greater move, $^+3$, is in the positive direction.
                    This is the difference of the number of units moved.
         $3 - 2 = 1$

$^+2 + {}^-5 = {}^-3$ The greater move, $^-5$, is in the negative direction.
                    This is the difference of the number of units moved.
         $5 - 2 = 3$

C. Add $^+17 + {}^-9$.

    8. Is the greater move in a positive or negative direction?
    9. Find the difference of the number of units moved, $17 - 9$.
    10. Complete. $^+17 + {}^-9 = \underline{\quad?\quad}$

D. Add.

    11. $^+5 + {}^-1$ **12.** $^-7 + {}^+4$ **13.** $^+3 + {}^-18$ **14.** $^+3 + {}^-4 + {}^-5$

Practice

Add.

1. $^+5 + ^-1$
2. $^+4 + ^-5$
3. $^+1 + ^-5$
4. $^+23 + ^-12$

5. $^+44 + ^-27$
6. $^+63 + ^-55$
7. $^+43 + ^-51$
8. $^+65 + ^-61$

9. $^+84 + ^-93$
10. $^-2 + ^+5$
11. $^-3 + ^+8$
12. $^-1 + ^+4$

13. $^-65 + ^+11$
14. $^-94 + ^+13$
15. $^-73 + ^+86$
16. $^-21 + ^+15$

17. $^-15 + ^+16$
18. $^-12 + ^+14$
19. $^+13 + ^-7$
20. $^+15 + ^-11$

21. $^-19 + ^+14$
22. $^+17 + ^-24$
23. $^-21 + ^+16$
24. $^-19 + ^+26$

25. $\begin{array}{r} ^+3 \\ + ^-1 \\ \hline \end{array}$
26. $\begin{array}{r} ^-7 \\ + ^+4 \\ \hline \end{array}$
27. $\begin{array}{r} ^-8 \\ + ^+9 \\ \hline \end{array}$
28. $\begin{array}{r} ^+6 \\ + ^-2 \\ \hline \end{array}$

29. $\begin{array}{r} ^+24 \\ + ^-19 \\ \hline \end{array}$
30. $\begin{array}{r} ^-18 \\ + ^+12 \\ \hline \end{array}$
31. $\begin{array}{r} ^-25 \\ + ^+19 \\ \hline \end{array}$
32. $\begin{array}{r} ^+34 \\ + ^-51 \\ \hline \end{array}$

33. $^-1 + ^+2 + ^+3$
34. $^-8 + ^+14 + ^-3$
35. $^-2 + ^+3 + ^-4$

36. $^+7 + ^-7 + ^+3$
37. $^+5 + ^+7 + ^-3$
38. $^-3 + ^-8 + ^+5$

★39. $^+3 + ^-9 + ^+2 + ^+3$
★40. $^-6 + ^+4 + ^+1 + ^+5$

Solve Problems

41. One week the price of a stock went up $2, then down $3, and then up $4. What was the net change in the price of the stock?

42. The temperature was $^-5°$. There was a change of temperature of $^+2°$, followed by a change of temperature of $^-3°$. What was the temperature then?

43. A helicopter pilot reported that he was 10 miles east ($^+10$) of the airport. He was told to search 7 miles west ($^-7$), and then 12 miles east ($^+12$). How far from the airport was he then?

★44. Lisa had 500 shares of stock worth $60 ($^+60$) a share. One month, each share rose $2 ($^+2$) in value. The next month, it dropped $6 ($^-6$) in value. What was the value of the 500 shares then?

SUBTRACTING INTEGERS

Examine these related subtraction and addition sentences.

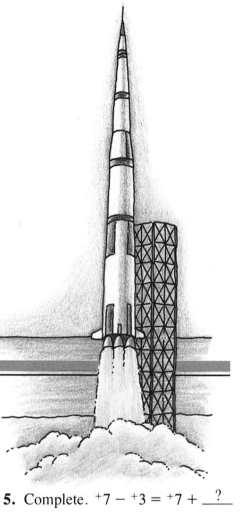

Subtraction	Addition
$^+6 - {}^+2 = {}^+4$	$^+6 + {}^-2 = {}^+4$
$^+6 - \phantom{{}^+}0 = {}^+6$	$^+6 + \phantom{{}^+}0 = {}^+6$
$^+6 - {}^-2 = {}^+8$	$^+6 + {}^+2 = {}^+8$
$^-6 - \phantom{{}^+}0 = {}^-6$	$^-6 + \phantom{{}^+}0 = {}^-6$
$^-6 - {}^-2 = {}^-4$	$^-6 + {}^+2 = {}^-4$

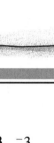

opposites
same numbers

▶ To subtract an integer, add its opposite.

Study and Learn

A. Give the opposites.

   1. $^+6$ **2.** $^-14$ **3.** $^-3$

B. Subtract $^+7 - {}^+3$.

   4. What is the opposite of $^+3$?   **5.** Complete. $^+7 - {}^+3 = {}^+7 + \underline{\ ?\ }$

   6. Add $^+7 + {}^-3$.   **7.** What is $^+7 - {}^+3$?

Subtract.

   8. $^+9 - {}^+1$ **9.** $^+4 - {}^+9$ **10.** $^+6 - {}^+14$ **11.** $^+15 - {}^+8$

C. Subtract $^+9 - {}^-2$.

   12. What is the opposite of $^-2$?   **13.** Complete. $^+9 - {}^-2 = {}^+9 + \underline{\ ?\ }$

   14. Add $^+9 + {}^+2$.   **15.** What is $^+9 - {}^-2$?

Subtract.

   16. $^+8 - {}^-3$ **17.** $^-8 - {}^-4$ **18.** $^-7 - {}^-14$ **19.** $^-6 - {}^+9$

Practice

Subtract.

1. $^+8 - {}^+3$ 2. $^+4 - {}^+1$ 3. $^+6 - {}^+2$ 4. $^+8 - {}^+6$

5. $^+14 - {}^+9$ 6. $^+16 - {}^+3$ 7. $^+14 - {}^+81$ 8. $^+92 - {}^+54$

9. $^+36 - {}^+62$ 10. $^+27 - {}^+58$ 11. $^+76 - {}^+14$ 12. $^+36 - {}^+10$

13. $^+6 - {}^-1$ 14. $^+4 - {}^-3$ 15. $^+8 - {}^-4$ 16. $^+7 - {}^-3$

17. $^+14 - {}^-3$ 18. $^+16 - {}^-7$ 19. $^+18 - {}^-96$ 20. $^+57 - {}^-38$

21. $^+51 - {}^-88$ 22. $^+79 - {}^-10$ 23. $^+64 - {}^-80$ 24. $^+32 - {}^-26$

25. $^-4 - {}^+1$ 26. $^-8 - {}^+5$ 27. $^-7 - {}^+6$ 28. $^-8 - {}^+2$

29. $^-41 - {}^+28$ 30. $^-91 - {}^+76$ 31. $^-82 - {}^+44$ 32. $^-92 - {}^+37$

33. $^-12 - {}^+81$ 34. $^-17 - {}^+32$ 35. $^-24 - {}^+78$ 36. $^-41 - {}^+20$

37. $^-5 - {}^-8$ 38. $^-3 - {}^-7$ 39. $^-4 - {}^-6$ 40. $^-3 - {}^-4$

41. $^-8 - {}^-92$ 42. $^-16 - {}^-20$ 43. $^-81 - {}^-31$ 44. $^-71 - {}^-29$

45. $^-81 - {}^-31$ 46. $^-42 - {}^-17$ 47. $^-86 - {}^-52$ 48. $^-30 - {}^-45$

49. $^-71 - {}^-12$ 50. $^-14 - {}^+92$ 51. $^+16 - {}^-78$ 52. $^-36 - {}^-54$

53. $^-3 - {}^+8$ 54. $^+41 - {}^-92$ 55. $^+16 - {}^+14$ 56. $^-58 - {}^-82$

Solve Problems

57. One day, the temperature was $^-5°C$. The next day, it was $^-11°C$. How many degrees less is $^-11°C$ than $^-5°C$?

58. Mt. McKinley is 20,320 feet above sea level ($^+20,320$) and Death Valley is 282 feet below sea level ($^-282$). What is the difference of the altitudes?

● 59. A submarine is 85 m below sea level ($^-85$). It fires a rocket which rises 215 m ($^+215$). How far above sea level is the rocket?

USING ABSOLUTE VALUE TO ADD INTEGERS

Here's a way to use absolute value when adding integers.

Use these steps to add 2 integers with the same signs.

Step 1   Find the sum of the absolute values of the integers.
Step 2   Use the sign of the integer with the greater absolute value.

Examples

$$^+5 \ + \ ^{\oplus}7$$
Step 1 $|^+5| + |^+7|$
Step 2 $\rightarrow ^+12$

$$^-6 \ + \ ^{\ominus}9$$
Step 1 $|^-6| + |^-9|$
Step 2 $\rightarrow ^-15$

Study and Learn

A. Add.

   1. $^+8 + ^+3$ **2.** $^+4 + ^+9$ **3.** $^-7 + ^-6$ **4.** $^-5 + ^-5$

Use these steps to add 2 integers with different signs.

Step 1   Find the difference of the absolute values of the integers.
Step 2   Use the sign of the integer with the greater absolute value.

Examples

$$^{\oplus}17 \ + \ ^-9$$
Step 1 $|^+17| - |^-9|$
Step 2 $\rightarrow ^+8$

$$^{\ominus}15 \ + \ ^+8$$
Step 1 $|^-15| - |^+8|$
Step 2 $\rightarrow ^-7$

B. Add.

   5. $^+11 + ^-3$ **6.** $^+9 + ^-12$ **7.** $^-6 + ^+10$ **8.** $^-14 + ^+7$

Practice

Add.

   1. $^+6 + ^+2$ **2.** $^+9 + ^+9$ **3.** $^-8 + ^-4$ **4.** $^-10 + ^-7$

   5. $^+8 + ^-5$ **6.** $^+6 + ^-12$ **7.** $^-4 + ^+13$ **8.** $^-15 + ^+12$

   9. $^+4 + ^+7 + ^+5$ **10.** $^-3 + ^-8 + ^-2$ **11.** $^+9 + ^-6 + ^-5$

Midchapter Review

Compare. Use $>$, $<$, or $=$. *(278)*

1. $^+2 \equiv ^+6$ **2.** $^-9 \equiv ^-5$ **3.** $^-4 \equiv ^+5$ **4.** $^+3 \equiv ^-8$

5. $|^+4| \equiv |^-7|$ **6.** $|^-3| \equiv |^+5|$ **7.** $|^+8| \equiv |^-8|$ **8.** $|^-3| \equiv |0|$

Add. *(280, 282)*

9. $^+7 + ^+8$ **10.** $^-4 + ^-6$ **11.** $^-8 + ^-9$ **12.** $^-7 + ^-4$

13. $^-7 + ^+9$ **14.** $^+5 + ^-1$ **15.** $^-6 + ^+3$ **16.** $^+4 + ^-8$

17. $^+4 + ^-4$ **18.** $^-12 + ^+12$ **19.** $^-2 + ^-3$ **20.** $^-6 + ^-4$

21. $^+6 + ^+2 + ^+4$ **22.** $^-3 + ^-2 + ^-9$ **23.** $^+5 + ^-7 + ^+8$ **24.** $^-10 + ^-4 + ^+7$

Subtract. *(284)*

25. $^+7 - ^+3$ **26.** $^+6 - ^+8$ **27.** $^+7 - ^-1$ **28.** $^+9 - ^-7$

29. $^+5 - ^-8$ **30.** $^-3 - ^-4$ **31.** $^-4 - ^-9$ **32.** $^-8 - ^-3$

33. $^-4 - ^+6$ **34.** $^-5 - ^+1$ **35.** $^-4 - ^-4$ **36.** $^+5 - ^-5$

Something Extra
Non-Routine Problems

1. Use only the numbers 1, 3, 9, and 27 with addition and subtraction to name the numbers from 1 to 25. Each number may be used at most once.

   Example $20 = 27 - 9 + 3 - 1$

2. Use only prime numbers with addition to name the positive even integers from 4 to 50.

   Example $24 = 11 + 13$

3. Use only the numbers 3, 4, 5, 6, and 7 with addition and multiplication to name the number 100.

Problem-Solving Skills

Circle Graphs

Jennifer interviewed 60 people. She asked each of them which of 3 television programs was his or her favorite. The results are shown by the circle graph.

The entire graph represents 100%, or all 60 choices.

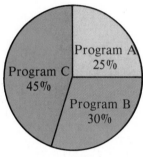

Study and Learn

A. Look at the graph above and answer these questions.

    1. Which program was selected as the favorite by the people?

    2. What percent of the people selected Program A?

    3. How many people selected Program A?
    Complete. 25% of 60 = ___?___

    4. How many people selected Program B?

    5. How many people selected Program C?

B. There are 400 workers in Middle Village.

Labor Force in Middle Village

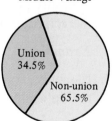

    6. What percent of the labor force is represented by the graph?

    7. Are most workers union or non-union?

    8. How many union workers are there?

    9. How many non-union workers are there?

Practice

The Alitos' weekly income is $400.

1. What percent of the budget is represented by the graph?

2. What takes up the largest part of the budget?

3. How much is spent on housing?

4. How much is spent on food?

5. How much is spent on each of the other items in the budget?

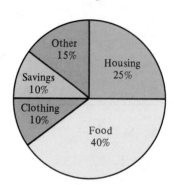

Alito Family Budget

The Field factory employs 1,200 people.

6. Which is the largest group of Field employees?

7. What fractional part of the employees are the office workers?

8. How many factory workers are employed?

9. How many supervisors are employed?

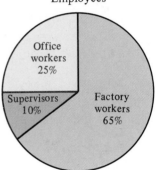

Field Factory Employees

Two thousand high school graduates were asked what they did after graduation.

10. What did most of the high school graduates do?

11. How many went to college?

12. How many went to work?

13. How many could not find jobs?

★ **14.** What fractional part of the graduates went to college or are working?

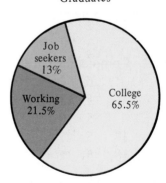

High School Graduates

MULTIPLYING INTEGERS

Multiplying whole numbers
$$3 \cdot 4 = 12$$
$$6 \cdot 9 = 54$$

Multiplying integers
$$^+3 \cdot {}^+4 = {}^+12$$
$$^+6 \cdot {}^+9 = {}^+54$$

▶ The product of two positive integers is a positive integer.

Study and Learn

A. Multiply.

1. $^+7 \cdot {}^+8$ **2.** $^+6 \cdot {}^+1$ **3.** $^+9 \cdot {}^+7$ **4.** $^+7 \cdot {}^+5$

B. Complete. Study the pattern.
$$^+6 \cdot {}^+2 = {}^+12$$
$$^+6 \cdot {}^+1 = {}^+6$$
$$^+6 \cdot \ 0 = 0$$

5. $^+6 \cdot {}^-1 = \underline{\ ?\ }$
6. $^+6 \cdot {}^-2 = \underline{\ ?\ }$

▶ The product of a positive integer and a negative integer is a negative integer.

Multiply.

7. $^+7 \cdot 0$ **8.** $^+6 \cdot {}^-3$ **9.** $^+7 \cdot {}^-4$ **10.** $^+9 \cdot {}^-6$

C. Complete.

11. $^+6 \cdot {}^-3 = {}^-18$; ${}^-3 \cdot {}^+6 = \underline{\ ?\ }$
12. $^+4 \cdot {}^-5 = {}^-20$; ${}^-5 \cdot {}^+4 = \underline{\ ?\ }$

▶ The product of a negative integer and a positive integer is a negative integer.

Multiply.

13. $^-2 \cdot {}^+4$ **14.** $^-1 \cdot {}^+8$ **15.** $^-7 \cdot {}^+6$ **16.** $^-9 \cdot {}^+6$

D. Complete. Study the pattern.
$$^-5 \cdot {}^+2 = {}^-10$$
$$^-5 \cdot {}^+1 = {}^-5$$
$$^-5 \cdot \ 0 = 0$$

17. $^-5 \cdot {}^-1 = \underline{\ ?\ }$
18. $^-5 \cdot {}^-2 = \underline{\ ?\ }$

▶ The product of two negative integers is a positive integer.

Multiply.

19. $^-3 \cdot {}^-9$ **20.** $^-4 \cdot {}^-8$ **21.** $^-7 \cdot {}^-9$ **22.** $^-8 \cdot {}^-7$

Practice

Multiply.

1. $^+3 \cdot {}^+9$	2. $^+7 \cdot {}^+6$	3. $^+8 \cdot {}^+3$	4. $^+7 \cdot {}^+9$
5. $^-6 \cdot {}^-7$	6. $^-4 \cdot {}^-5$	7. $^-8 \cdot {}^-9$	8. $^-6 \cdot {}^-4$
9. $^+3 \cdot {}^-4$	10. $^+8 \cdot {}^-3$	11. $^+7 \cdot {}^-5$	12. $^+9 \cdot {}^-8$
13. $^-4 \cdot {}^+8$	14. $^-7 \cdot {}^+7$	15. $^-9 \cdot {}^+5$	16. $^-6 \cdot {}^+7$
17. $0 \cdot {}^+8$	18. $0 \cdot {}^-7$	19. $^+3 \cdot 0$	20. $^-41 \cdot 0$
21. $^-1 \cdot {}^-18$	22. $^-1 \cdot {}^+1$	23. $^-1 \cdot {}^+8$	24. $^-11 \cdot {}^+10$
25. $^+1 \cdot {}^+96$	26. $^+11 \cdot {}^-38$	27. $^+1 \cdot 0$	28. $^+1 \cdot {}^-1$
29. $^+4 \cdot {}^+12$	30. $^-30 \cdot {}^+8$	31. $^-40 \cdot {}^-9$	32. $^+20 \cdot {}^-9$

Solve.

33. In the first quarter, the Lions lost 14 yd a minute for the first 4 minutes. What was the net change in yardage?

★ Multiply.

34. $^+3 \cdot {}^+4 \cdot {}^-1$	35. $^-4 \cdot {}^-8 \cdot {}^-1$	36. $^+2 \cdot {}^-3 \cdot {}^+4$
37. $^-4 \cdot {}^+2 \cdot {}^-8$	38. $^-1 \cdot {}^+5 \cdot {}^-6$	39. $^+1 \cdot {}^-1 \cdot {}^-3$

Something Extra
Non-Routine Problem

Is this a true or false conclusion? Give reasons.

Given: If a person is 17 years old, the person may obtain a driver's license.
Jeff has a driver's license.
Conclusion: Jeff is 17 years old.

PROPERTIES OF INTEGERS

A. Complete.

ADDITION PROPERTIES

	Property	Illustration check	Does the property hold? Yes	No
1.	Commutative	$^+7 + {}^-6 = {}^-6 + {}^+7$	?	?
2.	Associative	$(^+3 + {}^-2) + {}^-4 = {}^+3 + (^-2 + {}^-4)$	?	?
3.	of Zero	$^-7 + 0 = {}^-7$	?	?
4.	Opposite or Inverse	$^-6 + {}^+6 = 0$	?	?

MULTIPLICATION PROPERTIES

	Property	Illustration check	Does the property hold? Yes	No
5.	Commutative	$^-3 \cdot {}^+4 = {}^+4 \cdot {}^-3$	?	?
6.	Associative	$(^-3 \cdot {}^-4) \cdot {}^+2 = {}^-3 \cdot (^-4 \cdot {}^+2)$	?	?
7.	of One	$^-3 \cdot {}^+1 = {}^-3$	?	?
8.	Distributive	$^+2 \cdot (^-3 + {}^+4) = (^+2 \cdot {}^-3) + (^+2 \cdot {}^+4)$	?	?
9.	of Zero	$^-4 \cdot 0 = 0$	?	?

Practice

Solve. Then name the properties shown.

1. $(^-1 + {}^-1) + {}^-3 = n + (^-1 + {}^-3)$

2. $^+7 + n = {}^+9 + {}^+7$

3. $n + {}^-3 = {}^-3$

4. $^+5 + {}^-5 = n$

5. $^+6 \cdot {}^+2 = {}^+2 \cdot n$

6. $^-8 \cdot {}^+1 = n$

7. $(^+4 \cdot {}^-1) \cdot {}^-3 = n \cdot (^-1 \cdot {}^-3)$

8. $^-3 \cdot (^+7 + {}^-5) = (^-3 \cdot {}^+7) + (^-3 \cdot n)$

9. $^+7 \cdot n = 0$

Skills Review

Compare. Use $>$, $<$, or $=$. *(164)*

1. $\frac{3}{8} \equiv \frac{7}{8}$ **2.** $\frac{3}{5} \equiv \frac{12}{20}$ **3.** $\frac{3}{4} \equiv \frac{5}{6}$

Simplify. *(162)*

4. $\frac{6}{10}$ **5.** $\frac{12}{21}$ **6.** $\frac{8}{12}$

Write fractions. *(166)*

7. $2\frac{3}{4}$ **8.** $4\frac{1}{2}$ **9.** $6\frac{2}{3}$ **10.** $9\frac{3}{7}$ **11.** $8\frac{3}{10}$

Write mixed numbers. *(166)*

12. $\frac{7}{4}$ **13.** $\frac{3}{2}$ **14.** $\frac{13}{3}$ **15.** $\frac{26}{5}$ **16.** $\frac{54}{11}$

Add. Simplify when possible. *(168, 170)*

17. $\frac{3}{8}$ **18.** $\frac{5}{16}$ **19.** $\frac{3}{5}$ **20.** $\frac{5}{11}$ **21.** $\frac{5}{8}$
$+\frac{2}{8}$ $+\frac{1}{2}$ $+\frac{1}{4}$ $+\frac{2}{3}$ $+\frac{7}{12}$

22. $2\frac{3}{4}$ **23.** $3\frac{3}{4}$ **24.** $7\frac{1}{2}$ **25.** $7\frac{2}{3}$ **26.** $16\frac{3}{4}$
$+1\frac{5}{8}$ $+2\frac{5}{6}$ $+8\frac{7}{10}$ $+3\frac{1}{5}$ $+2\frac{1}{2}$

Subtract. Simplify when possible. *(173, 174)*

27. $12\frac{7}{8}$ **28.** $6\frac{3}{4}$ **29.** $5\frac{1}{3}$ **30.** $7\frac{4}{5}$ **31.** $6\frac{2}{5}$
$-6\frac{1}{8}$ $-3\frac{3}{5}$ $-2\frac{3}{4}$ $-2\frac{1}{2}$ $-3\frac{9}{10}$

32. $8\frac{1}{3} - 5$ **33.** $4 - 2\frac{2}{3}$ **34.** $7 - 3\frac{4}{5}$ **35.** $8 - 6\frac{1}{8}$

Multiply. Simplify when possible. *(178, 180)*

36. $\frac{3}{4} \times \frac{1}{5}$ **37.** $\frac{3}{5} \times \frac{2}{3}$ **38.** $\frac{3}{4} \times 12$ **39.** $\frac{7}{8} \times 24$

40. $1\frac{1}{2} \times 6$ **41.** $1\frac{1}{2} \times 2\frac{3}{4}$ **42.** $7\frac{2}{3} \times 6$ **43.** $2\frac{2}{5} \times 4\frac{3}{4}$

Divide. Simplify when possible. *(182)*

44. $\frac{3}{4} \div \frac{1}{2}$ **45.** $4 \div 1\frac{1}{2}$ **46.** $2\frac{2}{3} \div 1\frac{1}{4}$ **47.** $6 \div \frac{3}{4}$

DIVIDING INTEGERS

Multiplication and division are related.

Multiplication	Division
$^+6 \cdot {}^+3 = {}^+18$	$^+18 \div {}^+6 = {}^+3; \frac{^+18}{^+6} = {}^+3$
$^-5 \cdot {}^+4 = {}^-20$	$^-20 \div {}^-5 = {}^+4; \frac{^-20}{^-5} = {}^+4$
$^-2 \cdot {}^-7 = {}^+14$	$^+14 \div {}^-2 = {}^-7; \frac{^+14}{^-2} = {}^-7$
$^+8 \cdot {}^-9 = {}^-72$	$^-72 \div {}^+8 = {}^-9; \frac{^-72}{^+8} = {}^-9$

Division Rules for Integers

positive ÷ positive = positive
negative ÷ negative = positive
positive ÷ negative = negative
negative ÷ positive = negative

Study and Learn

A. Divide.

1. $^+12 \div {}^+3$ **2.** $^+4 \div {}^+1$ **3.** $\frac{^+9}{^+3}$

4. $^-24 \div {}^-3$ **5.** $^-8 \div {}^-1$ **6.** $\frac{^-12}{^-6}$

7. $^+36 \div {}^-6$ **8.** $^+42 \div {}^-7$ **9.** $\frac{^+18}{^-3}$

10. $^-24 \div {}^+8$ **11.** $^-17 \div {}^+1$ **12.** $\frac{^-42}{^+7}$

B. Complete.

13. $^+6 \cdot 0 = 0; 0 \div {}^+6 = \underline{?}$ ▶ Zero divided by a non-zero integer is zero.

14. $^-3 \cdot 0 = 0; 0 \div {}^-3 = \underline{?}$

15. $0 \cdot \underline{?} = {}^+3; {}^+3 \div 0 = \underline{?}$ ▶ Division by zero has no answer.

16. $0 \cdot \underline{?} = {}^-4; {}^-4 \div 0 = \underline{?}$

C. Divide if possible.

17. $0 \div {}^-1$ **18.** $^-1 \div {}^+1$ **19.** $^+5 \div 0$ **20.** $\frac{0}{^+8}$

Practice

Divide.

1. $^+8 \div {^+2}$	**2.** $^+12 \div {^+6}$	**3.** $^+15 \div {^+3}$	**4.** $^+20 \div {^+5}$
5. $^+36 \div {^+6}$	**6.** $^+45 \div {^+9}$	**7.** $^+50 \div {^+10}$	**8.** $^+80 \div {^+16}$
9. $^-36 \div {^-4}$	**10.** $^-36 \div {^-9}$	**11.** $^-56 \div {^-7}$	**12.** $^-56 \div {^-8}$
13. $^-27 \div {^-9}$	**14.** $^-36 \div {^-3}$	**15.** $^-36 \div {^-12}$	**16.** $^-64 \div {^-16}$
17. $^+24 \div {^-3}$	**18.** $^+40 \div {^-5}$	**19.** $^+72 \div {^-8}$	**20.** $^+36 \div {^-4}$
21. $^-42 \div {^+6}$	**22.** $^-28 \div {^+7}$	**23.** $^-48 \div {^+6}$	**24.** $^-40 \div {^+5}$
25. $0 \div {^-3}$	**26.** $0 \div {^+5}$	**27.** $0 \div {^-2}$	**28.** $^+9 \div {^-3}$
29. $^+16 \div {^-4}$	**30.** $^-8 \div {^+1}$	**31.** $^-30 \div {^-5}$	**32.** $^+45 \div {^-9}$
33. $^+30 \div {^-5}$	**34.** $^-63 \div {^-9}$	**35.** $^-42 \div {^-3}$	**36.** $^-60 \div {^-10}$
37. $^+80 \div {^-20}$	**38.** $^-45 \div {^-15}$	**39.** $^-60 \div {^+15}$	**40.** $^+80 \div {^-40}$

41. $\frac{^+18}{^+6}$	**42.** $\frac{^+56}{^+8}$	**43.** $\frac{^+6}{^+6}$	**44.** $\frac{^+55}{^+55}$
45. $\frac{^-27}{^-9}$	**46.** $\frac{^-48}{^-8}$	**47.** $\frac{^-24}{^-3}$	**48.** $\frac{^-64}{^-8}$
49. $\frac{^+36}{^-4}$	**50.** $\frac{^+40}{^-8}$	**51.** $\frac{^+30}{^-5}$	**52.** $\frac{^+40}{^-10}$
53. $\frac{^-21}{^+7}$	**54.** $\frac{^-35}{^+5}$	**55.** $\frac{^-42}{^+6}$	**56.** $\frac{^-42}{^+14}$
57. $\frac{0}{^-3}$	**58.** $\frac{0}{^+4}$	**59.** $\frac{0}{^-10}$	**60.** $\frac{^-7}{^-7}$
61. $\frac{^+17}{^-17}$	**62.** $\frac{^-56}{^-1}$	**63.** $\frac{^-60}{^+10}$	**64.** $\frac{^-46}{^+23}$

★ Solve.

65. The height of Mount Everest is $^+8{,}848$ m. The height of Lao Shan is $^+1{,}130$ m. About how many times higher is Mount Everest than Lao Shan?

EQUATIONS WITH INTEGERS

Solve and check.

$$x + {}^+3 = {}^+7$$
$$x + {}^+3 - {}^+3 = {}^+7 - {}^+3$$
$$x = {}^+4$$

Strategy: You want x alone.
You need to undo ${}^+3$ from x.
Use the subtraction property for equations.

Check:

$x + {}^+3$	${}^+7$
${}^+4 + {}^+3$	${}^+7$
${}^+7$	${}^+7$

Is ${}^+4$ a solution of $x + {}^+3 = {}^+7$?

So, ${}^+4$ is the solution of $x + {}^+3 = {}^+7$.

Study and Learn

A. Solve and check.

1. $x + {}^+4 = {}^-3$ **2.** $x - {}^+4 = {}^+7$ **3.** $x - {}^-3 = {}^-9$

Solve and check.

$$^+2x = {}^-18$$
$$\frac{^+2x}{^+2} = \frac{^-18}{^+2}$$
$$x = {}^-9$$

Strategy: You want x alone.
You need to undo ${}^+2$ from x.
Use the division property for equations.
The solution of $x = {}^-9$ is ${}^-9$.

Check:

^+2x	${}^-18$
$^+2 \cdot {}^-9$	${}^-18$
${}^-18$	${}^-18$

Is ${}^-9$ a solution of $^+2x = {}^-18$?

Yes, ${}^-9$ is the solution of $^+2x = {}^-18$.

B. Solve and check.

4. $^+3x = {}^+24$ **5.** $^-5x = {}^-40$ **6.** $^-7x = {}^+35$

C. Solve and check. Use the multiplication property for equations.

7. $\frac{x}{^+3} = {}^-7$ **8.** $\frac{x}{^-4} = {}^-8$ **9.** $\frac{x}{^-9} = {}^+6$

D. Solve. Use 2 properties of equations. Check.

10. $^+3x + {}^-3 = {}^+9$ **11.** $\frac{x}{^-2} + {}^+1 = {}^+3$ **12.** $^-5x - {}^-2 = {}^-13$

Practice

Solve and check.

1. $x + {}^+7 = {}^+9$

2. $x + {}^+8 = {}^+3$

3. $x + {}^-3 = {}^-5$

4. $x + {}^-1 = {}^+7$

5. $x + {}^-4 = {}^-8$

6. $x - {}^+1 = {}^+7$

7. $x - {}^+3 = {}^+1$

8. $x - {}^-5 = {}^-9$

9. $x - {}^-7 = {}^+3$

10. ${}^+3x = {}^+27$

11. ${}^+7x = {}^+42$

12. ${}^+5x = {}^-30$

13. ${}^+7x = {}^-35$

14. ${}^-3x = {}^+12$

15. ${}^-6x = {}^+48$

16. ${}^-7x = {}^-14$

17. ${}^-3x = {}^-15$

18. $\frac{x}{{}^+4} = {}^+7$

19. $\frac{x}{{}^+9} = {}^+6$

20. $\frac{x}{{}^+5} = {}^-4$

21. $\frac{x}{{}^+8} = {}^-2$

22. $\frac{x}{{}^-3} = {}^+8$

23. $\frac{x}{{}^-7} = {}^+6$

24. $\frac{x}{{}^-5} = {}^-6$

25. $\frac{x}{{}^-8} = {}^-8$

26. ${}^+2x + {}^+4 = {}^+12$

27. ${}^+3x + {}^+7 = {}^+19$

28. ${}^+3x + {}^-3 = {}^+12$

29. ${}^+4x + {}^-8 = {}^+4$

30. ${}^+4x + {}^-8 = {}^-20$

31. ${}^+7x + {}^-2 = {}^-23$

32. ${}^+3x - {}^-5 = {}^+23$

33. ${}^+2x - {}^+1 = {}^+1$

34. $\frac{x}{{}^+2} + {}^+3 = {}^+5$

35. $\frac{x}{{}^+3} + {}^+7 = {}^+8$

36. ${}^+2x = {}^+32$

37. $\frac{x}{{}^-3} + {}^+4 = {}^+6$

38. $\frac{x}{{}^+5} = {}^+8$

39. $\frac{x}{{}^-2} - {}^+3 = {}^-5$

★ Solve. Replacements for x: ${}^-10, {}^-9, \ldots, {}^+9, {}^+10$

40. $x + {}^+3 < {}^+6$

41. $x + {}^-4 > {}^-3$

42. ${}^+4x > {}^+12$

43. ${}^-3x > {}^-21$

44. ${}^+3x + {}^+4 > {}^+31$

45. ${}^+8x + {}^-4 > {}^+4$

Solve Problems

46. What number divided by ${}^-6$ equals ${}^-30$?

47. If ${}^+1$ is subtracted from twice a number, the result is ${}^-9$.
Find the number.

NEGATIVE EXPONENTS AND DECIMALS

Examine the pattern for the meaning of negative exponents.

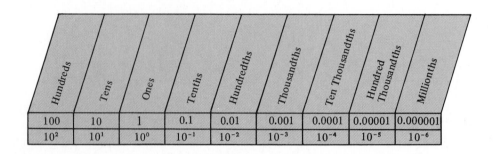

100	10	1	0.1	0.01	0.001	0.0001	0.00001	0.000001
10^2	10^1	10^0	10^{-1}	10^{-2}	10^{-3}	10^{-4}	10^{-5}	10^{-6}

Study and Learn

A. Complete.

1. $\frac{1}{10} = 0.1 = 10^{-\frac{?}{}}$ **2.** $\frac{1}{100} = 0.01 = 10^{-\frac{?}{}}$ **3.** $\frac{1}{10,000} = 0.0001 = 10^{-\frac{?}{}}$

4. $10^{-1} = 0.1 = \frac{1}{?}$ **5.** $10^{-2} = 0.01 = \frac{1}{?}$ **6.** $10^{-5} = 0.00001 = \frac{1}{?}$

B. Write standard numerals.

7. 10^2 **8.** 10^{-2} **9.** 10^1 **10.** 10^{-1} **11.** 10^6 **12.** 10^{-6}

C. Write in exponential notation.

13. 10   **14.** 0.1   **15.** 1,000   **16.** 0.001   **17.** 100   **18.** 0.01

Practice

Write standard numerals.

1. 10^4 **2.** 10^{-4} **3.** 10^5 **4.** 10^{-5} **5.** 10^3

6. 10^{-3} **7.** 10^{-1} **8.** 10^{-2} **9.** 10^{-6} ★**10.** 10^{-7}

Write in exponential notation.

11. 1,000   **12.** 0.001   **13.** 100,000   **14.** 0.00001   **15.** 1,000,000

16. 0.000001   **17.** 100   **18.** 10   **19.** 10,000   **20.** 10,000,000

EXPANDED NUMERALS WITH EXPONENTS

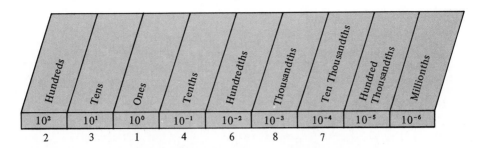

Standard numeral   Expanded numeral using exponents
231.4687 $= (2 \times 10^2) + (3 \times 10^1) + (1 \times 10^0) + (4 \times 10^{-1}) +$
$(6 \times 10^{-2}) + (8 \times 10^{-3}) + (7 \times 10^{-4})$

Study and Learn

A. Complete for the standard numeral 358.46127.

    1. The digit 3 has the value $3 \times 10^{\underline{?}}$.

    2. The digit 4 has the value $4 \times 10^{\underline{?}}$.

    3. The digit 1 has the value $1 \times 10^{\underline{?}}$.

    4. The digit 7 has the value $7 \times 10^{\underline{?}}$.

B. Write expanded numerals. Use exponents.

    5. 3.467   **6.** 28.5106   **7.** 374.691802

C. Write standard numerals.

    8. $(3 \times 10^2) + (4 \times 10^1) + (6 \times 10^0) + (8 \times 10^{-1}) + (7 \times 10^{-2}) + (8 \times 10^{-3})$

    9. $(4 \times 10^3) + (2 \times 10^2) + (0 \times 10^1) + (0 \times 10^0) + (6 \times 10^{-1}) +$
        $(3 \times 10^{-2}) + (4 \times 10^{-3})$

Practice

Write expanded numerals. Use exponents.

1. 64.37   **2.** 5,416.2743   **3.** 43.61371   **4.** 151.016341

Write standard numerals.

    5. $(6 \times 10^0) + (3 \times 10^{-1}) + (8 \times 10^{-2}) + (7 \times 10^{-3}) + (2 \times 10^{-4})$

    6. $(4 \times 10^1) + (0 \times 10^0) + (3 \times 10^{-1}) + (8 \times 10^{-2}) + (9 \times 10^{-3}) + (8 \times 10^{-4})$

Problem-Solving Applications

Career

The job of a wholesaler is to buy items from a manufacturer and then sell them to a retailer or store.

1. A salesperson from the Dale Wholesale Company sold 12 beds to the Most Attractive Furniture Store. The store paid $180 per bed. How much did the store pay for the beds? [HINT: multiply.]

2. The store listed the beds for $450 each and put them on sale at 40% off. Mrs. Rome bought a bed. What was the cost to her?

3. The Stanley Appliance Store paid a wholesale price of $203 for an air conditioner. The store sold it for $298 after a 20%-off sale. How much profit did the store make on the air conditioner?

4. Mr. Jackson paid $125 wholesale for a television. He sold it for $200. What was Mr. Jackson's percent of profit on the sale based on the cost of the television?

5. Ms. Klein sold ties to a store at the wholesale price of $50.40 a dozen. The store sold each tie for $7.50. How much did the store make on the dozen ties?

6. Mr. Johnson sold 8 television sets to a store at the wholesale price of $216 each. The storeowner sold the television sets at double the wholesale price. How much did the owner make on the sale of the 8 television sets?

★ 7. Chris sells books wholesale to a bookstore. She sells books marked $8.95 to the store at a 40% discount. If the owner pays for the books within 10 days, he gets a 2% discount. What does the owner pay for a book if payment is made within 10 days?

Chapter Review

Compare. Use >, <, or = *(278)*

1. $^+3 \equiv {^-4}$

2. $^-6 \equiv {^-1}$

Find. *(278)*

3. $|^-14|$

Add. *(280, 282)*

4. $^+6 + {^+8}$

5. $^-4 + {^-6}$

6. $^+3 + {^-5}$

Subtract. *(284)*

7. $^+6 - {^+2}$

8. $^-6 - {^-4}$

9. $^+4 - {^+9}$

Multiply. *(290)*

10. $^+4 \cdot {^+6}$

11. $^-8 \cdot {^-5}$

12. $^+8 \cdot {^-6}$

Divide. *(294)*

13. $^+16 \div {^+4}$

14. $^-56 \div {^-8}$

15. $\dfrac{^+56}{^-7}$

Solve. *(296)*

16. $x + {^-6} = {^+8}$

17. $\dfrac{x}{^-3} = {^-5}$

18. $^+2x + {^+3} = {^+27}$

Write in exponential notation. *(298)*

19. 10

20. 10,000

21. 0.001

Write expanded numerals. Use exponents. *(299)*

22. 78.21

23. 3,419.102

The Langs' yearly income is $30,000. *(288)*

24. How much is spent on housing?

25. How much is spent on food and other expenses combined?

Chapter Test

Compare. Use $>$, $<$, or $=$. *(278)*

1. $^+4 \equiv {^-5}$

2. $^-5 \equiv {^-2}$

Find. *(278)*

3. $|^+21|$

Add. *(280, 282)*

4. $^+6 + {^-3}$

5. $^-8 + {^-5}$

6. $^+6 + {^-8}$

Subtract. *(284)*

7. $^-3 - {^-7}$

8. $^+8 - {^-3}$

9. $^+3 - {^+8}$

Multiply. *(290)*

10. $^+3 \cdot {^+8}$

11. $^-9 \cdot {^-6}$

12. $^+7 \cdot {^-8}$

Divide. *(294)*

13. $^+24 \div {^+3}$

14. $^-56 \div {^-7}$

15. $\frac{0}{^-5}$

Solve. *(296)*

16. $x - {^+2} = {^-5}$

17. $^+4x = {^-20}$

18. $\frac{x}{2} + {^+1} = {^+4}$

Write in exponential notation. *(298)*

19. 100

20. 1,000

21. 0.0001

Write expanded numerals. Use exponents. *(299)*

22. 30.15

23. 526.1234

2,000 people selected their favorite television program. *(288)*

24. How many people selected Program A?

25. How many more people selected Program C than Program B?

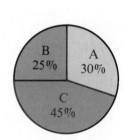

Skills Check

1. Which is not the same as 562 cm?

   A 56.2 m   B 5,620 mm

   C 5.62 m   D 0.00562 km

2. Which is the same as 42 in.?

   E $2\frac{1}{2}$ ft   F 3 ft

   G $3\frac{1}{2}$ ft   H none of the above

3. Which is the same as 659 g?

   A 659 kg   B 65.9 kg

   C 6.59 kg   D 0.659 kg

4. Which is the same as 56 oz?

   E $3\frac{1}{2}$ lb   F 4 lb

   G $4\frac{1}{2}$ lb   H 5 lb

5. Which is the same as 1,256 mL?

   A 125.6 L   B 12.56 L

   C 1.256 L   D 0.1256 L

6. Which is not a factor of 42?

   E 14   F 6

   G 4   H 2

7. What is the next number in this sequence?   2.1, 1.8, 1.5, . . .

   A 0.12   B 1.2

   C 1.4   D 1.6

8. Add.

$$4 \text{ lb } 9 \text{ oz}$$
$$+ 2 \text{ lb } 8 \text{ oz}$$

   E 9 lb 2 oz   F 8 lb 4 oz

   G 7 lb 1 oz   H 6 lb 3 oz

9. What is the length of $\overline{AB}$ to the nearest $\frac{1}{2}$ inch?

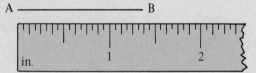

   A $2\frac{1}{2}$ in.   B 2 in.

   C $1\frac{1}{2}$ in.   D 1 in.

10. What is the length of $\overline{CD}$ to the nearest centimeter?

   E 4 cm   F 5 cm

   G 6 cm   H none of the above

12 REAL NUMBERS

The plane is landing at sunset at Los Angeles International Airport.

RATIONAL NUMBERS AND DECIMALS

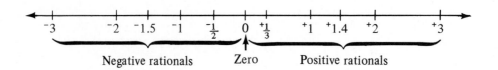

Negative rationals Zero Positive rationals

▶ A rational number is a number that can be written in
the form $\frac{a}{b}$ where a and b are integers and b is not zero.

$$^-3 = \frac{^-3}{^+1} \qquad\qquad ^+1.4 = \frac{^+14}{^+10}$$

Study and Learn

A. Write in the form $\frac{a}{b}$.

 1. 8   **2.** $^-8$ **3.** 0   **4.** $^-0.1$ **5.** 2.1

B. Compare. Use >, <, or =.

 6. $^-2 \equiv {}^-1\frac{1}{2}$ **7.** $\frac{^-2}{3} \equiv 0$ **8.** $^-1.4 \equiv 0.7$ **9.** $\frac{^-1}{2} \equiv \frac{^-3}{4}$

C. List in order from the least to the greatest. Use
 a number line.

 10. $^-6, \frac{2}{5}, \frac{7}{2}, {}^-1.1$ **11.** $^-4, {}^-1.4, \frac{3}{5}, \frac{^-3}{4}$

Every rational number can be shown as either a:

Terminating Decimal	or	*Repeating Decimal*

$$\frac{3}{4} \longrightarrow 4)\overline{3.00} \quad \begin{array}{r} 0.75 \\ \hline \end{array}$$

$$\begin{array}{r} 0.75 \\ 4)\overline{3.00} \\ \underline{2\ 8} \\ 20 \\ \underline{20} \\ \text{remainder} \rightarrow \quad 0 \end{array}$$

$$\frac{1}{3} \longrightarrow \begin{array}{r} 0.33\ldots \\ 3)\overline{1.00} \\ \underline{9} \\ 10 \\ \underline{9} \\ 1 \leftarrow \text{remainder} \end{array}$$

The division terminates (ends).   0.3333 . . . is a repeating decimal.
               0.3$\overline{3}$ means 3 repeats endlessly.

D. Write terminating decimals.

12. $\dfrac{-1}{2}$ **13.** $\dfrac{3}{25}$ **14.** $\dfrac{-1}{8}$ **15.** $\dfrac{2}{5}$ **16.** $^-1\dfrac{3}{4}$

E. Write repeating decimals.

17. $\dfrac{2}{3}$ **18.** $\dfrac{1}{9}$ **19.** $\dfrac{-4}{3}$ **20.** $^-3\dfrac{1}{7}$ **21.** $\dfrac{4}{9}$

Practice

Write in the form $\dfrac{a}{b}$.

1. 4 **2.** $^-3$ **3.** 6 **4.** $^-13$ **5.** 17

6. $^-0.3$ **7.** 2.64 **8.** $^-1.5$ **9.** 3.102 ★ **10.** $\dfrac{\frac{1}{2}}{3}$

Compare. Use >, <, or =.

11. $^-6 \equiv 5\dfrac{3}{4}$ **12.** $4 \equiv {}^-3.6$ **13.** $^-1.7 \equiv {}^-2.8$ **14.** $\dfrac{-1}{2} \equiv 0$

List in order from the least to the greatest.

15. $^-5, \dfrac{4}{5}, \dfrac{9}{2}, {}^-2.3$ **16.** $^-8, {}^-2.3, \dfrac{5}{6}, \dfrac{-3}{5}$

Write terminating decimals.

17. $\dfrac{1}{4}$ **18.** $\dfrac{1}{5}$ **19.** $\dfrac{3}{8}$ **20.** $\dfrac{7}{8}$ **21.** $\dfrac{7}{10}$

22. $\dfrac{-7}{10}$ **23.** $\dfrac{-3}{20}$ **24.** $\dfrac{-11}{25}$ **25.** $^-1\dfrac{3}{40}$ **26.** $\dfrac{-3}{2}$

Write repeating decimals.

27. $\dfrac{2}{9}$ **28.** $\dfrac{3}{11}$ **29.** $\dfrac{-1}{6}$ **30.** $\dfrac{-1}{7}$ **31.** $^-3\dfrac{5}{6}$

Skills Review

Compute. *(280, 290, 294)*

1. $^+6 + {}^+8$ **2.** $^-5 + {}^-9$ **3.** $^+7 + {}^-4$ **4.** $^+4 - {}^+2$

5. $^+8 - {}^-6$ **6.** $^-9 - {}^+3$ **7.** $^+6 \cdot {}^+2$ **8.** $^-8 \cdot {}^-7$

9. $^-8 \cdot {}^+9$ **10.** $^-81 \div {}^+9$ **11.** $^+50 \div {}^-5$ **12.** $^-72 \div {}^-8$

ADDING RATIONAL NUMBERS

Adding positive and negative rational numbers is like adding integers.

Integers	Rational Numbers
$5 + 2 = 7$	$\frac{5}{9} + \frac{2}{9} = \frac{7}{9}$
$^-3 + {}^-4 = {}^-7$	$\frac{^-3}{8} + \frac{^-4}{8} = \frac{^-7}{8}$
$^-14 + {}^-6 = {}^-20$	$^-1.4 + {}^-0.6 = {}^-2.0$

Study and Learn

A. Add and simplify.

1. $\frac{^-1}{7} + \frac{^-3}{7}$ **2.** $\frac{1}{10} + \frac{3}{10}$ **3.** $\frac{^-5}{8} + \frac{^-5}{8}$ **4.** $^-3.7 + {}^-2.1$

B. Adding a positive and a negative rational number is like adding positive and negative integers.

Integers	Rational Numbers
$3 + {}^-2 = 1$	$\frac{3}{7} + \frac{^-2}{7} = \frac{1}{7}$

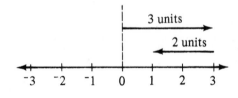   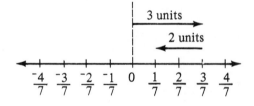

Add and simplify.

5. $\frac{^-5}{9} + \frac{2}{9}$ **6.** $\frac{^-3}{4} + \frac{1}{4}$ **7.** $\frac{^-7}{8} + \frac{3}{8}$ **8.** $^-2.6 + 4.0$

9. $\frac{3}{4} + \frac{^-1}{2}$ **10.** $\frac{2}{3} + \frac{^-1}{6}$ **11.** $\frac{3}{4} + \frac{^-5}{6}$ **12.** $3.1 + {}^-5.6$

C. Give the opposites.

13. $\frac{^-1}{2}$ **14.** $\frac{3}{4}$ **15.** $\frac{^-5}{6}$ **16.** 1.2

D. Add.

17. $\frac{1}{2} + \frac{^-1}{2}$ **18.** $\frac{3}{4} + \frac{^-3}{4}$ **19.** $\frac{^-5}{6} + \frac{5}{6}$ **20.** $1.2 + {}^-1.2$

Practice

Add and simplify.

1. $\frac{4}{9} + \frac{1}{9}$ **2.** $\frac{3}{8} + \frac{4}{8}$ **3.** $\frac{5}{10} + \frac{2}{10}$ **4.** $0.3 + 0.6$

5. $\frac{^-6}{12} + \frac{^-1}{12}$ **6.** $\frac{^-3}{7} + \frac{^-2}{7}$ **7.** $\frac{^-4}{12} + \frac{^-3}{12}$ **8.** $^-0.06 + {^-}0.09$

9. $\frac{3}{8} + \frac{^-3}{8}$ **10.** $\frac{1}{9} + \frac{^-2}{9}$ **11.** $\frac{3}{10} + \frac{^-3}{10}$ **12.** $6.2 + {^-}8.1$

13. $\frac{^-3}{10} + \frac{5}{10}$ **14.** $\frac{^-5}{6} + \frac{1}{6}$ **15.** $\frac{^-3}{4} + \frac{3}{4}$ **16.** $^-9.8 + 8.4$

17. $\frac{1}{6} + \frac{3}{8}$ **18.** $\frac{^-2}{5} + \frac{^-1}{4}$ **19.** $\frac{1}{3} + \frac{^-1}{2}$ **20.** $\frac{^-3}{4} + \frac{1}{8}$

21. $\frac{^-3}{8} + \frac{2}{8}$ **22.** $\frac{4}{5} + \frac{^-3}{5}$ **23.** $\frac{7}{12} + \frac{^-2}{12}$ **24.** $0.7 + {^-}0.3$

25. $\frac{7}{8} + \frac{^-3}{8}$ **26.** $\frac{^-3}{6} + \frac{1}{6}$ **27.** $\frac{5}{12} + \frac{^-1}{12}$ **28.** $1.6 + {^-}0.9$

29. $\frac{1}{2} + \frac{^-2}{6}$ **30.** $^-4.5 + 8.1$ **31.** $\frac{^-1}{4} + \frac{3}{8}$ **32.** $\frac{3}{4} + \frac{^-1}{2}$

33. $\frac{^-3}{10} + \frac{1}{4}$ **34.** $\frac{1}{2} + \frac{^-5}{6}$ **35.** $^-0.2 + {^-}4.1$ **36.** $\frac{5}{6} + \frac{^-5}{9}$

37. $\frac{3}{6} + \frac{5}{6}$ **38.** $6.9 + {^-}4.8$ **39.** $\frac{^-5}{8} + \frac{2}{3}$ **40.** $\frac{^-7}{10} + \frac{^-3}{10}$

★ **41.** $\frac{^-9}{10} + \frac{3}{5} + \frac{1}{2}$ ★ **42.** $\frac{^-5}{6} + \frac{3}{4} + \frac{^-1}{2}$ ★ **43.** $\frac{^-1}{3} + \frac{1}{6} + \frac{^-5}{6}$

Solve Problems

44. A stock decreased in value in January $\frac{3}{4}$ of a point $\left(\frac{^-3}{4}\right)$. In February, the stock increased in value $\frac{7}{8}$ of a point $\left(\frac{^+7}{8}\right)$. What was the total change in value for the 2 months?

SUBTRACTING RATIONAL NUMBERS

	Integers			*Rational Numbers*

Subtract.   Add opposites.

$4 - 3$ or $4 + {}^-3 = 1$

${}^-8 - {}^-6$ or ${}^-8 + 6 = {}^-2$

${}^-5 - 3$ or ${}^-5 + {}^-3 = {}^-8$

Subtract.   Add opposites.

$\frac{4}{7} - \frac{3}{7}$ or $\frac{4}{7} + \frac{{}^-3}{7} = \frac{1}{7}$

$\frac{{}^-8}{9} - \frac{{}^-6}{9}$ or $\frac{{}^-8}{9} + \frac{6}{9} = \frac{{}^-2}{9}$

${}^-0.5 - 0.3$ or ${}^-0.5 + {}^-0.3 = {}^-0.8$

▶ To subtract a rational number, add its opposite.

Study and Learn

A. Subtract $\frac{1}{8} - \frac{{}^-3}{8}$.

   1. Complete. $\frac{1}{8} - \frac{{}^-3}{8} = \frac{1}{8} + \underline{\quad?\quad}$

   2. What is $\frac{1}{8} - \frac{{}^-3}{8}$? Simplify.

B. Subtract and simplify.

   3. $\frac{{}^-3}{4} - \frac{1}{4}$ **4.** $\frac{{}^-9}{10} - \frac{{}^-3}{10}$ **5.** ${}^-0.8 - {}^-0.4$ **6.** $3.8 - {}^-1.9$

C. Subtract $\frac{1}{2} - \frac{{}^-3}{4}$.

   7. Complete. $\frac{1}{2} - \frac{{}^-3}{4} = \frac{?}{4} - \frac{{}^-3}{4}$ ⟵——— 4 is the least common denominator.

   8. Complete. $\frac{2}{4} - \frac{{}^-3}{4} = \frac{2}{4} + \underline{\quad?\quad}$

   9. What is $\frac{1}{2} - \frac{{}^-3}{4}$?

D. Subtract and simplify.

   10. $\frac{7}{8} - \frac{{}^-3}{4}$ **11.** $\frac{{}^-2}{3} - \frac{3}{4}$ **12.** $\frac{{}^-5}{6} - \frac{{}^-3}{8}$ **13.** ${}^-3\frac{1}{2} - 5\frac{2}{3}$

E. Subtract.

   14. $\frac{{}^-1}{2} - 0$ **15.** $0 - \frac{{}^-2}{3}$ **16.** $0 - \frac{1}{3}$ **17.** $0 - 0.4$

   18. $\frac{2}{3} - \frac{2}{3}$ **19.** $\frac{{}^-4}{5} - \frac{{}^-4}{5}$ **20.** ${}^-0.9 - {}^-0.9$ **21.** ${}^-0.8 - {}^-0.8$

Practice

Subtract and simplify.

1. $\frac{3}{5} - \frac{1}{5}$
2. $\frac{5}{8} - \frac{2}{8}$
3. $\frac{7}{10} - \frac{6}{10}$
4. $\frac{9}{12} - \frac{4}{12}$

5. $0.8 - 0.6$
6. $1.0 - 0.8$
7. $0.13 - 0.04$
8. $2.46 - 1.19$

9. $\frac{3}{5} - \frac{^-1}{5}$
10. $\frac{3}{8} - \frac{^-2}{8}$
11. $\frac{1}{6} - \frac{^-4}{6}$
12. $\frac{7}{12} - \frac{^-4}{12}$

13. $0.9 - {}^-0.6$
14. $0.7 - {}^-0.3$
15. $0.9 - {}^-0.3$
16. $1.6 - {}^-1.6$

17. $\frac{^-3}{8} - \frac{1}{8}$
18. $\frac{^-2}{5} - \frac{2}{5}$
19. $\frac{^-1}{6} - \frac{2}{6}$
20. $\frac{^-2}{10} - \frac{3}{10}$

21. $^-0.6 - 0.4$
22. $^-0.6 - 0.9$
23. $^-0.6 - 0.8$
24. $^-2.3 - 1.4$

25. $\frac{^-1}{8} - \frac{^-3}{8}$
26. $\frac{^-1}{12} - \frac{^-3}{12}$
27. $\frac{^-3}{4} - \frac{^-2}{4}$
28. $\frac{^-5}{6} - \frac{^-5}{6}$

29. $^-0.8 - {}^-0.3$
30. $^-1.9 - {}^-0.8$
31. $^-.3.4 - {}^-2.9$
32. $^-4.1 - {}^-2.1$

33. $\frac{1}{2} - \frac{1}{4}$
34. $\frac{3}{4} - \frac{2}{5}$
35. $\frac{5}{6} - \frac{^-3}{8}$
36. $\frac{2}{3} - \frac{^-3}{12}$

37. $\frac{^-3}{4} - \frac{5}{8}$
38. $\frac{^-3}{5} - \frac{3}{4}$
39. $\frac{^-1}{8} - \frac{^-3}{12}$
40. $\frac{^-5}{8} - \frac{^-4}{6}$

41. $\frac{^-3}{4} - 0$
42. $^-0.4 - 0$
43. $0 - \frac{3}{4}$
44. $0 - \frac{^-1}{2}$

45. $^-3.2 - {}^-4.8$
46. $\frac{^-2}{3} - \frac{1}{3}$
47. $\frac{2}{10} - \frac{^-4}{100}$
48. $5.6 - {}^-4.9$

Something Extra
Non-Routine Problems

1. is to as  is to

   a.  b.  c.  d.

2. S is to 2 as  is to

   a.   b.   c.   d.

MULTIPLYING AND DIVIDING

Integers	Rational Numbers
$2 \cdot 3 = 6$	$\frac{1}{3} \cdot \frac{5}{6} = \frac{5}{18}$
$^-2 \cdot {}^-3 = 6$	$\frac{^-2}{3} \cdot \frac{^-1}{5} = \frac{2}{15}$
$2 \cdot {}^-3 = {}^-6$	$\frac{5}{8} \cdot \frac{^-3}{4} = \frac{^-15}{32}$
$^-2 \cdot 3 = {}^-6$	$^-0.7 \cdot 0.8 = {}^-0.56$

▶ The product of 2 positive or of 2 negative numbers is positive.

▶ The product of a negative and a positive number is negative.

Study and Learn

A. Multiply and simplify.

1. $\frac{^-3}{5} \cdot \frac{2}{3}$ **2.** $\frac{^-3}{4} \cdot \frac{^-1}{2}$ **3.** $2\frac{1}{2} \cdot \frac{^-4}{5}$ **4.** $^-1.4 \cdot 2.3$

B. Multiply. Use a shortcut.

Example $\dfrac{^-3}{4} \cdot \dfrac{2}{3} = \dfrac{\overset{1}{^-\cancel{3}}}{\underset{2}{\cancel{4}}} \cdot \dfrac{\overset{1}{\cancel{2}}}{\underset{1}{\cancel{3}}} = \dfrac{^-1}{2}$

5. $\frac{^-3}{4} \cdot \frac{^-6}{9}$ **6.** $\frac{^-2}{5} \cdot \frac{7}{12}$ **7.** $\frac{8}{10} \cdot \frac{^-5}{6}$ **8.** $^-1\frac{1}{2} \cdot \frac{^-5}{6}$

▶ Two numbers are **reciprocals** if their product is 1.

C. Give the reciprocals.

9. $\frac{^-3}{4}$ **10.** $\frac{^-5}{8}$ **11.** $\frac{3}{4}$ **12.** $^-1\frac{1}{2}$

D. Divide and simplify.

13. $\frac{^-5}{8} \div \frac{^-2}{3}$ **14.** $10 \div {}^-2\frac{1}{2}$ **15.** $\frac{7}{8} \div {}^-3\frac{1}{2}$ **16.** $^-1.6 \div {}^-0.8$

E. Evaluate.

17. $\dfrac{6a + 4b}{4c}$ if $a = \frac{^-7}{3}$, $b = \frac{^-9}{2}$, $c = {}^-2$

Practice

Multiply and simplify.

1. $\frac{2}{3} \cdot \frac{1}{3}$ 2. $\frac{3}{4} \cdot \frac{1}{5}$ 3. $\frac{5}{6} \cdot \frac{1}{3}$ 4. $1\frac{1}{2} \cdot 1\frac{1}{3}$

5. $^-\frac{3}{4} \cdot {}^-\frac{1}{2}$ 6. $^-\frac{4}{5} \cdot {}^-\frac{7}{9}$ 7. $^-\frac{4}{5} \cdot {}^-\frac{2}{3}$ 8. $^-\frac{6}{7} \cdot {}^-\frac{3}{5}$

9. $^-\frac{5}{6} \cdot \frac{2}{3}$ 10. $^-\frac{3}{4} \cdot \frac{5}{8}$ 11. $^-\frac{6}{7} \cdot \frac{3}{7}$ 12. $^-\frac{5}{8} \cdot \frac{7}{12}$

13. $\frac{7}{8} \cdot {}^-\frac{3}{5}$ 14. $\frac{5}{12} \cdot {}^-\frac{5}{6}$ 15. $\frac{7}{10} \cdot {}^-\frac{3}{4}$ 16. $\frac{2}{3} \cdot {}^-\frac{4}{5}$

17. $^-0.4 \cdot {}^-0.8$ 18. $^-0.8 \cdot 0.3$ 19. $^-1.6 \cdot {}^-0.7$ 20. $8.4 \cdot {}^-3.4$

21. $\frac{3}{4} \cdot {}^-\frac{8}{9}$ 22. $^-\frac{5}{6} \cdot \frac{3}{4}$ 23. $^-\frac{3}{10} \cdot {}^-\frac{2}{3}$ 24. $\frac{5}{6} \cdot {}^-\frac{3}{10}$

25. $^-1\frac{1}{2} \cdot 3$ 26. $2\frac{1}{6} \cdot {}^-8$ 27. $^-\frac{3}{4} \cdot {}^-4$ 28. $^-1\frac{2}{5} \cdot 2\frac{1}{3}$

★ 29. $^-\frac{1}{2} \cdot \frac{2}{3} \cdot {}^-6$ ★ 30. $^-0.1 \cdot {}^-0.2 \cdot {}^-0.4$ ★ 31. $\frac{3}{4} \cdot \frac{1}{3} \cdot {}^-\frac{2}{3}$

Divide and simplify.

32. $0.6 \div 0.3$ 33. $^-0.09 \div {}^-0.03$ 34. $^-1.6 \div 0.4$ 35. $0.72 \div {}^-0.9$

36. $\frac{5}{6} \div \frac{3}{4}$ 37. $\frac{3}{8} \div \frac{3}{10}$ 38. $^-\frac{2}{3} \div {}^-\frac{1}{3}$ 39. $^-\frac{4}{5} \div {}^-\frac{7}{9}$

40. $^-\frac{3}{4} \div \frac{1}{2}$ 41. $^-\frac{5}{8} \div \frac{3}{4}$ 42. $\frac{1}{4} \div {}^-\frac{5}{12}$ 43. $\frac{5}{6} \div {}^-\frac{2}{3}$

44. $\frac{3}{4} \div 2$ 45. $^-\frac{8}{10} \div {}^-4$ 46. $\frac{7}{8} \div {}^-4$ 47. $^-\frac{15}{24} \div 8$

48. $8 \div \frac{3}{4}$ 49. $^-6 \div {}^-\frac{1}{2}$ 50. $7 \div {}^-\frac{1}{3}$ 51. $^-4 \div \frac{5}{6}$

52. $^-2\frac{1}{3} \div {}^-\frac{1}{4}$ 53. $3\frac{2}{5} \div {}^-\frac{3}{5}$ 54. $2\frac{1}{2} \div 1\frac{1}{4}$ 55. $5\frac{1}{2} \div {}^-1\frac{3}{4}$

56. Evaluate $\frac{4x + 3y}{2z}$ if $x = {}^-\frac{3}{2}$, $y = {}^-\frac{4}{3}$, $z = {}^-3$

IRRATIONAL NUMBERS

Decimals

Terminating	Non-terminating	
	Repeating	*Non-repeating*
0.1	$0.3\overline{3}$	0.141596535 . . .
0.27	$0.0\overline{101}$	0.12112111211112 . . .
0.125		There is a pattern here, but no block of numbers repeats.

Rational numbers

Irrational numbers

▶ A non-terminating, non-repeating decimal is an irrational number.

Study and Learn

A. Rational or irrational?

1. 0.164

2. 0.414214 . . .

3. $0.142857\overline{142857}$

4. 0.123123312333123333 . . .

> There are two ways to make up irrational numbers.
>   **1.** Write numbers at random after the decimal points as in 0.17283156 . . .
>   **2.** Use a pattern as in 0.117111711117111117 . . .

B. **5.** Make up 4 irrational numbers.

Practice

Rational or irrational?

1. 0.124

2. $0.123\overline{123}$

3. 0.3164893141967 . . .

4. 3.0100100010000100001 . . .

5. $0.058823\overline{58823}$

6. 0.919119111911119 . . .

★ **7.** 0.401401401401 . . .

★ **8.** π

Midchapter Review

Repeating or terminating? *(306)*

1. $\frac{1}{4}$ **2.** $^-\frac{1}{3}$ **3.** $\frac{2}{11}$ **4.** $4\frac{1}{5}$

Add and simplify. *(308)*

5. $^-\frac{3}{8} + ^-\frac{3}{8}$ **6.** $\frac{5}{6} + \frac{1}{6}$ **7.** $\frac{1}{2} + ^-\frac{3}{5}$ **8.** $^-\frac{3}{4} + ^-\frac{5}{8}$

9. $^-\frac{5}{6} + ^-\frac{7}{8}$ **10.** $^-0.06 + 0.09$ **11.** $^-1.2 + ^-3.4$ **12.** $^-3.41 + 2.61$

Subtract and simplify. *(310)*

13. $0.8 - ^-0.6$ **14.** $^-\frac{1}{6} - \frac{5}{6}$ **15.** $^-\frac{3}{4} - \frac{1}{2}$ **16.** $\frac{2}{3} - \frac{4}{5}$

17. $0 - ^-\frac{2}{3}$ **18.** $^-\frac{3}{5} - ^-\frac{3}{5}$ **19.** $^-0.9 - 1.4$ **20.** $^-2.36 - 4.49$

Multiply and simplify. *(312)*

21. $\frac{3}{4} \cdot \frac{1}{3}$ **22.** $^-\frac{4}{5} \cdot ^-\frac{3}{4}$ **23.** $^-1\frac{1}{2} \cdot \frac{3}{4}$ **24.** $^-6.1 \cdot ^-2.4$

Divide and simplify. *(312)*

25. $\frac{3}{8} \div ^-\frac{1}{2}$ **26.** $^-\frac{4}{5} \div ^-2$ **27.** $^-8 \div 2\frac{1}{2}$ **28.** $^-0.64 \div ^-0.8$

Rational or irrational? *(314)*

29. 0.3167 **30.** $2.41424344454\ldots$

31. $0.566666\ldots$ **32.** $0.31764859087156\ldots$

Something Extra
Calculator Activity

Use your calculator to check this formula for finding the sum of consecutive whole numbers.

$$\text{For the sum of whole numbers 1 to } n, \quad S = \frac{n \cdot (n + 1)}{2}$$

Example $1 + 2 + 3 + 4 = \frac{4 \cdot 5}{2}$, or 10

1. Check the example by adding.
2. Use the formula and check for the sum of the first 100 whole numbers.

Problem-Solving Skills

Bar Graphs

Information can be pictured by a bar graph.

This bar graph shows that in September the 8th grade borrowed more books than the 7th grade borrowed.

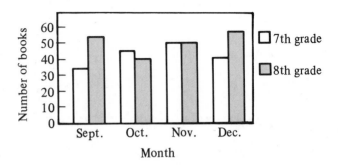

Study and Learn

A. Look at the bar graph above.

   1. How many books did the 7th grade borrow in September?

   2. How many books did the 8th grade borrow in September?

   3. In which month did the 7th grade borrow more books than the 8th grade?

   4. In which month did the 2 grades borrow the same number of books?

   5. How many more books did the 8th grade borrow than the 7th grade in December?

B. Sometimes a bar may be used to give 2 pieces of information.

   Example   In September, 15 fiction and 10 (25-15) biographies were borrowed.

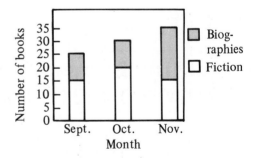

   6. How many fiction and how many biographies were borrowed in November?

   7. In which month were the most fiction books borrowed?

   8. In which month was the total number of fiction and biographies borrowed greatest?

Practice

Use this bar graph to answer Exercises 1 to 4.

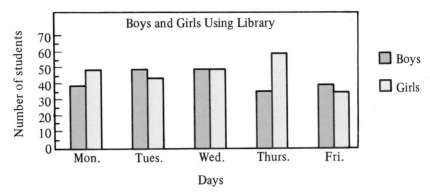

1. On what day did the most girls use the library?

2. On what days did more boys than girls use the library?

3. On what day did the same number of boys and girls use the library?

4. How many more girls than boys used the library on Monday?

Use this bar graph to answer Exercises 5 to 8.

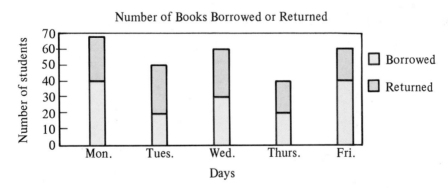

5. Which was the busiest day for borrowing and returning books?

6. On which 3 days were the greatest number of books returned?

7. How many more books were borrowed than returned on Monday?

8. How many more books were returned than borrowed on Tuesday?

REPEATING DECIMALS AND FRACTIONS

Observe these patterns.

$0.\overline{1} = 0.111111\ldots$ $10 \times 0.\overline{1} = 1.111111\ldots = 1.\overline{1}$

$0.\overline{12} = 0.121212\ldots$ $100 \times 0.\overline{12} = 12.121212\ldots = 12.\overline{12}$

A. Use this way to change a 1-digit repeating decimal to a fraction:

$0.\overline{1}$ is the repeating decimal. ▶ $N = 0.\overline{1}$

Multiply both sides by 10. ▶ $10\,N = 1.\overline{1}$

Subtract the original equation. ▶ $\underline{N = 0.\overline{1}}$

$9\,N = 1$

Divide both sides by 9. ▶ $\dfrac{9\,N}{9} = \dfrac{1}{9}$

$N = \dfrac{1}{9}$ So, $0.\overline{1} = \dfrac{1}{9}$

Change to fractions:

1. $0.\overline{3}$ **2.** $0.\overline{5}$ **3.** $0.\overline{6}$ **4.** $0.\overline{7}$ **5.** $0.\overline{8}$

B. Use this way to change a 2-digit repeating decimal to a fraction:

$0.\overline{12}$ is the repeating decimal. ▶ $N = 0.\overline{12}$

Multiply both sides by 100. ▶ $100\,N = 12.\overline{12}$

Subtract the original equation. ▶ $\underline{N = 0.\overline{12}}$

$99\,N = 12$

Divide both sides by 99. ▶ $\dfrac{99\,N}{99} = \dfrac{12}{99}$

$N = \dfrac{12}{99} = \dfrac{4}{33}$ So, $0.\overline{12} = \dfrac{4}{33}$

Change to fractions:

6. $0.\overline{36}$ **7.** $0.\overline{45}$ **8.** $0.\overline{06}$ **9.** $0.\overline{78}$ **10.** $0.\overline{30}$

Computer

To change $\frac{2}{3}$ to a decimal on the computer, enter **PRINT 2/3**.
The output is .666666667. This is the computer's display for $0.\overline{6}$. (Since most microcomputers have a 9-digit limit, the 9th digit is rounded).

For each computer output: (a) What repeating decimal is represented?
(b) What fraction input gave the output?

1. .333333333   **2.** .121212121   **3.** .575757576

SQUARES OF NUMBERS

To square a number means to multiply it by itself.
Read 7^2 : 7 squared or second power of 7
$7^2 = 7 \cdot 7 = 49$ $(^-7)^2 = ^-7 \cdot {}^-7 = 49$
$\left(\frac{3}{4}\right)^2 = \frac{3}{4} \cdot \frac{3}{4} = \frac{9}{16}$ $\left(^-\frac{3}{4}\right)^2 = ^-\frac{3}{4} \cdot {}^-\frac{3}{4} = \frac{9}{16}$
Squares of rational numbers, except 0, are positive.

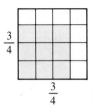

Study and Learn

A. Compute.

   1. 4^2 **2.** 8^2 **3.** $(^-3)^2$ **4.** $\left(\frac{2}{3}\right)^2$ **5.** $\left(^-\frac{2}{3}\right)^2$

   6. 0.4^2 **7.** 2.3^2 **8.** 1.9^2 **9.** 3.1^2 **10.** $(^-4.2)^2$

B. The square of 5 is 25.

   11. What other number is there whose square is 25?

   12. Give 2 numbers whose square is 100.

C. Some numbers are squares of whole numbers. 36 is the square of 6. 36 is called a **perfect square.** Which of these numbers are perfect squares?

   13. 2   **14.** 4   **15.** 16   **16.** 49   **17.** 80

Practice

Compute.

   1. 5^2 **2.** 9^2 **3.** 12^2 **4.** 20^2 **5.** $(^-9)^2$

   6. $(^-10)^2$ **7.** $(^-15)^2$ **8.** $(^-100)^2$ **9.** $\left(\frac{2}{5}\right)^2$ **10.** $\left(^-\frac{2}{5}\right)^2$

   11. $\left(\frac{3}{5}\right)^2$ **12.** $\left(^-\frac{3}{5}\right)^2$ **13.** 0.3^2 **14.** 1.1^2 **15.** $(^-2.1)^2$

Give 2 numbers whose square is the given number.

   16. 16   **17.** 9   **18.** 4   **19.** 49   ★ **20.** 2.25

Which of these numbers are perfect squares?

   21. 25   **22.** 50   **23.** 64   **24.** 100   **25.** 500

FINDING SQUARE ROOTS

A square root of 9 is 3 because $3 \cdot 3 = 9$ or $3^2 = 9$.
A square root of 9 is $^-3$ because $^-3 \cdot {}^-3 = 9$ or $(^-3)^2 = 9$.

▶ A positive rational number has 2 square roots.
$$\sqrt{9} = 3 \text{ and } {}^-\sqrt{9} = {}^-3.$$
$\sqrt{}$ means positive square root $^-\sqrt{}$ means negative square root

Study and Learn

A. Give 2 square roots.

    1. 25   **2.** 100   **3.** $\frac{4}{9}$ **4.** 0.09   **5.** 0.64

B. Find the square roots.

    6. $\sqrt{36}$ **7.** $\sqrt{81}$ **8.** $\sqrt{\frac{1}{36}}$ **9.** $^-\sqrt{0.16}$ **10.** $^-\sqrt{400}$

C. Estimate $\sqrt{289}$, and then find the exact square root.

    THINK: $\sqrt{100} < \sqrt{289}$ and $\sqrt{289} < \sqrt{400}$
    So, $10 < \sqrt{289}$ and $\sqrt{289} < 20$, or
          $\sqrt{289}$ is between 10 and 20 $10 < \sqrt{289} < 20$ ⟵   Estimate

    11. Try whole numbers between 10 and 20 to find one whose
         square is 289. What is $\sqrt{289}$?

D. Estimate, and then find the exact square root.

    12. $\sqrt{225}$ **13.** $\sqrt{529}$ **14.** $\sqrt{961}$ **15.** $\sqrt{2,809}$ **16.** $\sqrt{8,464}$

E. Estimate $\sqrt{5}$, and then find the square root to the nearest tenth.

    THINK: $\sqrt{4} < \sqrt{5}$ and $\sqrt{5} < \sqrt{9}$
    So, $2 < \sqrt{5}$ and $\sqrt{5} < 3$, or
          $\sqrt{5}$ is between 2 and 3 $2 < \sqrt{5} < 3$ ⟵   Estimate

    17. Try tenths between 2 and 3. Try $2.1 \cdot 2.1$, $2.2 \cdot 2.2$,
         $2.3 \cdot 2.3$, and so on. What is $\sqrt{5}$ to the nearest tenth?

F. Estimate, and then find the square root to the nearest tenth.

    18. $\sqrt{7}$ **19.** $\sqrt{11}$ **20.** $\sqrt{24}$ **21.** $\sqrt{73}$ **22.** $\sqrt{95}$

Practice

Give 2 square roots.

1. 4   **2.** 16   **3.** 1   **4.** 25

5. 81   **6.** 64   **7.** 100   **8.** 49

9. $\frac{1}{9}$ **10.** $\frac{4}{25}$ **11.** $\frac{36}{49}$ **12.** $\frac{81}{100}$

13. 0.04   **14.** 0.01   **15.** 0.49   **16.** 0.0025

Find the square roots.

17. $\sqrt{49}$ **18.** $\sqrt{100}$ **19.** $\sqrt{1}$ **20.** $\sqrt{4}$

21. $^-\sqrt{16}$ **22.** $^-\sqrt{36}$ **23.** $^-\sqrt{100}$ **24.** $^-\sqrt{4}$

25. $\sqrt{\frac{4}{81}}$ **26.** $\sqrt{\frac{1}{100}}$ **27.** $^-\sqrt{\frac{9}{25}}$ **28.** $^-\sqrt{\frac{49}{64}}$

29. $\sqrt{0.36}$ **30.** $\sqrt{0.81}$ **31.** $^-\sqrt{0.0064}$ **32.** $^-\sqrt{0.01}$

33. $\sqrt{2,500}$ **34.** $\sqrt{3,600}$ **35.** $^-\sqrt{4,900}$ **36.** $^-\sqrt{6,400}$

Estimate, and then find the exact square root.

37. $\sqrt{196}$ **38.** $\sqrt{361}$ **39.** $\sqrt{625}$ **40.** $\sqrt{784}$

41. $\sqrt{1,024}$ **42.** $\sqrt{2,209}$ **43.** $\sqrt{4,624}$ **44.** $\sqrt{5,041}$

Estimate, and then find the square root to the nearest tenth.

45. $\sqrt{2}$ **46.** $\sqrt{8}$ **47.** $\sqrt{12}$ **48.** $\sqrt{20}$

Solve Problems

49. A square board has an area of 91 cm^2. How long is each side to the nearest tenth?

50. A brace for a 3-foot by 5-foot rectangular gate is $\sqrt{34}$ feet long. Find the length of the brace to the nearest tenth.

★ **51.** The length of a diagonal of a square is $\sqrt{2}$ times the length of a side. A square has a side 8 cm long. Find the length of a diagonal to the nearest tenth.

THE SQUARE ROOT TABLE

$(1.4)^2 = 1.96$ $(1.5)^2 = 2.25$

So, $\sqrt{2} = 1.4$ to the nearest tenth.

$\sqrt{2}$ may be found to the nearest hundredth.

$(1.41)^2 = 1.9881$ $(1.42)^2 = 2.0164$

So, $\sqrt{2} = 1.41$ to the nearest hundredth.

You can continue this process endlessly. You will not find a number whose square is exactly 2.

$\sqrt{2}$ is a non-terminating, non-repeating decimal.

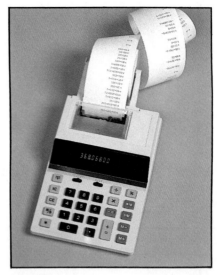

Study and Learn

A. The square root table below gives square roots to 3 decimal places.

Example $\sqrt{5} = 2.236$

Find the square roots. Use the table.

1. $\sqrt{7}$ **2.** $\sqrt{10}$ **3.** $\sqrt{29}$ **4.** $\sqrt{41}$

TABLE OF SQUARE ROOTS

Number	Square Root	Number	Square Root	Number	Square Root	Number	Square Root
1	1	13	3.606	26	5.099	38	6.164
2	1.414	14	3.742	27	5.196	39	6.245
3	1.732	15	3.873	28	5.292	40	6.325
4	2	16	4	29	5.385	41	6.403
5	2.236	17	4.123	30	5.477	42	6.481
6	2.449	18	4.243	31	5.568	43	6.557
7	2.646	19	4.359	32	5.657	44	6.633
8	2.828	20	4.472	33	5.745	45	6.708
9	3	21	4.583	34	5.831	46	6.782
10	3.162	22	4.690	35	5.916	47	6.856
11	3.317	23	4.796	36	6	48	6.928
12	3.464	24	4.899	37	6.083	49	7
		25	5			50	7.071

Some square roots are rational numbers, such as $\sqrt{1}$, $\sqrt{4}$, and $\sqrt{9}$.
Some square roots are irrational numbers. $\sqrt{2}$, $\sqrt{3}$, and $\sqrt{5}$.

▶ If the square root of a whole number is not a whole number, then it is an irrational number.

B. Rational or irrational?

    5. $\sqrt{16}$ **6.** $\sqrt{10}$ **7.** $\sqrt{24}$ **8.** $\sqrt{25}$

Practice

Find the square roots. Use the table on page 322.

  1. $\sqrt{8}$ **2.** $\sqrt{12}$ **3.** $\sqrt{23}$ **4.** $\sqrt{25}$ **5.** $\sqrt{37}$

  6. $\sqrt{48}$ **7.** $\sqrt{31}$ **8.** $\sqrt{15}$ **9.** $\sqrt{35}$ **10.** $\sqrt{43}$

Rational or irrational?

  11. $\sqrt{6}$ **12.** $\sqrt{20}$ **13.** $\sqrt{31}$ **14.** $\sqrt{36}$ **15.** $\sqrt{64}$

Skills Review

Write percents. *(250, 252)*

  1. 0.7   **2.** 0.23   **3.** 0.148

  4. 3.6   **5.** $\frac{47}{100}$ **6.** $\frac{6}{100}$

  7. $\frac{3}{4}$ **8.** $\frac{1}{25}$ **9.** $\frac{1}{8}$

Compute. *(254)*

  10. 10% of 80   **11.** 50% of 36   **12.** 25% of 96   **13.** 45% of 75

  14. 0.5% of 60   **15.** 6.8% of 24   **16.** 13.2% of 15   **17.** 52.3% of 40

  18. $1\frac{1}{2}$% of 34   **19.** $2\frac{1}{4}$% of 43   **20.** $4\frac{3}{4}$% of 24   **21.** $9\frac{1}{2}$% of 61

Compute. *(256)*

  22. What percent of 10 is 6?   **23.** What percent of 50 is 28?

  24. What percent of 80 is 46?   **25.** What percent of 4 is 8?

THE PYTHAGOREAN RELATIONSHIP

A famous Greek philosopher and mathematician named Pythagoras discovered this relationship about the measures of the sides of a right triangle:

$$a^2 + b^2 = c^2$$

measures of legs   measure of hypotenuse

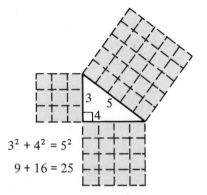

$3^2 + 4^2 = 5^2$

$9 + 16 = 25$

Study and Learn

A. Find the length of the hypotenuse (the longest side).

Example
$$a^2 + b^2 = c^2$$
$$6^2 + 8^2 = c^2$$
$$36 + 64 = c^2$$
$$100 = c^2$$
$$10 = c$$

a = 6, b = 8, c

1.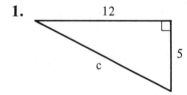

12, 5, c

2.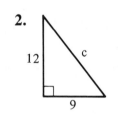

12, c, 9

B. Find the missing lengths.

Example
$$a^2 + b^2 = c^2$$
$$3^2 + b^2 = 5^2$$
$$9 + b^2 = 25$$
$$b^2 = 16$$
$$b = 4$$

a = 3, c = 5, b

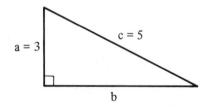

3.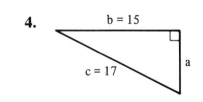

c = 15, a = 12, b

4.

b = 15, c = 17, a

324

Practice

Find the missing lengths to the nearest whole number.

1.

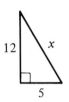

2.

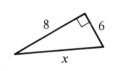

3.

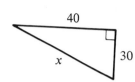

4.

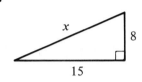

5.

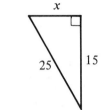

6.

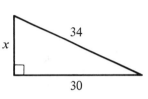

7.

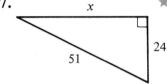

★ **8.**

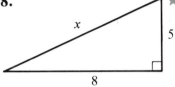

★ **9.**

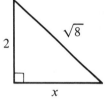

Solve Problems

10. Find the distance across the pond.

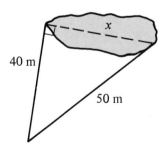

★ **11.** Find the length of the brace for the gate to the nearest tenth.

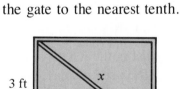

12. A 24-m television antenna is erected. A 26-m cable is connected to the top of the antenna and anchored to the ground. How far from the antenna is the cable anchored?

★ **13.** The length of one side of a right triangle is 14 cm, and the length of the other side is 18 cm. Find the length of the hypotenuse to the nearest tenth.

325

THE REAL NUMBER LINE

The length of a side can be rational or irrational.

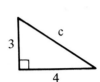

$$a^2 + b^2 = c^2$$
$$3^2 + 4^2 = c^2$$
$$9 + 16 = c^2$$
$$25 = c^2$$
$$5 = c$$

$$a^2 + b^2 = c^2$$
$$1^2 + 1^2 = c^2$$
$$1 + 1 = c^2$$
$$2 = c^2$$
$$\sqrt{2} = c$$

▶ The real numbers are all the rational numbers and all the irrational numbers.

Study and Learn

A. Each point on a line may be associated with a rational or an irrational number. This is the real number line. To associate the irrational number $\sqrt{2}$ with a point on a number line, you can use the Pythagorean relationship.

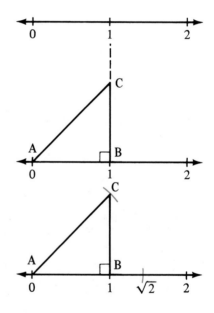

1. Draw a number line. Choose a unit and label 0, 1, and 2. Label 0 as point A and 1 as point B.

2. Construct a perpendicular to $\overline{AB}$ at B.

3. Mark point C on the perpendicular so that $BC = 1$ unit.

4. Draw $\overline{AC}$.

5. What is the length of $\overline{AC}$? [HINT: see display.]

6. Use your compass and mark off $\overline{AC}$ on the number line. (Put the compass point on 0.)

B. Look at the numbers shown on this number line.

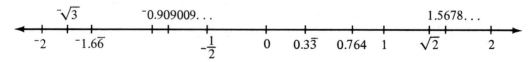

7. Which are rational?

8. Which are irrational?

326

Practice

Find the missing lengths. Write the answers as square roots.

1. 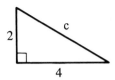   **2.** 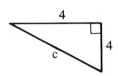   **3.**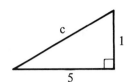

Draw a number line. Use the method in Item A on page 326 to find the points associated with these real numbers.

4. $\sqrt{8}$ **5.** $\sqrt{13}$ **6.** $\sqrt{17}$ ★ **7.** $^-\sqrt{8}$

Match the points on the number line with the numbers listed below.

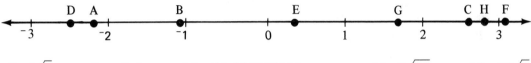

8. $\sqrt{8}$ **9.** $^-2.5$ **10.** $^-1.010010001\ldots$ **11.** $\sqrt{10}$ **12.** $^-\sqrt{5}$

13. In **8–12** above, which numbers are rational? irrational? real?

Computer

This computer program calculates the length of the hypotenuse of a right triangle when the lengths of the sides are entered.

```
10 INPUT "LENGTH OF SIDE A ";A

20 INPUT "LENGTH OF SIDE B ";B

30 PRINT "C= THE SQUARE ROOT
     OF " A^2 + B^2

40 PRINT "C= " SQR(A^2 + B^2)
```

Line 40 prints the hypotenuse as a decimal.
SQR is the computer's square root function.

H A N D S   O N

Use this computer program to check **1–3** in the Practice. Type in the lines; then type **RUN.** Remember ⌈ RETURN ⌉.

Problem-Solving Applications

Career

Astronomers are people who study the universe.

1. Astronomers have estimated that there are 200 billion stars in our galaxy. They believe that only 1 star in 25 might have planets suitable for life. About how many stars in our galaxy might have planets suitable for life? [HINT: divide.]

2. Light travels about 3×10^5 km per second. It takes light about $1\frac{1}{3}$ seconds to reach the earth from the moon. About how far is the Earth from the moon?

3. Neptune rotates on its axis in 15 hours 8 minutes. The Earth rotates on its axis once every 24 hours. How much longer does it take Earth to rotate than Neptune?

4. Light travels about 9 trillion km per year. The distance is called a light year by astronomers. The star, Alpha Centauri, is $4\frac{1}{3}$ light years from us. How many kilometers from Earth is the star?

5. Astronomers believe that the nearest star which might have planets suitable for life is 10.8 light years away. How long would it take for a signal from such a planet to be received on Earth and then an answer to be sent back to the planet?

6. One year there were about 2,000 astronomers. About 85% of these were involved in research. How many were not involved in research?

★ 7. The distance from Pluto to the sun is about 3.67×10^9 mi. The distance from Neptune to the sun is about 2.79×10^9 mi. How much farther away from the sun is Pluto?

Chapter Review

Compare. Use $<$ or $>$. *(306)*

Write as a decimal. *(306)*

1. $\frac{^-2}{3} \equiv 0$ **2.** $^-3.4 \equiv ^-4.1$ **3.** $\frac{1}{2}$

Add or subtract and simplify. *(308, 310)*

4. $\frac{^-5}{8} + \frac{^-5}{8}$ **5.** $\frac{3}{4} + \frac{^-5}{8}$ **6.** $^-3.46 + 2.19$

7. $\frac{^-2}{3} - \frac{1}{2}$ **8.** $0 - \frac{3}{4}$ **9.** $^-3.16 - 5.87$

Multiply or divide and simplify. *(312)*

10. $\frac{^-3}{4} \cdot \frac{1}{3}$ **11.** $^-1\frac{1}{2} \cdot \frac{3}{8}$ **12.** $^-6.3 \cdot {}^-1.9$

13. $\frac{5}{8} \div \frac{^-1}{2}$ **14.** $\frac{^-2}{3} \div 2$ **15.** $^-0.72 \div {}^-0.9$

Compute. *(319)*

16. 8^2 **17.** $(^-5)^2$ **18.** $\left(\frac{^-4}{5}\right)^2$ **19.** 1.3^2

Find square roots. *(320)*

20. $\sqrt{64}$ **21.** $^-\sqrt{4}$ **22.** $\sqrt{\frac{9}{49}}$ **23.** $\sqrt{0.36}$

24. $\sqrt{1,600}$ **25.** $\sqrt{4,900}$ **26.** $\sqrt{196}$ **27.** $\sqrt{576}$

Rational or irrational? *(314, 322)*

28. $0.34\overline{4}$ **29.** $0.4040040004\ldots$ **30.** $\sqrt{10}$ **31.** $\sqrt{81}$

32. Two sides of a right triangle are 5 cm and 12 cm. How long *(324)* is the hypotenuse?

33. How many more boys than girls *(316)* bowled on Monday?

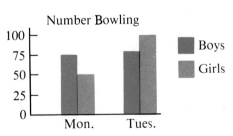

Number Bowling

Chapter Test

Compare. Use $<$ or $>$. *(306)*

1. $\frac{^-1}{4} \equiv 0$

2. $^-2.9 \equiv \ ^-3.1$

Write as a decimal. *(306)*

3. $\frac{2}{3}$

Add or subtract and simplify. *(308, 310)*

4. $\frac{^-3}{4} + \frac{^-3}{4}$

5. $\frac{2}{3} + \frac{^-5}{6}$

6. $^-5.36 + 4.08$

7. $\frac{^-3}{4} - \frac{2}{3}$

8. $0 - \frac{^-1}{2}$

9. $^-4.08 - 6.73$

Multiply or divide and simplify. *(312)*

10. $\frac{^-5}{6} \cdot \frac{1}{5}$

11. $^-2\frac{1}{2} \cdot \frac{2}{3}$

12. $^-3.9 \cdot \ ^-2.5$

13. $\frac{5}{6} \div \frac{^-1}{3}$

14. $\frac{^-3}{4} \div 3$

15. $^-0.81 \div 0.9$

Compute. *(319)*

16. 6^2

17. $(^-3)^2$

18. $\left(\frac{^-2}{3}\right)^2$

19. 2.1^2

Find the square roots. *(320)*

20. $\sqrt{36}$

21. $^-\sqrt{16}$

22. $\sqrt{\frac{9}{64}}$

23. $\sqrt{0.04}$

24. $\sqrt{900}$

25. $\sqrt{6,400}$

26. $\sqrt{256}$

27. $\sqrt{441}$

Rational or irrational? *(314, 322)*

28. $0.45\overline{5}$

29. $\sqrt{8}$

30. $\sqrt{64}$

31. $0.3131131113\ldots$

32. The hypotenuse of a right triangle is 15 m. One side is
(324) 12 m. What is the length of the other side?

33. In October, how many more eighth-grade
(316) students had perfect attendance than
seventh-grade students?

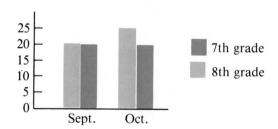

330

Skills Check

1. What is the least number of bills and coins to make change from a 20 dollar bill for a purchase of $13.40?

    A one $5, one $1, one 50¢, two 5¢

    B one $5, one $1, two 25¢, one 10¢

    C six $1, one 50¢, one 10¢

    D one $5, one $1, one 50¢, one 10¢

2. Sue Ellen bought clothing for $146. The sales tax was 5%. How much tax did she pay?

    E $7.30   F $73

    G $153.30   H $219

3. Sneakers marked $21.25 are sold at a 20% discount during a sale. How much is saved during the sale?

    A $0.43   B $4.25

    C $5.31   D $17.00

4. What is the area of a rectangle with a length of 23 in. and a width of 14 in.?

    E 37 in.²   F 74 in.²

    G 322 in.²   H 422 in.²

5. Which program received the most votes?

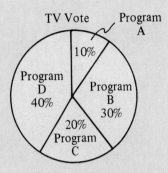

    A program A   B program B

    C program C   D program D

6. A 6 lb 8 oz piece of meat became 4 lb 14 oz after cooking. How much was lost during cooking?

    E 2 lb 6 oz   F 1 lb 10 oz

    G 1 lb 8 oz   H 1 lb 4 oz

7. A machine produces 6,000 radio parts an hour. On the average, 5% of the parts are defective. How many are defective?

    A 3   B 30

    C 300   D 3,000

8. Vegetable soup is selling at 2 cans for $0.39. What is the cost of a dozen cans of vegetable soup?

    E $4.68   F $2.48

    G $2.34   H $1.17

13 GEOMETRY AND MEASUREMENT

Silicon chips contain many integrated circuits.

CONGRUENT TRIANGLES

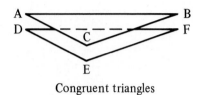

Congruent triangles

Corresponding Sides	Corresponding Angles
$\overline{AB} \cong \overline{DF}$	$\angle C \cong \angle E$
$\overline{AC} \cong \overline{DE}$	$\angle B \cong \angle F$
$\overline{CB} \cong \overline{EF}$	$\angle A \cong \angle D$

▶ Congruent triangles are triangles that can be matched so that corresponding sides are congruent and corresponding angles are congruent.

Study and Learn

A. $\triangle LMN$ and $\triangle PQR$ are congruent.

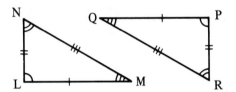

   1. Name the corresponding sides.

   2. Name the corresponding angles.

B. Construct a triangle congruent to $\triangle ABC$ by copying its sides.

   3. Draw a triangle and label it $\triangle ABC$.

   4. Draw a working segment and copy $\overline{AB}$. Label it $\overline{DE}$.

   5. Measure $\overline{AC}$ with your compass. With your compass point on D, make an arc the measure of $\overline{AC}$.

   6. Measure $\overline{BC}$ with your compass. With your compass point on E, make an arc the measure of $\overline{BC}$. Label the point where the 2 arcs meet point F.

   7. Draw $\overline{DF}$ and $\overline{EF}$.

   8. Cut out $\triangle DEF$ and place it on $\triangle ABC$. Are all the parts congruent?

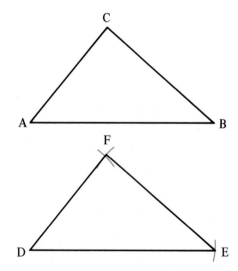

▶ Two triangles are congruent if their corresponding sides are congruent. This principle is called SSS (side-side-side).

C. Construct a triangle congruent to $\triangle ABC$ by copying 2 sides and the included angle ($\angle A$ is the included angle of sides $\overline{AB}$ and $\overline{AC}$).

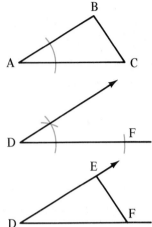

    9. Draw a triangle and label it $\triangle ABC$.

   10. Draw a working segment and copy $\overline{AC}$. Label it $\overline{DF}$.

   11. Copy $\angle A$ using point D as the vertex.

   12. On the ray you drew to copy $\angle A$, copy $\overline{AB}$. Label it $\overline{DE}$.

   13. Draw $\overline{EF}$.

   14. Cut out $\triangle DEF$ and place it on $\triangle ABC$. Are all the parts congruent?

D. Find x.

   15. $\triangle ABC \cong \triangle DEF$

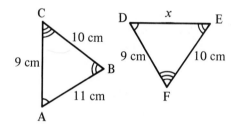

   16. $\triangle TUV \cong \triangle XYZ$

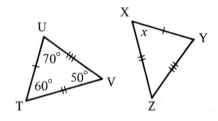

Practice

Name the corresponding sides and corresponding angles. Find x.

1. $\triangle ABC \cong \triangle DEF$

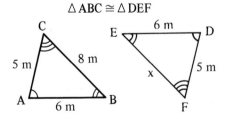

2. $\triangle LMN \cong \triangle PQR$

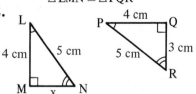

★ **3.** Draw a triangle and label it $\triangle DEF$. Construct a triangle congruent to $\triangle DEF$ by copying 2 angles and the included side. ($\overline{DE}$ is the included side of angles D and E.)

SIMILAR TRIANGLES

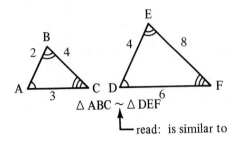

B
2 △ 4
A 3 C

E
4 8
D 6 F

△ ABC ~ △ DEF

└─ read: is similar to

Corresponding Angles

$\angle A \cong \angle D$

$\angle B \cong \angle E$

$\angle C \cong \angle F$

Corresponding Sides

$\dfrac{BC}{EF} = \dfrac{4}{8}$ or $\dfrac{1}{2}$

$\dfrac{AC}{DF} = \dfrac{3}{6}$ or $\dfrac{1}{2}$

$\dfrac{AB}{DE} = \dfrac{2}{4}$ or $\dfrac{1}{2}$

$\dfrac{BC}{EF} = \dfrac{AC}{DF} = \dfrac{AB}{DE}$

▶ Similar triangles have:
   1. corresponding angles congruent, and
   2. corresponding sides in proportion.

Study and Learn

A. Decide if the triangles are similar.

   1. Are the corresponding angles congruent?

   2. Are the corresponding sides in proportion?

   3. Are the triangles similar?

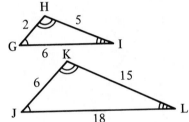

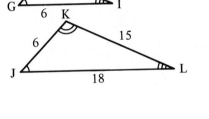

B. $\triangle MNO \sim \triangle PQR$. Find x. Complete.

   4. $\dfrac{NM}{QP} = \dfrac{MO}{?}$

   5. $\dfrac{4}{2} = \dfrac{?}{x}$

   6. $4x = \underline{\quad ? \quad}$

   7. $x = \underline{\quad ? \quad}$

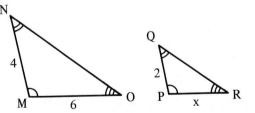

C. $\triangle STU \sim \triangle WVU$.

   8. Find x.

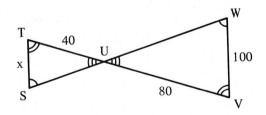

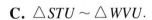

336

Practice

Are the triangles similar? Why?

1.

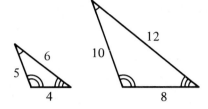

2.

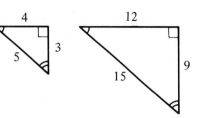

Each pair of triangles is similar. Find x.

3.

4.

5.

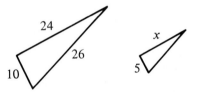

6.

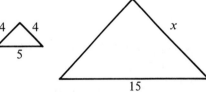

7.

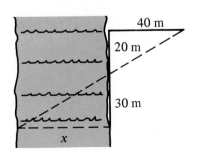

8.

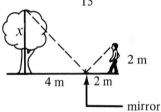

Solve Problems

9. Find the distance across the river.

★ **10.** Assume that the sun's rays form similar triangles. Find the height of the school building.

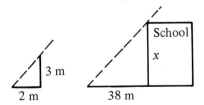

SCALE DRAWINGS

A scale drawing is similar to the actual object. The scale may be shown in different ways.

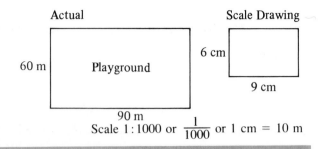

Actual

Scale Drawing

60 m | Playground

90 m

6 cm

9 cm

Scale 1:1000 or $\frac{1}{1000}$ or 1 cm = 10 m

Study and Learn

A. Find the actual length of the living room.

1. Measure the length of the scale drawing.

2. Write a proportion and solve.

   Scale Length   Actual Length

   $$\frac{0.5 \text{ cm}}{3 \text{ cm}} = \frac{1 \text{ m}}{x}$$

Living Room

Scale: 0.5 cm = 1 m

3. What is the actual length of the living room?

B. Make a scale ruler by copying, on a strip of paper, the scale shown on the map.

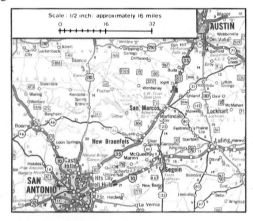

4. Use your scale ruler to find the land distance between San Antonio and Austin, Texas.

C. Use the scale 0.5 cm represents 1 m to make a scale drawing of a room 4.8 m by 6.6 m.

5. What length is represented by 1 cm?

6. How many centimeters on the drawing will represent 4.8 m?
   Solve. $\frac{1 \text{ cm}}{2 \text{ m}} = \frac{x}{4.8 \text{ m}}$

7. How many centimeters on the drawing will represent 6.6 m?

8. What are the dimensions of the drawing?

9. Draw the scale drawing of the room.

Practice

Find the actual length and width. Measure to the nearest centimeter.

1. Living room

2. Kitchen

3. Bedroom 1

4. Bedroom 2

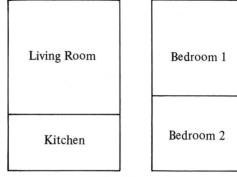

Scale 0.5 cm = 1 m

Make a scale ruler from the scale at the bottom of the map.

5. Find the land distance
   between Tallahassee
   and Lake City, Florida.

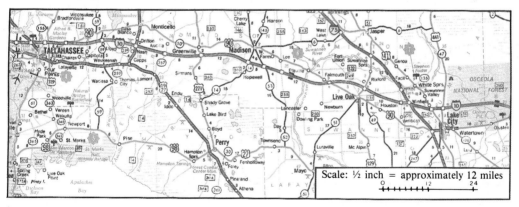

Scale: ½ inch = approximately 12 miles

Make scale drawings.

	Item	Dimensions	Scale
6.	Table top	1.5 m by 2.5 m	6 cm = 1 m
7.	Door	1 m by 2 m	0.5 cm = 1 m
8.	Playground	160 m by 300 m	1 cm = 40 m
9.	Room	2.3 m by 3 m	1 : 100

TRIGONOMETRIC RATIOS

In right $\triangle ABC$,

the ratio $\frac{a}{c}$ is the sine of m$\angle A$,

the ratio $\frac{b}{c}$ is the cosine of m$\angle A$, and

the ratio $\frac{a}{b}$ is the tangent of m$\angle A$.

These abbreviations are used:

$\sin A = \frac{a}{c}$ $\qquad$ $\cos A = \frac{b}{c}$ $\qquad$ $\tan A = \frac{a}{b}$

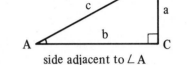

▶ The sine, cosine, and tangent are trigonometric ratios. The word **trigonometry** means triangle measurement.

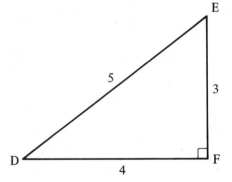

Study and Learn

A. Find the lengths of these sides of $\triangle DEF$.

   1. The hypotenuse

   2. The side opposite $\angle D$

   3. The side adjacent to $\angle D$

B. Find the ratios. Use $\triangle DEF$ at the right.

   4. $\tan D$ **5.** $\sin D$

   6. $\cos D$ **7.** $\tan E$

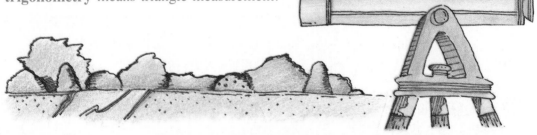

C. The trigonometric ratios are the same for a certain angle, no matter what the size of the triangle. The trigonometric ratios for angles measuring 0° through 90° are given in the table on page 416.

Find the ratios. Use the table on page 416.

   8. $\sin 38°$ **9.** $\cos 21°$ **10.** $\tan 74°$

340

D. You can use trigonometric ratios to find the height of a building.

11. m∠A is given. Which length is given, opposite, adjacent, or hypotenuse?

12. Which length is *x*, opposite, adjacent, or hypotenuse?

13. Which ratio involves adjacent and opposite?

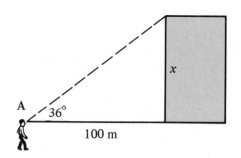

14. Complete.

$$\tan 36° = \frac{x}{100}$$

$$0.727 = \frac{x}{100}$$

$$x = \underline{\ \ ?\ \ }$$

15. What is the height of the building?

Practice

Find the ratios. Use the table on page 416.

1. sin 54°	**2.** sin 16°	**3.** sin 87°
4. cos 27°	**5.** cos 45°	**6.** cos 71°
7. tan 38°	**8.** tan 4°	**9.** tan 69°

Find *x*.

10.

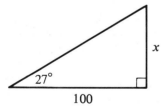

11.

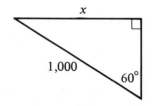

12.

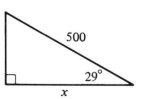

13.

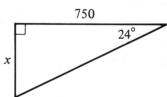

341

AREA

Rectangle

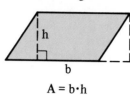

$A = b \cdot h$

Parallelogram

$A = b \cdot h$

Triangles

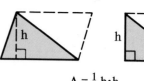

$A = \frac{1}{2} b \cdot h$

Study and Learn

A. Find the areas.

1.
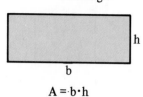
3 m
4 m

2.
6 cm
22 cm

3.
8 mm
15 mm

4.
5 m
3 m
4 m

B. A square is a rectangle. Find the areas of these squares.

5.
4 cm
4 cm

6.
11 m
11 m

C. Here's how to find a formula for the area of a trapezoid.

7. Copy this trapezoid on square-ruled paper.

8. Double the trapezoid as shown.

9. What figure results?

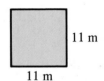

a = 3
h = 2
b = 5

10. The base of the parallelogram is $a + b$, or ___?___ .

11. What is the area of the parallelogram?

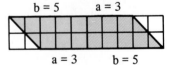

b = 5   a = 3

a = 3   b = 5

12. How does the area of the trapezoid compare with that of the parallelogram?

13. What is the area of the trapezoid?

▶ The area of a trapezoid is $\frac{1}{2}$ the height times the sum of the bases.
$A = \frac{1}{2} h \cdot (a + b)$

D. Find the areas of the trapezoids.

14.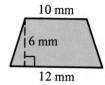
10 mm
6 mm
12 mm

15.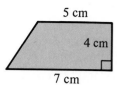
5 cm
4 cm
7 cm

Practice

Find the areas.

1.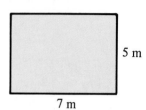
5 m
7 m

2.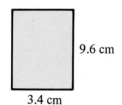
9.6 cm
3.4 cm

3.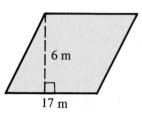
6 m
17 m

4.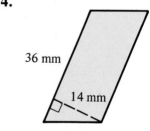
36 mm
14 mm

5.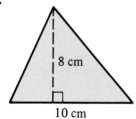
8 cm
10 cm

6.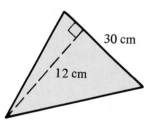
30 cm
12 cm

7.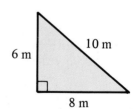
6 m
10 m
8 m

8.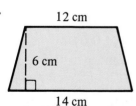
12 cm
6 cm
14 cm

9.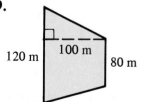
120 m
100 m
80 m

Solve Problems

10. Ms. Stone bought a carpet for a rectangular living room that is 6 m by 4 m. The carpet sold for $14.95/m². How much did she pay to carpet her living room?

 11. What happens to the area of a square when a side is doubled?

343

AREA IN METRIC SYSTEM

A square centimeter (cm²) is a unit of area.

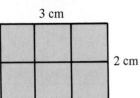

The area of this rectangle is 6 cm².

Study and Learn

A. You can change between metric units of area.

Examples

$1 \text{ m}^2 = \underline{\ \ ?\ \ } \text{ cm}^2$
$1 \text{ m}^2 = 1 \text{ m} \times 1 \text{ m}$
$\qquad = 100 \text{ cm} \times 100 \text{ cm}$
$\qquad = 10{,}000 \text{ cm}^2$

$5 \text{ km}^2 = \underline{\ \ ?\ \ } \text{ m}^2$
$5 \text{ km}^2 = 5 \text{ km} \times 1 \text{ km}$
$\qquad = 5{,}000 \text{ m} \times 1{,}000 \text{ m}$
$\qquad = 5{,}000{,}000 \text{ m}^2$

Complete.

1. $1 \text{ cm}^2 = \underline{\ \ ?\ \ } \text{ mm}^2$ **2.** $1 \text{ km}^2 = \underline{\ \ ?\ \ } \text{ m}^2$ **3.** $2 \text{ m}^2 = \underline{\ \ ?\ \ } \text{ cm}^2$

4. $4 \text{ cm}^2 = \underline{\ \ ?\ \ } \text{ mm}^2$ **5.** $3 \text{ km}^2 = \underline{\ \ ?\ \ } \text{ m}^2$ **6.** $1 \text{ m}^2 = \underline{\ \ ?\ \ } \text{ mm}^2$

B. Use decimals to change from a smaller unit to a larger unit.

Examples

$1 \text{ cm}^2 = \underline{\ \ ?\ \ } \text{ m}^2$
$1 \text{ cm}^2 = 1 \text{ cm} \times 1 \text{ cm}$
$\qquad = 0.01 \text{ m} \times 0.01 \text{ m}$
$\qquad = 0.0001 \text{ m}^2$

$5 \text{ mm}^2 = \underline{\ \ ?\ \ } \text{ cm}^2$
$5 \text{ mm}^2 = 5 \text{ mm} \times 1 \text{ mm}$
$\qquad = 0.5 \text{ cm} \times 0.1 \text{ cm}$
$\qquad = 0.05 \text{ cm}^2$

Complete.

7. $1 \text{ mm}^2 = \underline{\ \ ?\ \ } \text{ cm}^2$ **8.** $1 \text{ m}^2 = \underline{\ \ ?\ \ } \text{ km}^2$ **9.** $1 \text{ mm}^2 = \underline{\ \ ?\ \ } \text{ m}^2$

10. $25 \text{ cm}^2 = \underline{\ \ ?\ \ } \text{ m}^2$ **11.** $500 \text{ m}^2 = \underline{\ \ ?\ \ } \text{ km}^2$ **12.** $400 \text{ mm}^2 = \underline{\ \ ?\ \ } \text{ cm}^2$

C. The units of area commonly used for measuring land are the are (a) and hectare (ha).

$1 \text{ are (a)} = 100 \text{ m}^2$ $1 \text{ hectare (ha)} = 10{,}000 \text{ m}^2$

Complete.

13. $2 \text{ ha} = \underline{\ \ ?\ \ } \text{ m}^2$ **14.** $40{,}000 \text{ m}^2 = \underline{\ \ ?\ \ } \text{ ha}$ **15.** $4 \text{ a} = \underline{\ \ ?\ \ } \text{ m}^2$

344

D. Which unit of area would you use to measure the following?
Choose km², ha, m², cm², or mm².

16. Alaska   **17.** Fabric for a suit   **18.** The school property

19. An arm bandage   **20.** A shirt button   **21.** A rug

Practice

Complete.

1. 2 m² = __?__ cm²   **2.** 6 m² = __?__ mm²

3. 900 m² = __?__ km²   **4.** 4 km² = __?__ m²

5. 6 cm² = __?__ mm²   **6.** 400 cm² = __?__ m²

7. 30,000 cm² = __?__ m²   **8.** __?__ m² = 7,000,000 mm²

9. 2 cm² = __?__ m²   **10.** 4 m² = __?__ km²

11. 3 mm² = __?__ cm²   **12.** 53 mm² = __?__ m²

13. 3 ha = __?__ m²   **14.** 70,000 m² = __?__ ha

15. 5 a = __?__ m²   **16.** 700 m² = __?__ a

Which unit of area would you use? Choose km², ha, m², cm², or mm².

17. Draperies   **18.** New York City   **19.** The playground

Solve Problems

● **20.** A rectangular piece of land is 300 m long and 200 m wide. Express its area in hectares.

● **21.** A triangular piece of land has a base of 1,000 m and a height of 500 m. Express its area in hectares.

● **22.** The side of a square measures 50 cm. Express its area in m².

★ **23.** A rug 100 cm by 200 cm is placed on a floor 3 m by 4 m. Find the floor area not covered by the rug.

AREA OF A CIRCLE

A goat is tied to a stake with a rope 7 m long. Over how many
square meters can the goat graze?

Area of circle $= \pi \cdot$ (radius)^2

$$\pi \doteq 3.14$$
$$A = \pi r^2$$
$$= 3.14 \cdot (7)^2$$
$$= 3.14 \cdot 49$$
$$= 153.86 \text{ m}^2$$

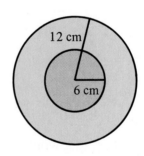

The goat can graze over about 153.9 m^2 rounded to the nearest tenth.

Study and Learn

A. Find the areas. Use $\pi \doteq 3.14$. Round the answer to the nearest tenth.

     1. $r = 10$ cm   **2.** $r = 4$ m   **3.** $d = 10$ cm   **4.** $d = 18$ cm

B. Here's how to find the area of the
shaded portion (ring). Use
$\pi \doteq 3.14$.

     5. What is the area of the large circle?

     6. What is the area of the small circle?

     7. Subtract the area of the small circle
     from the area of the large circle.

     8. What is the area of the ring?

Practice

Find the areas. Use $\pi \doteq 3.14$. Round the answer to the nearest tenth.

     1. $r = 8$ mm   **2.** $r = 14$ m   **3.** $d = 12$ cm   **4.** $d = 24$ cm

Find the area of the shaded region.

5.

12 cm

6.

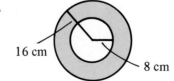

16 cm   8 cm

7.

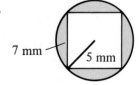

7 mm   5 mm

Midchapter Review

Find x. *(334, 336)*

1. $\triangle ABC \cong \triangle DEF$

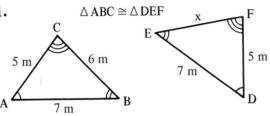

2. $\triangle XYU \cong \triangle STU$

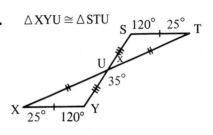

3. $\triangle LMN \sim \triangle PQR$

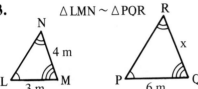

4. $\triangle GHI \sim \triangle JKI$

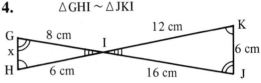

Find the actual lengths. Measure to the nearest centimeter. *(338)*

5.

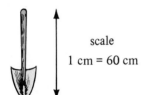

scale
1 cm = 60 cm

6.

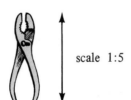

scale 1:5

7.

scale 1:300

Complete. *(344)*

8. $3 \text{ m}^2 = \underline{} \text{ cm}^2$

9. $4 \text{ m}^2 = \underline{} \text{ mm}^2$

10. $8 \text{ cm}^2 = \underline{} \text{ m}^2$

11. $8 \text{ ha} = \underline{} \text{ m}^2$

Find the areas. Use $\pi \doteq 3.14$. *(342, 346)*

12.

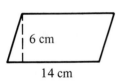

6 cm

14 cm

13.

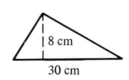

8 cm

30 cm

14.

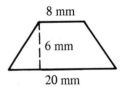

8 mm

6 mm

20 mm

15.

12 mm

16.

160 cm

17.

20 mm

Problem-Solving Skills

Using Rounded Numbers

Rounded numbers can be used in order to find a method to solve problems and check if your answer is reasonable.

READ   Mr. Valenti owns a sporting goods store. He paid $328.44 for a dozen jogging suits. What is the cost of 1 jogging suit?

PLAN   1 dozen = 12
12 rounded to the nearest ten is 10.
$328.44 rounded to the nearest hundred dollars is $300.

Think of a method to solve the problem.
Divide $300 by 10 to find the approximate cost of 1 jogging suit. $300 ÷ 10 = $30

SOLVE   Use the original numbers to find the exact cost.
$328.44 ÷ 12 = $27.37

CHECK $27.37 is close to $30, so the answer is reasonable.

Study and Learn

A can of 3 tennis balls sells for $1.99. Sue bought 24 cans for her tennis club. What was the total cost of the tennis balls?

PLAN   Complete.

1. $1.99 rounded to the nearest dollar is __?__ .

2. 24 rounded to the nearest ten is __?__ .

3. To find the approximate total cost, use the operation of __?__ .

4. $2.00 × 20 = __?__

SOLVE   Complete.

5. Use the original numbers to find the exact cost.
$1.99 × 24 = __?__

CHECK

6. Is $47.76 close to $40.00?

348

Practice

Solve. First use rounded numbers.

1. A barbell set of weights is on sale for $68.99. After the sale the set will sell for $84.25. How much more will the set cost after the sale?

2. A box of a dozen golf balls costs a store owner $4.92. What is the cost to the store owner of 75 boxes of these golf balls?

3. The school basketball team bought a new basketball on sale for $13.99. They saved $2.98 off the regular price. What was the regular price of the basketball?

4. Mr. Hawthorne bought a tennis racket for $24.90, a pair of tennis shorts for $19.95, sneakers for $24.25, and a warm-up suit for $26.95. How much did he spend in all?

5. A soccer ball costs $17.49 during a sale. A soccer club bought 18 balls at the beginning of the soccer season. How much did they pay for the soccer balls?

6. A store owner pays $46.19 for a basketball backboard. She sells it for $61.25 during a sale. How much does the store owner make if a backboard is sold during the sale?

7. A rod and reel combination is on sale for $29.99. When the sale ends, it will cost $41.50. How much is saved by buying during the sale?

8. The jogging club has 17 members. They want to buy jogging shoes, which sell for $24.99 a pair. If they all buy their shoes at once, the price will be $21.25 a pair. How much will they save by buying the shoes together?

USING CUSTOMARY UNITS OF AREA

Dion wants to carpet a hall that measures 4 ft by 9 ft. Carpeting costs $5.95 per square yard. What will carpeting for the hall cost?

$$1 \text{ yd} = 3 \text{ ft}$$
$$1 \text{ yd}^2 = 9 \text{ ft}^2$$

$$\text{Area of the hall} = 4 \text{ ft} \times 9 \text{ ft}$$
$$= 36 \text{ ft}^2$$
$$= 4 \text{ yd}^2 \longleftarrow (36 \text{ ft} \div 9 \text{ ft})$$

Carpeting will cost $4 \times \$5.95$, or $23.80.

Study and Learn

A. Complete. Use these relationships.

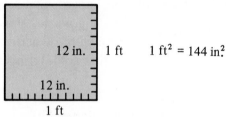

$1 \text{ ft}^2 = 144 \text{ in.}^2$

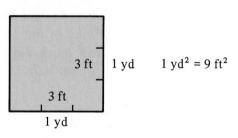

$1 \text{ yd}^2 = 9 \text{ ft}^2$

1. $3 \text{ ft}^2 = \underline{\;\;?\;\;} \text{ in.}^2$

2. $4 \text{ yd}^2 = \underline{\;\;?\;\;} \text{ ft}^2$

3. $72 \text{ ft}^2 = \underline{\;\;?\;\;} \text{ yd}^2$

4. $288 \text{ in.}^2 = \underline{\;\;?\;\;} \text{ ft}^2$

B. Give the areas in square yards.

5.

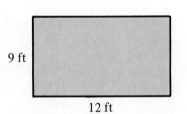

9 ft

12 ft

6.

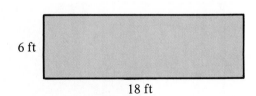

6 ft

18 ft

C. Complete. Use these relationships.

$1 \text{ acre} = 43,560 \text{ ft}^2$ $640 \text{ acres} = 1 \text{ mi}^2$

7. $2 \text{ acres} = \underline{\;\;?\;\;} \text{ ft}^2$

8. $2 \text{ mi}^2 = \underline{\;\;?\;\;} \text{ acres}$

350

Practice

Complete.

1. $5 \text{ ft}^2 = \underline{\quad ? \quad} \text{ in.}^2$ **2.** $7 \text{ ft}^2 = \underline{\quad ? \quad} \text{ in.}^2$

3. $10 \text{ ft}^2 = \underline{\quad ? \quad} \text{ in.}^2$ **4.** $20 \text{ ft}^2 = \underline{\quad ? \quad} \text{ in.}^2$

5. $6 \text{ yd}^2 = \underline{\quad ? \quad} \text{ ft}^2$ **6.** $10 \text{ yd}^2 = \underline{\quad ? \quad} \text{ ft}^2$

7. $12 \text{ yd}^2 = \underline{\quad ? \quad} \text{ ft}^2$ **8.** $20 \text{ yd}^2 = \underline{\quad ? \quad} \text{ ft}^2$

9. $432 \text{ in.}^2 = \underline{\quad ? \quad} \text{ ft}^2$ **10.** $1{,}440 \text{ in.}^2 = \underline{\quad ? \quad} \text{ ft}^2$

11. $18 \text{ ft}^2 = \underline{\quad ? \quad} \text{ yd}^2$ **12.** $45 \text{ ft}^2 = \underline{\quad ? \quad} \text{ yd}^2$

13. $81 \text{ ft}^2 = \underline{\quad ? \quad} \text{ yd}^2$ **14.** $270 \text{ ft}^2 = \underline{\quad ? \quad} \text{ yd}^2$

15. $3 \text{ acres} = \underline{\quad ? \quad} \text{ ft}^2$ **16.** $10 \text{ acres} = \underline{\quad ? \quad} \text{ ft}^2$

17. $6 \text{ mi}^2 = \underline{\quad ? \quad} \text{ acres}$ **18.** $10 \text{ mi}^2 = \underline{\quad ? \quad} \text{ acres}$

Give the areas in square yards.

19.

12 ft

15 ft

20.

6 ft

36 ft

21.

5 ft

18 ft

22.

12 ft

27 ft

23.

9 ft 6 in.

18 ft

24.

6 ft 3 in.

36 ft

Solve Problems

25. Carpeting sells for $17.65/yd². Find the cost of carpeting for a room 12 ft by 18 ft.

SURFACE AREA

The surface area of a solid is the sum of the areas of its faces.

Area of bottom = 10 · 5 or 50 cm²
Area of top   = 10 · 5 or 50 cm²
Area of side   = 5 · 6 or 30 cm²
Area of side   = 5 · 6 or 30 cm²
Area of front = 10 · 6 or 60 cm²
Area of back   = 10 · 6 or 60 cm²
Surface Area   280 cm²

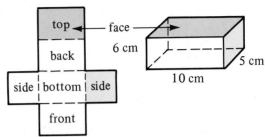

Study and Learn

A. Find the surface areas.

1.

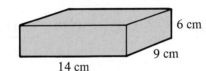

6 cm
9 cm
14 cm

2.

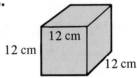

12 cm
12 cm
12 cm

B. Here's a way to find the surface area of the cylinder.

3. What kind of figure is each base?

4. Find the area of one base. Use $\pi \doteq 3.14$.

5. What is the area of both bases?

6. When the lateral surface is unfolded, what figure is formed?

7. What is the width of this rectangle?

8. Find the length of this rectangle. (It's the same as the circumference of the circle. Use $\pi \doteq 3.14$.)

9. Find the area of the lateral surface.

10. What is the surface area of the cylinder?

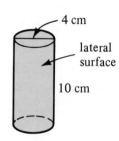

4 cm
lateral surface
10 cm

C. Here's a way to find the surface area of the pyramid.

11. The base is a square. Find its area.

12. What is the area of one face? Of the 4 faces?

13. What is the surface area of the pyramid?

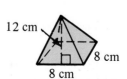

12 cm
8 cm
8 cm

D. Here is how to find the surface area of a cone.

   14. Find the area of the circular base?

   15. Find the area of the lateral surface.
Use $\frac{1}{2} \cdot c \cdot l$, where c is the circumference
of the circular base and l is the slant height.

   16. What is the surface area of the cone?

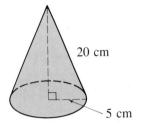

E. Use $S = 4\pi r^2$ to find the surface area
of a sphere.

   17. Find the surface area of the
sphere whose radius is 4 cm long.

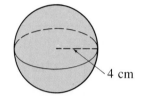

Practice

Find the surface areas.

1.

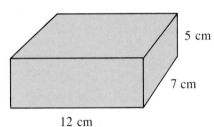

2.

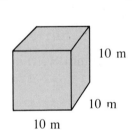

3.

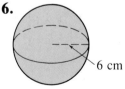

4.

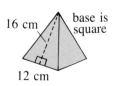

5.

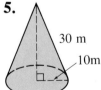

6.

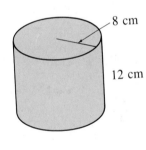

7. A liter of paint will cover about 12 m². How much paint is needed
to paint the walls and ceiling of a room that is 7 m long, 6 m wide,
and 3 m high?

VOLUME IN THE METRIC SYSTEM

A cubic centimeter (cm³) is a unit of volume.

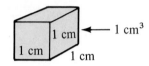

The volume of this solid is 8 cm³.

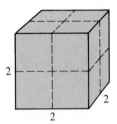

Study and Learn

A. You can change between metric units of volume.

Examples $2 \text{ cm}^3 = \underline{\quad ? \quad} \text{ mm}^3$

$2 \text{ cm}^3 = 2 \text{ cm} \times 1 \text{ cm} \times 1 \text{ cm}$

$\qquad\qquad \downarrow \qquad\quad \downarrow \qquad\quad \downarrow$

$\qquad\quad = 20 \text{ mm} \times 10 \text{ mm} \times 10 \text{ mm}$

$\qquad\quad = 2{,}000 \text{ mm}^3$

Complete.

   1. $4 \text{ m}^3 = \underline{\quad\quad} \text{ cm}^3$ **2.** $7 \text{ cm}^3 = \underline{\quad ? \quad} \text{ mm}^3$ **3.** $1 \text{ m}^3 = \underline{\quad ? \quad} \text{ mm}^3$

B. Use decimals to change from a smaller unit of volume to a larger unit of volume.

Examples $3 \text{ mm}^3 = \underline{\quad ? \quad} \text{ cm}^3$

$3 \text{ mm}^3 = 3 \text{ mm} \times 1 \text{ mm} \times 1 \text{ mm}$

$\qquad\qquad \downarrow \qquad\quad \downarrow \qquad\quad \downarrow$

$\qquad\quad = 0.3 \text{ cm} \times 0.1 \text{ cm} \times 0.1 \text{ cm}$

$\qquad\quad = 0.003 \text{ cm}^3$

Complete.

   4. $8 \text{ cm}^3 = \underline{\quad ? \quad} \text{ m}^3$ **5.** $500 \text{ cm}^3 = \underline{\quad ? \quad} \text{ m}^3$ **6.** $9 \text{ mm}^3 = \underline{\quad ? \quad} \text{ cm}^3$

   7. $5{,}000 \text{ mm}^3 = \underline{\quad ? \quad} \text{ cm}^3$ **8.** $7{,}000{,}000 \text{ mm}^3 = \underline{\quad ? \quad} \text{ m}^3$

Practice

Complete.

1. $2 \text{ m}^3 = \underline{\ ?\ } \text{ cm}^3$

2. $80 \text{ m}^3 = \underline{\ ?\ } \text{ cm}^3$

3. $8 \text{ cm}^3 = \underline{\ ?\ } \text{ mm}^3$

4. $60 \text{ cm}^3 = \underline{\ ?\ } \text{ mm}^3$

5. $5{,}000 \text{ cm}^3 = \underline{\ ?\ } \text{ mm}^3$

6. $3 \text{ m}^3 = \underline{\ ?\ } \text{ mm}^3$

7. $7 \text{ m}^3 = \underline{\ ?\ } \text{ mm}^3$

8. $2 \text{ cm}^3 = \underline{\ ?\ } \text{ m}^3$

9. $8{,}000 \text{ cm}^3 = \underline{\ ?\ } \text{ m}^3$

10. $6 \text{ mm}^3 = \underline{\ ?\ } \text{ cm}^3$

11. $4{,}000 \text{ mm}^3 = \underline{\ ?\ } \text{ cm}^3$

12. $5{,}000{,}000 \text{ mm}^3 = \underline{\ ?\ } \text{ m}^3$

13. $9{,}000 \text{ mm}^3 = \underline{\ ?\ } \text{ m}^3$

14. $20 \text{ cm}^3 = \underline{\ ?\ } \text{ mm}^3$

15. $30{,}000 \text{ mm}^3 = \underline{\ ?\ } \text{ m}^3$

16. $5 \text{ m}^3 = \underline{\ ?\ } \text{ cm}^3$

★ 17. $25 \text{ m}^3 = \underline{\ ?\ } \text{ mm}^3$

★ 18. $1 \text{ km}^3 = \underline{\ ?\ } \text{ m}^3$

Solve.

19. A tank has a volume of 4,000 cm³. Express its volume in cubic millimeters.

20. In excavating a cellar, 70 m³ of dirt were removed. How many cubic centimeters of dirt is this?

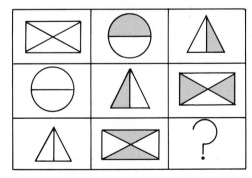

VOLUMES OF PRISMS AND CYLINDERS

The volume of a solid is the number of cubic units it contains.

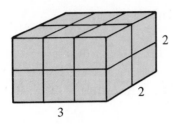

Volume is
12 cubic units

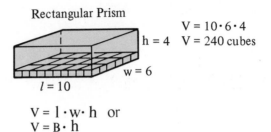

Rectangular Prism

$V = 10 \cdot 6 \cdot 4$
$V = 240$ cubes
$h = 4$
$w = 6$
$l = 10$

$V = l \cdot w \cdot h$ or
$V = B \cdot h$

Study and Learn

A. Find the volumes. Use $V = l \cdot w \cdot h$ or $V = B \cdot h$.

1.

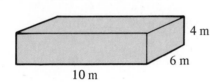

4 m
6 m
10 m

2.

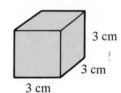

3 cm
3 cm
3 cm

B. Find the volume. Use $V = B \cdot h$.

3. What kind of figure is the base?

4. What is B (area of the base)?

5. What is the volume?

18 cm
6 cm
10 cm

C. Find the volume. Use $V = B \cdot h$.

6. What kind of figure is the base?

7. Find B. Use $\pi \doteq 3.14$.

8. Find the volume.

$r = 10$ cm
30 cm

D. Find the volumes.

9.

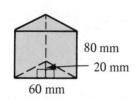

80 mm
20 mm
60 mm

10.

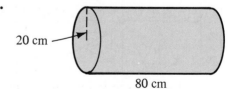

20 cm
80 cm

★ **E.** The volume of a sphere is found by the formula $V = \frac{4}{3} \cdot \pi \cdot r^3$.

11. Use the formula to find the volume of the sphere.

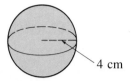
4 cm

Practice

Find the volumes.

1.

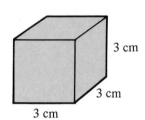
3 cm
3 cm
3 cm

2.

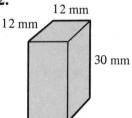
12 mm
12 mm
30 mm

3.

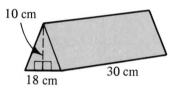
10 cm
18 cm
30 cm

4.

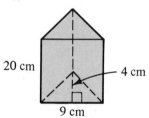
20 cm
4 cm
9 cm

5.

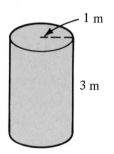
1 m
3 m

★**6.**

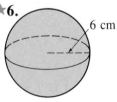
6 cm

Solve Problems

★ **7.** 1 cm³ of lead is 11.3 g. What is the mass of a lead bar 20 cm long, 5 cm wide, and 5 cm high?

★ **8.** 1 cm³ of iron is 7.8 g. What is the mass of an iron bar 300 cm long with a diameter of 20 cm?

RELATING METRIC MEASURES

In the metric system, a liter is defined in terms of cubic centimeters.

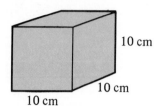

10 cm

10 cm

10 cm

This cube holds 1 liter,
so 1 L = 1,000 cm³

Study and Learn

A. Change to liters. Complete.

1. 5,000 cm³ = 5 × 1,000 cm³
        = 5 × 1 L
        = __?__ L

2. 2,500 cm³ = 2.5 × 1,000 cm³
        = 2.5 × 1 L
        = __?__ L

3. 10,000 cm³ = __?__ L

4. 650 cm³ = __?__ L

B. Change to cubic centimeters. Complete.

5. 8 L = 8 × 1 L
      = 8 × 1,000 cm³
      = __?__ cm³

6. 7.3 L = 7.3 × 1 L
        = 7.3 × 1,000 cm³
        = __?__ cm³

7. 3 L = __?__ cm³

8. 2.5 L = __?__ cm³

C. 1 L = 1,000 cm³, so 1 mL = 1 cm³. Complete.

9. 3 mL = __?__ cm³

10. 400 mL = __?__ cm³

11. 240 cm³ = __?__ mL

12. 7.4 cm³ = __?__ mL

▶ 1,000 cm³ of water is 1 kg.
1 cm³ of water is 1 g.

D. How many grams of water?

13. 7 cm³   **14.** 34.1 cm³   **15.** 2,300 cm³   **16.** 3,400 cm³

E. How many cubic centimeters of water?

17. 3 kg   **18.** 8 g   **19.** 341.7 g   **20.** 4,600 g

Practice

Complete.

1. 2,000 cm³ = __?__ L   **2.** 7,000 cm³ = __?__ L

3. 3,500 cm³ = __?__ L   **4.** 4,800 cm³ = __?__ L

5. 6,140 cm³ = __?__ L   **6.** 8,345 cm³ = __?__ L

7. 350 cm³ = __?__ L   **8.** 413 cm³ = __?__ L

9. 7 L = __?__ cm³   **10.** 4 L = __?__ cm³

11. 12 L = __?__ cm³   **12.** 40 L = __?__ cm³

13. 5.7 L = __?__ cm³   **14.** 8.1 L = __?__ cm³

15. 4 mL = __?__ cm³   **16.** 30 mL = __?__ cm³

17. 600 mL = __?__ cm³   **18.** 7.4 mL = __?__ cm³

19. 350 cm³ = __?__ mL   **20.** 17.6 cm³ = __?__ mL

21. 3,700 cm³ = __?__ L   **22.** 18 L = __?__ cm³

23. 850 cm³ = __?__ mL   **24.** 8.2 mL = __?__ cm³

How many grams of water?

25. 8 cm³   **26.** 50 cm³   **27.** 300 cm³

28. 4,100 cm³   **29.** 42.3 cm³   **30.** 117.9 cm³

How many cubic centimeters of water?

31. 12 kg   **32.** 6 g   **33.** 53 g

34. 2,400 g   **35.** 46.7 g   **36.** 131.7 g

Solve Problems

37. What is the mass in kilograms of the water in the fish tank?

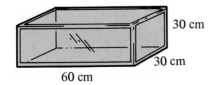

38. How many liters of water can this tank hold?

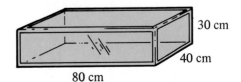

VOLUMES OF CONES AND PYRAMIDS

Study and Learn

A. The formula for the volume of a rectangular prism can be used to find a formula for the volume of a pyramid.

1. Make a model of a rectangular prism out of cardboard.

2. Make a model of a pyramid out of cardboard. Use the same base and height for both figures.

3. Fill the pyramid with sand or salt.

4. Pour the contents of the pyramid into the rectangular prism.

5. Appproximately how full is the rectangular prism?

6. The volume of the pyramid is what part of the volume of the rectangular prism?

▶ Volume of a pyramid: $V = \frac{1}{3}B \cdot h$

B. Find the volumes.

7.

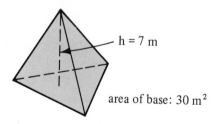

h = 7 m

area of base: 30 m²

8.

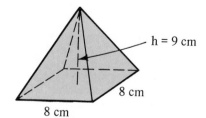

h = 9 cm

8 cm

8 cm

C. Examine these diagrams.

9. The volume of the pyramid is what part of the volume of the rectangular prism?

10. The volume of the cone appears to be what part of the volume of the cylinder?

▶ Volume of a cone: $V = \frac{1}{3}B \cdot h$

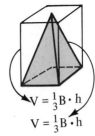

$V = \frac{1}{3}B \cdot h$

$V = \frac{1}{3}B \cdot h$

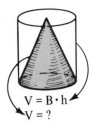

$V = B \cdot h$

$V = ?$

D. Find the volumes. Use $\pi \doteq 3.14$. Give the answers to the nearest whole number.

11.

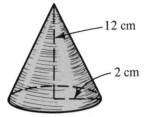

12 cm

2 cm

12.

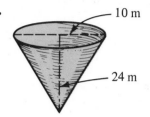

10 m

24 m

Practice

Find the volumes.

1.

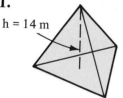

h = 14 m

area of base: 72 m²

2.

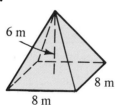

6 m

8 m

8 m

3.

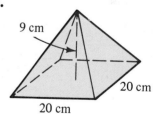

9 cm

20 cm

20 cm

Find the volumes. Use $\pi \doteq 3.14$. Give the answers to the nearest whole number.

4.

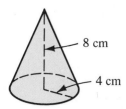

8 cm

4 cm

5.

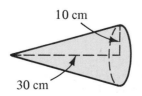

10 cm

30 cm

6.

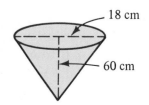

18 cm

60 cm

361

USING CUSTOMARY UNITS OF VOLUME

Find the number of cubic yards of coal removed in digging out a hole
9 ft by 15 ft by 21 ft in a mine.

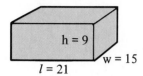

$V = l \cdot w \cdot h$
$\quad = 21 \cdot 15 \cdot 9$
$\quad = 2{,}835 \text{ ft}^3$
$\quad = 105 \text{ yd}^3 \longleftarrow$ 1 yd^3 = 27 ft^3 and 2,835 ÷ 27 = 105

105 yd^3 of coal are removed.

Study and Learn

A. Complete. Use these relationships.

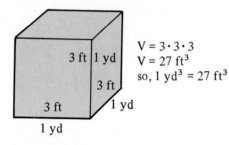

$V = 3 \cdot 3 \cdot 3$
$V = 27 \text{ ft}^3$
so, 1 yd^3 = 27 ft^3

$V = 12 \cdot 12 \cdot 12$
$V = 1{,}728 \text{ in.}^3$
so, 1 ft^3 = 1,728 in.^{3}

1. 2 yd^3 = ___?___ ft^3

2. 10 yd^3 = ___?___ ft^3

3. 3 ft^3 = ___?___ in.^{3}

4. 10 ft^3 = ___?___ in.^{3}

5. 54 ft^3 = ___?___ yd^3

6. 8,640 in.^{3} = ___?___ ft^3

Practice

Complete.

1. 4 yd^3 = ___?___ ft^3

2. 8 yd^3 = ___?___ ft^3

3. 6 ft^3 = ___?___ in.^{3}

4. 20 ft^3 = ___?___ in.^{3}

5. 270 ft^3 = ___?___ yd^3

6. 5,184 in.^{3} = ___?___ ft^3

Solve.

7. A bin 8 ft wide and 7 ft 6 in. long is filled to a depth of 5 ft
   with coal. One ton of coal occupies about 50 ft^3. How many tons
   of coal are in the bin?

362

Skills Review

Multiply. *(110, 112)*

1. 2.16
　　× 7

2. 0.34
　　× 0.9

3. 1.46
　　× 2.4

4. 0.37
　　× 0.09

5. 2.57
　　× 0.60

6. 34.1
　　× 2.40

Divide. *(116, 118)*

7. $6\overline{)0.6}$

8. $3\overline{)15.3}$

9. $4\overline{)0.0016}$

10. $33\overline{)4.95}$

11. $0.6\overline{)2.4}$

12. $0.03\overline{)9}$

13. $2.4\overline{)0.744}$

14. $0.87\overline{)78.3}$

Multiply. Simplify when possible. *(178, 180)*

15. $\frac{2}{5} \times \frac{5}{6}$

16. $\frac{4}{9} \times \frac{3}{16}$

17. $\frac{3}{4} \times 20$

18. $7\frac{1}{2} \times 8$

19. $1\frac{1}{2} \times 4$

20. $\frac{2}{3} \times 1\frac{1}{2}$

21. $1\frac{1}{2} \times 1\frac{1}{2}$

22. $3\frac{3}{4} \times \frac{8}{9} \times 6$

Divide. Simplify when possible. *(182)*

23. $\frac{3}{4} \div \frac{1}{2}$

24. $4 \div 1\frac{1}{2}$

25. $\frac{2}{3} \div 4$

26. $4\frac{2}{3} \div 1\frac{1}{6}$

Write percents. *(250, 252)*

27. $\frac{7}{100}$

28. 0.38

29. $\frac{3}{5}$

30. 0.069

Write simplest fractions or whole numbers. *(250)*

31. 6%

32. 25%

33. 200%

34. $12\frac{1}{2}$%

Write decimals. *(252)*

35. 7%

36. 150%

37. 6.7%

38. $\frac{1}{2}$%

Compute. *(254, 256, 258)*

39. 12% of 60

40. $\frac{1}{2}$% of 400

41. 0.7% of 6,000

42. 20% of what number is 75?

43. What percent of 12 is 9?

44. 0.3 is what percent of 25?

Solve. *(264)*

45. At a sale, a dress marked $90 was sold at a discount of 30%. What was the sale price?

Problem-Solving Applications
Career

1. Mr. Engel, an architect, made a scale drawing of a house that he designed. The living room is to be 7 m long. The scale on the drawing is 0.5 cm = 1 m. What is the scale length of the living room? [HINT: Write a proportion and solve.]

2. Mr. Bendler used a scale of 1 cm for 10 m on a scale drawing of a playground. The distance of the playground to the school building is 14.5 cm. What is the actual distance of the playground to the school building?

3. Of all the architects that work in a certain city, 30% are women. There are 150 architects in the city. How many of the architects are men?

4. Ms. Koch designs schools. A classroom is to have 35 students. Each student needs 3 m² of floor space. What should the area of the classroom be?

5. An architect designed a house to cost $80,000 to build. The actual cost of building the house was $90,000. What was the percent increase in the cost of building the house?

6. Mrs. Gonzalez made a scale drawing of a house. She used the scale 0.5 cm = 1 m. The scale length of the bedroom is 2.5 cm and the width is 2 cm. What is the actual area of the bedroom?

Chapter Review

Find x. *(334, 336)*

1. $\triangle GHI \cong \triangle JKL$

6 cm, x, 8 cm, 10 cm, 6 cm, 8 cm

2. $\triangle ABC \cong \triangle DEC$

75° 30° 30° 75° x 75°

3. $\triangle XYZ \sim \triangle UVW$

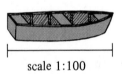

5 cm, 8 cm, 4 cm, 2 cm, x

4. $\triangle PQR \sim \triangle STU$

5 m, 3 m, x, 6 m

Find the actual lengths. Measure the scale lengths to the nearest centimeter. *(338)*

5.

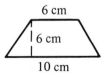

scale 1:27

6.

scale 1:9

7.

scale 1:100

Complete. *(344, 350)*

8. $4 \text{ m}^2 = \underline{\ ?\ } \text{ cm}^2$

9. $50 \text{ mm}^2 = \underline{\ ?\ } \text{ cm}^2$

10. $2 \text{ ft}^2 = \underline{\ ?\ } \text{ in.}^2$

11. $8 \text{ yd}^2 = \underline{\ ?\ } \text{ ft}^2$

Find the areas. Use $\pi \doteq 3.14$. *(342, 346)*

12.

12 m, 25 m

13.

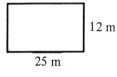

9 mm, 15 mm

14.

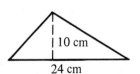

10 cm, 24 cm

15.

6 cm, 6 cm, 10 cm

16.

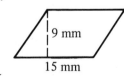

12 cm

17.

20 mm

Chapter Review continues

Find the surface areas. *(352)*

18.

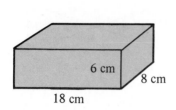

6 cm
8 cm
18 cm

19.

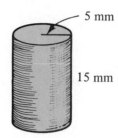

5 mm
15 mm

20.

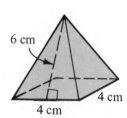

6 cm
4 cm
4 cm

Complete. *(354, 362)*

21. $2 \text{ m}^3 = \underline{\quad?\quad} \text{ cm}^3$

22. $4 \text{ cm}^3 = \underline{\quad?\quad} \text{ m}^3$

23. $3 \text{ yd}^3 = \underline{\quad?\quad} \text{ ft}^3$

24. $4 \text{ ft}^3 = \underline{\quad?\quad} \text{ in.}^3$

Find the volumes. *(356)*

25.

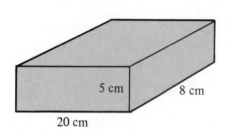

5 cm
8 cm
20 cm

26.

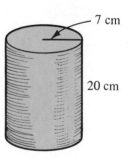

7 cm
20 cm

Complete. *(358)*

27. $4{,}000 \text{ cm}^3 = \underline{\quad?\quad} \text{ L}$

28. $6.2 \text{ L} = \underline{\quad?\quad} \text{ cm}^3$

29. $7.5 \text{ mL} = \underline{\quad?\quad} \text{ cm}^3$

30. How many grams of water in 10.5 cm^3? *(358)*

31. How many cubic centimeters of water in 3 kg? *(358)*

Solve. *(348, 364)*

32. A dozen fielder's gloves cost a store owner $145.44. What is the cost of 1 glove?

33. Jody used a scale of 0.4 cm = 1 m. What is the actual length of a room which is 1.6 cm on the scale drawing?

Chapter Test

Find *x*. *(334, 336)*

1. △ABC ≅ △DEF

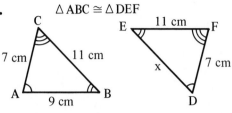

2. △LMP ≅ △NMP

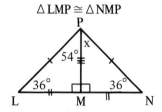

3. △QRS ~ △TUV

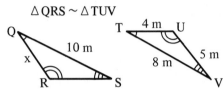

4. △GHI ~ △JHK

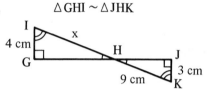

Find the actual lengths. Measure the scale lengths to the nearest centimeter. *(338)*

5.

scale
1 cm = 5 cm

6.

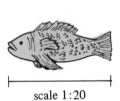

scale 1:20

7.

scale $\frac{1}{33}$

Complete. *(344, 350)*

8. 8 m² = __?__ cm²

9. 30 cm² = __?__ m²

10. 3 ft² = __?__ in.²

11. 5 yd² = __?__ ft²

Find the areas. Use $\pi \doteq 3.14$. *(342, 346)*

12.

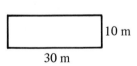

10 m

30 m

13.

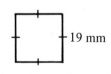

19 mm

14.

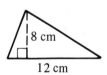

8 cm

12 cm

15.

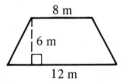

8 m

6 m

12 m

16.

2 m

17.

15 mm

Chapter Test continues

Chapter Test continued

Find the surface areas. *(352)*

18.

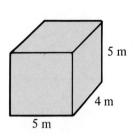

5 m
4 m
5 m

19.

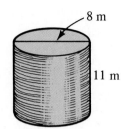

8 m
11 m

20.

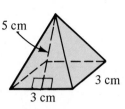

5 cm
3 cm
3 cm

Complete. *(354, 362)*

21. $3 \text{ m}^3 = \underline{\ ?\ } \text{ cm}^3$

22. $7{,}000 \text{ cm}^3 = \underline{\ ?\ } \text{ m}^3$

23. $5 \text{ yd}^3 = \underline{\ ?\ } \text{ ft}^3$

24. $5 \text{ ft}^3 = \underline{\ ?\ } \text{ in.}^3$

Find the volumes. *(356)*

25.

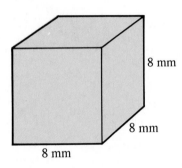

8 mm
8 mm
8 mm

26.

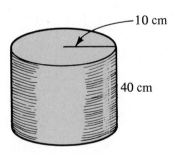

10 cm
40 cm

Complete. *(358)*

27. $4{,}150 \text{ cm}^3 = \underline{\ ?\ } \text{ L}$ **28.** $19 \text{ L} = \underline{\ ?\ } \text{ cm}^3$ **29.** $8.7 \text{ mL} = \underline{\ ?\ } \text{ cm}^3$

30. How many grams of water in 14 cm^3?
(358)

31. How many cubic centimeters of water in 4.5 kg?
(358)

Solve. *(348, 364)*

32. A jump rope sells for $4.99. A gym bought 55 jump ropes. How much did they pay for the ropes?

33. Jim used a scale of 1.5 cm = 1 m on a drawing. What is the scale length of a distance of 9 m?

Skills Check

1. The people working in Ann's department earn $134, $256, $176, and $158 per week. What is their average weekly earnings?

   A $181   B $191

   C $201   D $211

2. A radio marked $64 is sold at a discount of 25% during a sale. How much does it cost during the sale?

   E $12.80   F $16

   G $48   H $51.20

3. Mr. Green drove his car a distance of 484 km at an average speed of 88 km/h. How long did he drive?

   A 4 hours   B 5 hours

   C $5\frac{1}{2}$ hours   D 6 hours

4. What is the volume of the rectangular solid?

   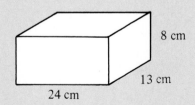

   E 2,496 cm³   F 312 cm³

   G 104 cm³   H 45 cm³

5. What is the area of the right triangle?

   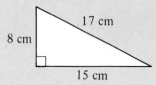

   A 40 cm²   B 60 cm²

   C 120 cm²   D 136 cm²

6. Ms. Levy bought 2 bags of onions. One bag was marked 1.3 kg. The other bag was marked 820 g. How much heavier was the first bag?

   E 480 g   F 818.7 g

   G 480 kg   H 818.7 kg

7. There are 5 black checkers and 6 red checkers in a box. A checker is drawn. What is the probability that it is red?

   A $\frac{1}{11}$ B $\frac{5}{11}$

   C $\frac{6}{11}$ D $\frac{11}{11}$

8. Which class has the best attendance record?

   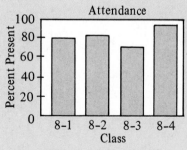

   E 8–1   F 8–2

   G 8–3   H 8–4

14 STATISTICS AND PROBABILITY

This solar furnace is located at Odeillo, France, in the Pyrenees.

ORDERED PAIRS

A point may be located by an ordered pair of numbers. Two perpendicular number lines called axes are used.

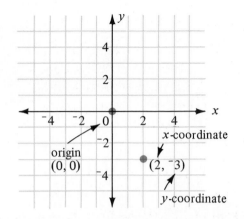

The horizontal number line is called the **x-axis.** The vertical number line is called the **y-axis.** The point where the axes intersect is called the **origin.**

Study and Learn

A. Give the coordinates.

 1. *A*   **2.** *B*   **3.** *C*

 4. *D*   **5.** *E*   **6.** *F*

 7. *G*   **8.** *H*   **9.** *I*

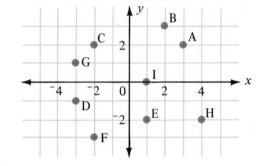

B. To plot the point $(4, {}^-2)$, count 4 units to the right of the origin, then 2 units down.

Plot these points.

 10. $(1, 2)$ **11.** $({}^-2, 3)$ **12.** $({}^-1, {}^-1)$ **13.** $(4, {}^-5)$ **14.** $(0, {}^-4)$

C. The axes divide the plane into 4 parts called **quadrants.**

In which quadrant is each point?

 15. $(1, 2)$ **16.** $(1, {}^-2)$

 17. $({}^-1, 2)$ **18.** $({}^-1, {}^-2)$

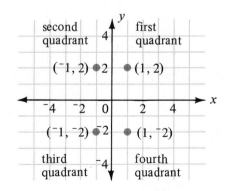

372

D. You can **slide** a figure on a graph by changing the ordered pairs of its vertices.

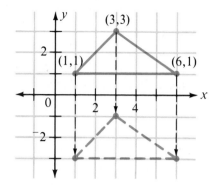

19. Slide the triangle to the fourth quadrant by changing the *y*-coordinates 4 units. Give the new coordinates of the vertices.

Practice

Give the coordinates of these points.

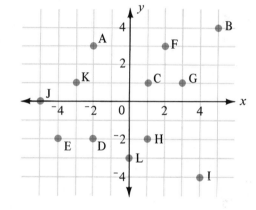

1. *A*   **2.** *B*   **3.** *C*

4. *D*   **5.** *E*   **6.** *F*

7. *G*   **8.** *H*   **9.** *I*

10. *J*   **11.** *K*   **12.** *L*

Plot these points.

13. $(3, 1)$ **14.** $(4, 0)$ **15.** $(3, {}^-1)$ **16.** $({}^-2, 2)$

17. $({}^-2, {}^-2)$ **18.** $(0, 2)$ **19.** $(5, {}^-3)$ **20.** $(2, 2)$

In which quadrant is each point?

21. $({}^-1, 2)$ **22.** $(3, 4)$ **23.** $({}^-4, 2)$ **24.** $(4, {}^-1)$

25. Plot three vertices and draw a triangle in the first quadrant. Slide the triangle into the second and then into the third quadrant. Give the coordinates of the vertices of the triangle in each quadrant.

373

GRAPHING INTEGER PAIRS FOR EQUATIONS

The sum of two numbers is 6. What are the numbers?

Let x = one number
y = other number
Equation: $x + y = 6$

There are many number pairs which are solutions of the equation. Three of them are (0, 6), (1, 5), and (2, 4).

Study and Learn

A. Find the number pair solutions of $y = 2x$. Complete the table.

Select any value for x. $y = 2x$
Let $x = 1$.
Then: $y = 2x$
$y = 2 \cdot 1$
$y = 2$

x	y
1	2
2	
3	

1. Let $x = 2$. Find the corresponding value for y.

2. Let $x = 3$. Find the corresponding value for y.

B. Complete this table for the equation $y = x + 4$.

	x	y
3.	0	
4.	1	
5.	2	
6.	3	

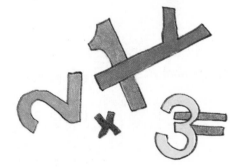

C. Graph integer pairs for the equation $y = 2x + 1$.

7. Make a table of x and y values. Let $x = {}^-2, {}^-1, 0,$ and 1.

8. Write each pair as an ordered pair.   **9.** Graph the 4 pairs on a set of axes.

Example

x	y
⁻2	⁻3

$\longrightarrow ({}^-2, {}^-3)$

Practice

Find integer pair solutions for each equation. Let $x = 0, 1, 2,$ 3, 4, and 5. Place the integer pairs in table form.

1. $y = 3x$ **2.** $y = x + 1$ **3.** $y = x + 2$

4. $y = x - 1$ **5.** $y = x - 2$ **6.** $y = 2x + 2$

7. $y = 2x + 4$ **8.** $y = 3x - 2$ **9.** $y = 3x - 3$

Graph integer pairs for each equation. Let $x = {}^-2, {}^-1, 0, 1,$ and 2.

10. $y = 2x$ **11.** $y = 3x$ **12.** $y = x + 3$

13. $y = x + 5$ **14.** $y = 3x - 1$ ★ **15.** $y = x^2$

Something Extra

Activity

★ Graph. Replacements for x: integers

Example   This is the graph of $x > {}^-2$.

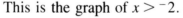

indicates that points continue without end

1. $x < 3$ **2.** $x > {}^-4$ **3.** $2x > 10$ **4.** $x + 1 < {}^-2$

The inequalities $x > 2$ and $x < 7$ can be combined.

         $2 < x < 7$ ⟵ Read: x is greater than 2 and less than 7.

Graph of $2 < x < 7$ Replacements for x: integers

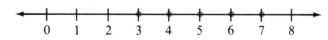

Give a combined inequality for each graph.

5. 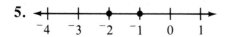   **6.**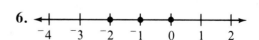

GRAPHING EQUATIONS

Equation: $y = x + 1$

Step 1 Make a table of values.

x	0	1	2	$^-1$	$^-2$	$2\frac{1}{2}$
y	1	2	3	0	$^-1$	$3\frac{1}{2}$

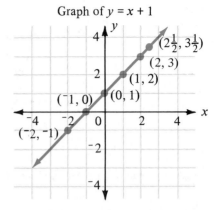

Graph of $y = x + 1$

Step 2 Plot the number pairs.
Step 3 Draw a line through the plotted points.

Study and Learn

A. Look at the graph of $y = x + 1$.

  1. The point with coordinates (4, 5) is on the graph of $y = x + 1$. Is (4, 5) a solution of $y = x + 1$?

  2. The point with coordinate (3, 1) is not on the graph of $y = x + 1$. Is (3, 1) a solution of $y = x + 1$?

B. Graph the equation $y = 2x$.

  3. Make a table of x and y values for $y = 2x$.

  4. Write ordered pairs from the table of values.

  5. Plot the ordered pairs.

  6. Draw a line through the plotted points.

C. Complete the tables. Then graph the equations.

7. $y = 2x - 1$

x	y
2	
1	
0	
$^-1$	
$^-2$	

8. $y = \frac{1}{2}x$

x	y
2	
0	
$^-2$	
$^-4$	
$^-6$	

9. $y = {}^-2x + 1$

x	y
2	
1	
0	
$^-1$	
$^-2$	

376

D. Look at the graph.

    10. Write a table of ordered pairs for the graph.

    11. Write an equation for the graph.

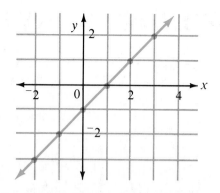

Practice

Complete the tables. Then graph the equations.

1. $y = x + 2$

x	y
3	
2	
1	
0	
$^{-}1$	
$^{-}2$	

2. $y = 3x - 1$

x	y
3	
2	
1	
0	
$^{-}1$	
$^{-}2$	

3. $y = \frac{1}{3}x$

x	y
9	
6	
3	
0	
$^{-}3$	
$^{-}6$	

Graph the equations.

    4. $y = x + 3$ **5.** $y = 5x$ **6.** $y = x - 2$ **7.** $y = 2x - 3$

    8. $y = \frac{1}{2}x + 1$ **9.** $y = {}^{-}x + 7$ **10.** $y = {}^{-}x - 1$ **11.** $y = \frac{1}{3}x + 1$

★ **12.** $x + y = 9$ ★ **13.** $x = y - 4$ ★ **14.** $y = x^2$ ★ **15.** $y = x^2 - 1$

Look at the graph.

16. Write a table of ordered pairs for the graph.

17. Write an equation for the graph.

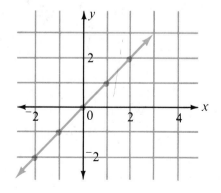

377

GRAPHING INEQUALITIES

Graph $y > x + 1$.

Step 1 Make a table of values for the related equation $y = x + 1$.

Step 2 Graph $y = x + 1$ as a dashed line.

Step 3 Test a point on one side of the line.
$$A(2, 1) \quad y > x + 1$$
$$1 > 2 + 1$$
$$1 > 3 \quad \text{False}$$

Step 4 Test a point on the other side of the line.
$$B(1, 3) \quad y > x + 1$$
$$3 > 1 + 1$$
$$3 > 2 \quad \text{True}$$

Step 5 Shade the side containing the point that tested true.

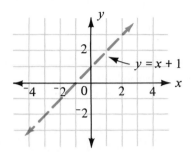

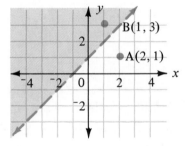

Study and Learn

A. Graph $y > 2x - 1$.

1. Make up a table of values for the related equation $y = 2x - 1$.

2. Use square-ruled paper and plot $y = 2x - 1$ with a dashed line.

3. Test a point on one side of the line. $A(3, 1)$ Is the inequality true?

4. Test a point on the other side of the line. $B(^-2, 3)$ Is the inequality true?

5. Shade the side containing point B. This is the graph of $y > 2x - 1$.

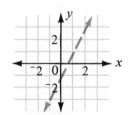

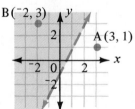

B. Graph these inequalities.

    6. $y > x + 2$ **7.** $y > 3x$ **8.** $y < x$ **9.** $y < x - 1$

Practice

Test a point on each side of the line. Complete the graph.

1. $y > 2x + 1$

2. $y < 2x + 1$

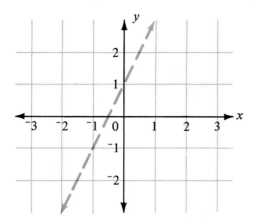 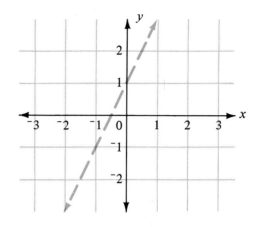

Graph these inequalities.

3. $y > x + 3$ **4.** $y > 2x$ **5.** $y < x + 3$ **6.** $y < 2x$

Skills Review

Compare. Use $>$, $<$, or $=$. *(164)*

1. $\frac{5}{8} \equiv \frac{3}{8}$ **2.** $\frac{3}{4} \equiv \frac{9}{12}$ **3.** $\frac{5}{6} \equiv \frac{7}{8}$

Simplify. *(162)*

4. $\frac{6}{10}$ **5.** $\frac{9}{12}$ **6.** $\frac{12}{16}$

Add. Simplify when possible. *(168, 170)*

7. $\begin{array}{r} \frac{5}{8} \\ + \frac{1}{4} \\ \hline \end{array}$ **8.** $\begin{array}{r} \frac{9}{10} \\ + 1\frac{4}{5} \\ \hline \end{array}$ **9.** $\begin{array}{r} 2\frac{3}{4} \\ + 1\frac{2}{5} \\ \hline \end{array}$ **10.** $\begin{array}{r} 3\frac{5}{6} \\ + 2\frac{2}{3} \\ \hline \end{array}$ **11.** $\begin{array}{r} 4\frac{3}{8} \\ + 5\frac{5}{6} \\ \hline \end{array}$

Subtract. Simplify when possible. *(173, 174)*

12. $\begin{array}{r} \frac{7}{12} \\ - \frac{1}{3} \\ \hline \end{array}$ **13.** $\begin{array}{r} 3\frac{5}{8} \\ - 2 \\ \hline \end{array}$ **14.** $\begin{array}{r} 4\frac{5}{8} \\ - 1\frac{1}{2} \\ \hline \end{array}$ **15.** $\begin{array}{r} 6\frac{1}{4} \\ - 2\frac{3}{8} \\ \hline \end{array}$ **16.** $\begin{array}{r} 7 \\ - 3\frac{5}{6} \\ \hline \end{array}$

Midchapter Review

Plot these points. *(372)*

1. $A(3, 0)$ **2.** $B(1, {}^-2)$ **3.** $C({}^-1, {}^-1)$ **4.** $D({}^-4, 5)$

Graph. *(376)*

5. $y = x + 4$ **6.** $y = 2x$ **7.** $y = 3x - 2$

Computer—Artificial Intelligence

Some people believe that computers can be given artificial intelligence, the ability to think like humans. Others argue that a computer does what it is programmed to do, and nothing more. The scientists doing artificial intelligence research say that a program called **Caduceus** makes a computer imitate the thought processes of a doctor.

A. Caduceus checks a patient's symptoms against those of the thousands of diseases stored in its memory. If the symptoms are not clearly those of one disease, Caduceus asks for more information about the patient and may even call for medical tests. By matching all the input with the information stored in its memory, Caduceus makes the same diagnosis and prescribes the same treatment that human doctors might. Since Caduceus has a large memory and works at computer speed, it makes its decisions much more quickly than human doctors.

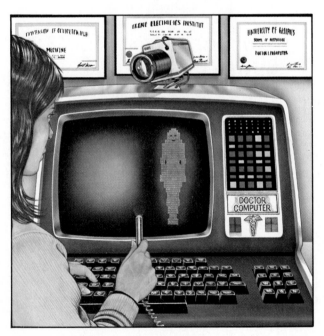

B. Critics of artificial intelligence argue that although it makes its decisions in the same way a human doctor might, the computer is not really able to think or intelligent. They say that since the program was written by humans, the computer is just doing what it is told to do, and not thinking for itself. What do you think?

SYMMETRY AND COORDINATES

A figure is **symmetric** about a line if the corresponding parts of the figure on opposite sides of the line are mirror images.

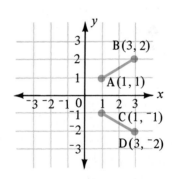

axis of symmetry

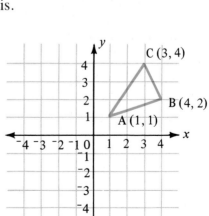

axis of symmetry

Practice

$\overline{CD}$ is a mirror image of $\overline{AB}$ about the x-axis.

1. What are the coordinates of point C?

2. How do the coordinates of A and C compare?

3. What are the coordinates of point D?

4. How do the coordinates of B and D compare?

5. What is true of the distances of points A and C from the x-axis?

6. What is true of the distances of points B and D from the x-axis?

Find the mirror image of $\triangle ABC$ about the y-axis.

7. What are the coordinates of the point that is directly across from point A and the same distance from the y-axis?

8. What is the mirror image of point B about the y-axis?

9. What is the mirror image of point C about the y-axis?

10. Copy the graph and draw in the mirror image of $\triangle ABC$.

381

Problem-Solving Skills

Using Graphs

Under Plan 1 a salesperson can earn $100 per week plus 10% of sales. Under Plan 2 the salesperson can earn 20% of sales. How much must be sold so that both plans give the same earnings?

Answer: If sales are $1,000, the earnings under either plan are the same.

Study and Learn

A. Mr. Leff left his house at 8:00 am and drove 60 km/h on a business trip. His son left at 9:00 am. To overtake his father, the son drove at 80 km/h. At what time will the son overtake his father?

   1. Copy the graph at the right.

   2. Plot Mr. Leff's trip.

Time	Distance
8:00 am	0 km
9:00 am	60 km
10:00 am	120 km
11:00 am	180 km
12:00 noon	240 km

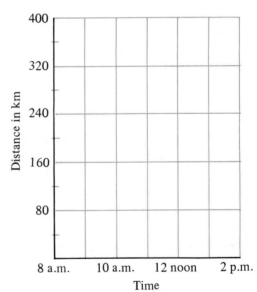

   3. Plot the son's trip.

   4. Where do the lines intersect?

   5. Answer the problem.

B. Draw a graph and solve.

   6. José has 24 records. Each week he adds 2 records to his collection. Susan has no records, but she is beginning to collect them. She is collecting 4 records a week. In how many weeks will she have as many as José?

A ship leaves New York harbor at 6:00 am. It goes at an average speed of 10 knots per hour. Another ship leaves New York at 9:00 am. It follows the first ship at an average speed of 15 knots per hour.

1. Draw a graph.

2. At what time will the second ship overtake the first ship?

3. How far from New York will they meet?

Joan has $60 in the bank. Each week she deposits $5. Bill has no money in the bank, but he is working part time. He deposits $10 a week.

4. Draw a graph.

5. In how many weeks will Joan and Bill have deposited the same amount of money?

6. How much money will each have saved?

Draw a graph and solve.

7. A company has 2 pay plans for salespersons. Under Plan 1, a salesperson earns $100 a week plus 25% of sales. Under Plan 2, a salesperson earns 50% of sales. How much must be sold so that a salesperson would earn the same amount under either plan?

INTERPRETING DATA

The scores made by the school basketball team were 47, 68, 50, 34, 64, 47, and 54. The scores are arranged below from highest to lowest.

highest ⟶ 68
64
54
50 ——— *Median* (50), or middle number
47
47 ⟩ *Mode* (47), or number that occurs most often
lowest ⟶ 34
364

Mean, or average: $7\overline{)364}$ ... 52

Range: 34 to 68

Study and Learn

A. Here are the test scores of 9 students on a math test:
85, 70, 95, 90, 80, 75, 85, 85, 100.

1. Arrange the scores in order from highest to lowest.

2. What is the range of the scores?

3. What is the mode of the scores?

4. What is the median of the scores?

5. What is the mean of the scores?

B. Here are Jeremy's bowling scores: 100, 120, 114, 118, 94, 102.

6. Find the mean of the bowling scores.

7. Jeremy's average for 7 bowling games is 112. What was his total for the 7 games?

C. Find the median of these salaries: $75, $105, $246, $156, $182, $98.

8. How many salaries are there? Arrange them in order.

9. Is there a middle salary?

10. What are the 2 middle salaries?

11. What is their mean? This is the median.

384

Practice

Find the ranges.

1. 32, 52, 72, 92, 13

2. 20, 14, 23, 35, 7, 6

3. 34, 41, 33, 12, 18, 31, 29, 28, 13, 16, 18, 31, 49, 18

Find the modes.

4. 34, 21, 22, 34, 29

5. 85, 90, 85, 90, 75, 65

6. 86, 92, 81, 90, 59, 85, 69, 81, 39, 64, 100, 99, 79

Find the means.

7. 81, 88, 87, 89, 69

8. 135, 146, 119, 124, 138, 148

9. 34, 22, 36, 49, 38, 29, 40, 51, 48, 38, 43, 22

Find the medians.

10. 8, 5, 13, 16, 9

11. 22, 31, 41, 29, 34, 38

12. 81, 96, 87, 75, 68, 78, 90, 90, 100, 75, 87, 61, 58, 60

These are the salaries of 9 people working in a store:
$150, $110, $210, $180, $220, $185, $180, $180, $250.

13. Arrange the salaries in order. What is the range?

14. What is the mode of the salaries?

15. What is the median of the salaries?

16. What is the mean of the salaries?

Solve Problems

17. The ages of the workers in an office are 48, 29, 32, 41, 37, 39, 43, 23, and 26. What is the mean of the ages of these office workers?

★ **18.** In 7 games, a team averaged 6 runs. In 6 games, the team scored 39 runs. How many runs did they score in the seventh game?

GRAPHING DATA

A survey was made of the number of people using the tennis courts on Brown Boulevard. The data may be shown in table form or by a graph.

PEOPLE USING TENNIS COURTS

Week	Number of People
1	300
2	150
3	250
4	200

Step 1 Select axes and scales.

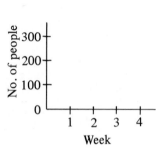

Step 2 Draw bars. Use information in the table.

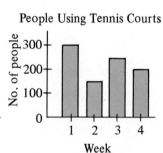

Study and Learn

A. On square-ruled paper construct a bar graph of the data in the table.

1. Draw a vertical and a horizontal axis.

2. From numbers in the table, decide upon a scale for each axis.

3. Construct each bar equally wide and equally spaced.

4. Label the graph.

VOTES FOR CLASS PRESIDENT

Name	Votes
Gene	70
Maria	60
Randy	80

B. Construct bar graphs from the tables.

5. TICKETS SOLD TO SCHOOL DANCE

Grade	Number of Tickets
7th	110
8th	150
9th	180

6. TELEPHONES PER 100 PEOPLE

Country	Number of Telephones
United States	45
Canada	35
Norway	25

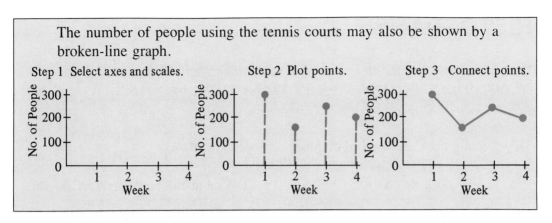

The number of people using the tennis courts may also be shown by a broken-line graph.

Step 1 Select axes and scales. Step 2 Plot points. Step 3 Connect points.

C. Construct broken-line graphs from the tables.

7. BETH'S SCHOOL MARKS

Test	Mark
1	90
2	95
3	80

8. TEMPERATURE RECORD

Time	Temperature
8 am	20°C
10 am	20°C
noon	23°C

Practice

Construct bar graphs.

1. TED'S PAPER ROUTE

Day	Papers sold
1	40
2	35
3	60

★ **2.** POPULATION 100 YEARS AGO

State	Population
California	864,694
Georgia	1,542,180
Illinois	3,077,871

Construct broken-line graphs.

3. HAROLD'S TEST SCORES

Test	Score
1	90
2	75
3	80

4. SIZE OF SENIOR CLASS

Year	Number
1960	350
1970	300
1980	500

HISTOGRAMS

Here are the test scores of a class on a mathematics test: 100, 91, 83, 70, 100, 95, 82, 70, 83, 100, 90, 85, 72, 77, 81, 76, 64, 91, 82, 77, 87, 66, 82, 70, 50, 82, 85, 89, 73, 87.

Tables and graphs are used to organize and visualize data.

FREQUENCY TABLE

Score	Tally	Frequency
100					3
95			1
91				2
90			1
89			1
87				2
85				2
83				2
82						4
81			1
77				2
76			1
73			1
72			1
70					3
66			1
64			1
50			1
	Total	30

The table can be organized like this.

Scores	Frequency
50–60	1
61–70	5
71–80	5
81–90	13
91–100	6
Total	30

A histogram can be made from this table.

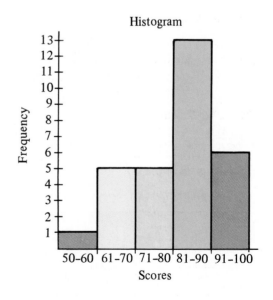

388

A. Answer these questions about the test scores. Use the frequency tables and the histogram.

  1. What is the mode of the test scores?

  2. Did more students receive scores between 70 and 85 than between 85 and 100?

B. Use the frequency table on the left to find the mean of the test scores.

  3. Multiply each score by its frequency.

  4. Add the products.   **5.** Divide by 30.

Practice

Here are the number of people using Steve's Racquetball Club each day for a month: 110, 110, 125, 130, 150, 100, 115, 130, 125, 130, 150, 100, 125, 150, 130, 110, 120, 140, 135, 145, 130, 105, 130, 145, 115, 130, 140, 130, 145, 100.

1. Make frequency tables.

2. What is the mode?

3. What is the mean?

4. Complete the histogram.

5. From the histogram, the range of the number of people using Steve's Racquetball Club most often is between what 2 numbers?

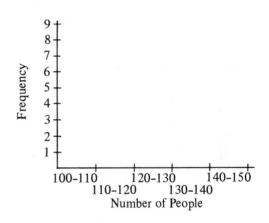

Make histograms.

6.

Salaries	Frequency
$276–300	2
$251–275	3
$226–250	4
$201–225	6
$176–200	12
$150–175	8

7.

Years of service	Frequency
1–5	9
6–10	10
11–15	6
16–20	8
21–25	2
26–30	3

CIRCLE GRAPHS

Ilga interviewed 40 students. She asked each which of 4 popular records was his or her favorite. The results are in the first 2 columns of the table. The next 2 columns are used to construct a circle graph of the results.

Record	Number	Fractional Part of Circle	Measure of Central Angle
A	10	$\frac{10}{40}$ or $\frac{1}{4}$	$\frac{1}{4} \times 360° = 90°$
B	20	$\frac{20}{40}$ or $\frac{1}{2}$	$\frac{1}{2} \times 360° = 180°$
C	4	$\frac{4}{40}$ or $\frac{1}{10}$	$\frac{1}{10} \times 360° = 36°$
D	6	$\frac{6}{40}$ or $\frac{3}{20}$	$\frac{3}{20} \times 360° = 54°$
Total	40		360°

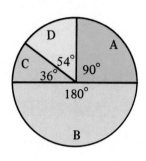

Study and Learn

A. A central angle is formed by 2 radii of a circle.

    1. How many central angles are shown?

    2. What is the sum of the central angles?

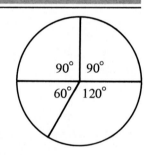

    ▶ The sum of the central angles of a circle is 360°.

B. Complete the table.

	Days Absent	No. of Students	Fractional Part of Circle	Measure of Central Angle
Example	0	10	$\frac{10}{30}$ or $\frac{1}{3}$	$\frac{1}{3} \times 360° = 120°$
3.	1	4		
4.	2	8		
5.	3	5		
6.	4	3		
	Total	30		360°

    7. Make a circle graph.

Practice

Complete the tables. Make circle graphs.

1. CEREAL PREFERENCE SURVEY

Cereal	Number	Fractional Part of Circle	Measure of Central Angle
A	4		
B	3		
C	10		
D	8		
E	9		
F	2		
Total	36		

2. STUDENT QUIZ

Score	Number	Fractional Part of Circle	Measure of Central Angle
100	10		
80	24		
60	15		
40	6		
20	5		
Total	60		

Make circle graphs.

3. The Jackson sisters sold 1,200 boxes of shoes. Ann sold 200 boxes, Peg sold 350 boxes, Ruby sold 250 boxes, and Kay sold 400 boxes.

4. In Mrs. Spencer's mathematics class, 30% received A's, 40% received B's, 20% received C's, and 10% just passed.

5. Approximately 40% of the families contain 2 persons, 25% contain 3 persons, 20% contain 4 persons, and 15% contain more than 4 persons.

PROBABILITY

Lee is going to draw 1 marble from the box. It is equally likely that any of the 4 marbles will be drawn. The probability of drawing a white marble is $\frac{1}{4}$.

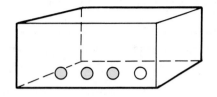

Probability of drawing a white marble $= \dfrac{\text{number of white marbles}}{\text{number of marbles}}$

Study and Learn

A. Find the probability of spinning a 2.

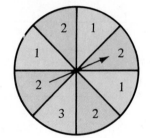

1. How many 2's are there? This is the number of *favorable outcomes*.

2. How many *possible outcomes* are there?

3. What is the ratio $\dfrac{\text{number of favorable outcomes}}{\text{number of possible outcomes}}$?

4. What is the probability of spinning a 2?

   Write $P(2) = \frac{4}{8}$, or $\frac{1}{2}$.

B. This box contains 5 checkers. One checker is black, 3 are red, and 1 is white.

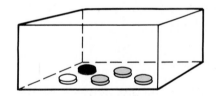

5. What is $P(\text{black})$?

6. What is the probability of *not* getting black?

C. Look at the spinner.

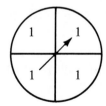

7. What is $P(1)$?

8. What is $P(2)$?

▶ A probability of 1 indicates certainty.
A probability of 0 indicates impossibility.

392

Practice

Assume that all outcomes are equally likely.

The box shown at the right contains
numbered blocks. Find the probabilities.

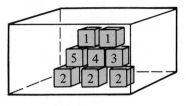

   1. P(2)   **2.** P(3)   **3.** P(4)

   4. P(5)   **5.** P(6)   **6.** P(1)

Use the spinner at the right.
Find the probabilities.

   7. P(6)   **8.** P(5)   **9.** P(4)

An envelope contains six $1 bills, eight $5 bills, and three $10 bills.

10. What is the probability of drawing a $1 bill?

11. What is the probability of drawing a $5 bill?

12. What is the probability of drawing a $10 bill?

A penny is tossed.

13. What is the probability that it
will fall heads up?

14. What is the probability that it
will fall tails up?

A die is tossed.

15. What is the probability the face with 2 dots
will be on top?

★ **16.** What is the probability that 3 or fewer
dots will be on top?

★ **17.** What is the probability that an odd number
of dots will be on top?

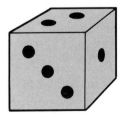

★ **18.** What is the probability that an even number
of dots will be on top?

SAMPLE SPACE

The sample space of an experiment consists of all the possible outcomes. For tossing a die, the sample space consists of 1, 2, 3, 4, 5, 6. P(1 or 2) means the probability of getting a number less than 3.

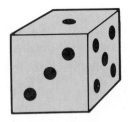

$$P(1 \text{ or } 2) = \frac{2}{6} \begin{array}{l} \longleftarrow \text{ number of favorable outcomes} \\ \longleftarrow \text{ number of possible outcomes} \end{array}$$

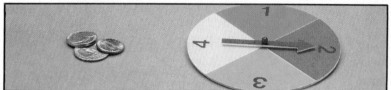

Study and Learn

A. For tossing a die, the sample space consists of 1, 2, 3, 4, 5, 6. Find the probabilities.

    1. P(1)   **2.** P(2 or 4 or 6)   **3.** P(number > 3)

B. Find the sample space for tossing 2 coins. Use *H* for heads and *T* for tails. Complete the table of possible outcomes.

	1st coin	2nd coin	Ordered Pair
	H	H	(H, H)
4.	H		
5.	T		
6.		H	

    ▶ The sample space for tossing 2 coins consists of (H, H), (H, T), (T, T), (T, H).

C. Two coins are tossed 100 times. Predict how many times the outcome will be (H, H).

    7. What is P(H, H)?

    8. Multiply $\frac{1}{4} \times 100$.

Practice

A slip of paper is drawn at random from the box.

1. What is the sample space?

Find the probabilities.

2. P(3)   **3.** P(3 or 4)

4. P(even number)   **5.** P(number > 8)

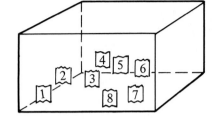

6. Predict the number of times 3 would be drawn in 40 draws.

The pointers on the red and white spinners are spun at the same time.

7. Give the ordered pairs in the sample space. Use (red, white).

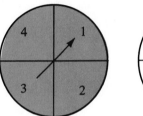

Find the probabilities.

8. P(2, 3)   **9.** P(4, 4)

10. P(sum of both numbers is 3)

11. Predict how many times (1, 2) will occur in 400 spins.

Skills Review

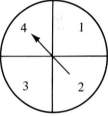

Compute. *(254, 256, 258)*

1. 10% of 80   **2.** 25% of 12   **3.** 30% of 70   **4.** 68% of 97

5. 45% of 75   **6.** 15% of 50   **7.** 20% of 150   **8.** 175% of 300

9. 25% of what is 8?   **10.** 30% of what is 12?   **11.** 8% of what is 16?

12. 3% of what is 60?   **13.** 150% of what is 72?   **14.** 160% of what is 32?

15. What percent of 20 is 5?   **16.** What percent of 100 is 84?   **17.** What percent of 40 is 24?

18. What percent of 16 is 2?   **19.** What percent of 230 is 69?   **20.** What percent of 32 is 48?

COMPOUND PROBABILITY

The probability that an M or an E is drawn is $\frac{2}{11} + \frac{1}{11}$, or $\frac{3}{11}$.

$\boxed{M}\boxed{A}\boxed{T}\boxed{H}\boxed{E}\boxed{M}\boxed{A}\boxed{T}\boxed{I}\boxed{C}\boxed{S}$

$P(M \text{ or } E) = P(M) + P(E)$

Study and Learn

A. A bag contains 4 red marbles, 7 blue marbles, and 1 yellow marble. One marble is drawn. Find the probabilities.

   1. P(blue)   **2.** P(yellow)   **3.** P(blue or yellow)

▶ If 2 events cannot occur at the same time, then the probability that one or the other of them will occur is the *sum* of their probabilities.

B. You can discuss the probability of events which occur one after another.

Here is how to find the probability that on 2 successive tosses of a coin 2 heads result.

   4. What is $P(H)$ on the first toss?

   5. What is $P(H)$ on the second toss?

   6. Give the ordered pairs for the sample space for 2 successive tosses. Use (1st toss, 2nd toss).

   7. What is $P(H, H)$?

   8. True or false? $P(H, H) = P(H) \cdot P(H)$

▶ If an event can occur and then afterward a second event can occur, then the probability that both events can occur is the product of their probabilities.

C. Do an experiment to test the probability of getting 2 heads on 2 successive tosses of a coin. Do 100 trials and record your results. How many times would you expect to get 2 heads? How many times did you get 2 heads?

Practice

A jar contains seven $1 bills, eight $5 bills, and six $10 bills. A bill is drawn from the jar. Find the probabilities.

1. P($1)   **2.** P($5)   **3.** P($10)   **4.** P($1 or $10)

Two bills are drawn; the first is replaced before the second bill is drawn. Find the probabilities.

5. P(Two $1 bills)   **6.** P(Two $5 bills)   **7.** P($1, then $10)

8. P($5, then $10)   ★ **9.** P($1, then $5, then $10)

Computer

```
10 LET C = 0
20 FOR N = 1 TO 100 STEP 1
30 LET A = INT(RND(1)*2)+1:LET B = INT(RND(1)*2)+1
40 IF A = 1 THEN A$ = "HEADS":GOTO 60
50 A$ = "TAILS"
60 IF B = 1 THEN B$ = "HEADS":GOTO 80
70 B$ = "TAILS"
80 IF A$ = "HEADS" AND B$ = "HEADS" THEN C = C + 1
90 PRINT A$" , "B$
100 NEXT N
110 PRINT "2 HEADS CAME UP " C " TIMES."
```

This computer program simulates the 100-trial experiment in **C**, page 396. You may want to try it on your computer. The FOR/STEP/NEXT loop (lines 20 and 100) makes the computer loop through lines 30 to 90. A and B in line 30 each randomly generate 1 and 2 to identify HEADS or TAILS. C is used as a counter, and each time that HEADS is stored in both A$ and B$, C becomes C+1.

Problem-Solving Applications

Using Venn Diagrams

30 students in a class.
20 of them like to watch football on TV.
17 of them like to watch baseball on TV.
8 of them like to watch both football and baseball on TV.

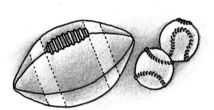 

How many students like to watch football, but not baseball on TV?
How many students like to watch baseball, but not football on TV?
How many students do not like to watch either football or baseball on TV?

To answer these questions, you can use a **Venn diagram.**

The rectangle *R* represents the 30 students in the class. The circle *F* represents the 20 students who like to watch football. The circle *B* represents the 17 students who like to watch baseball. The *intersection* of the two circles represents the 8 students who like to watch both football and baseball.

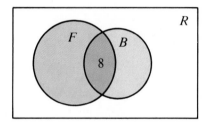

12 like football, but not baseball.
9 like baseball, but not football.
1 likes neither football nor baseball.

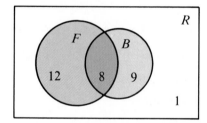

Draw a Venn diagram. Solve.

250 students in Grade 8.
200 went to the school play on Friday.
180 went on the school trip to Washington on Saturday.
140 went to the school play and on the trip to Washington.

1. How many students went only to the school play?

2. How many students went only on the trip to Washington?

3. How many students went to neither the play nor took the trip?

Chapter Review

Plot these points. *(372)*

1. $A(4, 2)$ **2.** $B(^-1, 0)$ **3.** $C(^-2, ^-1)$

Graph. *(376)*

4. $y = x + 3$ **5.** $y = 2x + 2$ **6.** $y = 3x - 1$

Daily attendance at the handball court one week was 31, 29, 36, 48, 48, 71, 76. *(384)*

7. Find the range.   **8.** Find the mean.

9. Find the median.   **10.** Find the mode.

11. Construct a bar graph.
(386) Sales: Mon. $300; Tues. $400; Wed. $215; Thurs. $525; Fri. $809; Sat. $1,200

12. Construct a broken-line graph.
(386) Attendance: August, 23,000; September, 40,000; October, 35,000; November, 24,000; December, 18,000.

13. Lena asked 20 students which sport they like to play most,
(390) baseball, football or soccer. 5 people said baseball, 3 people said football and 12 people said soccer. Make a circle graph.

A box contains 4 red marbles, 2 green marbles, 1 white marble and 5 black marbles. Find the probabilities. *(392)*

14. P(red)   **15.** P(not green)   **16.** P(yellow)

Two spinners are spun at the same time. *(394)*

17. Give the ordered pairs in the sample space.

18. Predict how many times (1, 4) will occur in 300 spins.

Draw a graph or Venn diagram. Solve. *(382, 398)*

19. Opi has $50 in the bank. Each week she deposits $7. Steve has no money in the bank, but plans to deposit $12 a week. In how many weeks will they have the same amount of money in the bank?

20. 38 scouts on hiking trip.
32 had breakfast. 35 had lunch.
30 had both breakfast and lunch.
How many had only breakfast?
How many had only lunch?
How many had neither?

Chapter Test

Plot these points. *(372)*

1. $A(3, 1)$ **2.** $B(0, 3)$ **3.** $C(^-3, 2)$

Graph. *(376)*

4. $y = x - 1$ **5.** $y = x + 4$ **6.** $y = 4x - 3$

The average daily temperatures for 5 days were 23°C, 21°C, 23°C, 27°C, 26°C. *(384)*

7. Find the range. **8.** Find the mean.

9. Find the median. **10.** Find the mode.

11. Construct a bar graph. Sales: *(386)* Howard, $600; Mimi, $720; Daniel, $525; Isaac, $410; Kate, $475

12. Construct a broken-line graph. *(386)* Quiz scores: Mon., 86%; Tues., 85%; Wed., 90%; Thurs., 94%; Fri., 97%

13. Jorge noted the colors of the first 100 cars to drive past his *(390)* house. 25 were blue, 15 were black, 20 were yellow, 10 were white, 15 were green, 15 were red. Make a circle graph.

A box contains 7 blue blocks, 3 white blocks, and 6 red blocks. Find the probabilities. *(392)*

14. P(white) **15.** P(not green) **16.** P(red)

Two spinners are spun at the same time. *(394)*

17. Give the ordered pairs in the sample space.

18. Predict how many times (1, 1) will occur in 400 spins.

Draw a graph or Venn diagram. Solve. *(382, 398)*

19. Mr. James left his home at 7:00 am driving at a speed of 60 km/h. Mrs. James left at 8:00 am and followed him driving at a speed of 80 km/h. At what time will she overtake him?

20. 300 students in 11th grade. 240 take math. 220 take science. 200 take both math and science. How many take math, but not science? How many take science, but not math? How many take neither?

Skills Check

1. Which is the best buy?

   A 2 rolls for 25¢ B 3 rolls for 36¢

   C 4 rolls for 46¢ D 5 rolls for 55¢

2. Bill borrowed $500 from his uncle for 2 years at 6% interest. What was the interest at the end of the 2 years?

   E $30   F $60

   G $300   H $600

3. For which of the following times will the hands of a clock form right angles?

   A noon   B 6:00 am

   C 3:00 am   D 2:00 am

4. The radius of a magnifying glass is 30 mm. What is the area of the magnifying glass? Use 3.14 for π.

   E 282,600 mm²   F 2,826 mm²

   G 188.400 mm²   H 94.20 mm²

5. What point is located at (3, 4)?

   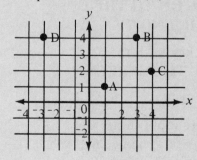

   A A   B B

   C C   D D

6. Weekly income is $400. How much is budgeted for food?

   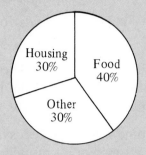

   E $40   F $160

   G $200   H $400

7. The scale on a map is 1 cm = 50 km. Two cities are 3 cm apart on the map. How many kilometers apart are the cities?

   A 50 km   B 150 km

   C 175 km   D 200 km

8. A sports car gets 34 miles per gallon of gasoline. How far can it go on 9 gal of gasoline?

   E 340 mi   F 306 mi

   G 296 mi   H 25 mi

9. The Blass family drove for 3 hours at an average speed of 65 km/h to their cousins' house. How far is it to their cousins' house?

   A 68 km   B 185 km

   C 195 km   D 245 km

BASIC COMMANDS AND PUNCTUATION

The keyboard is used to enter *input* into the computer. The input is entered with BASIC language commands.

▶ **BASIC** stands for **B**eginner's **A**ll-purpose **S**ymbolic **I**nstructional **C**ode.

A. Turn on the computer. The TV screen is one of the computer's output units. The blinking light is called a cursor. It is the computer's signal that it is ready to accept input.

1. Some keys have two symbols. Use the ⸢ SHIFT ⸣ key to get the top symbol printed. Type a few of the symbols.

2. The space bar is the long key at the front of the keyboard. It is used to put a space between words. Type your name.

3. The ⸢ RETURN ⸣ key signals the end of an input. It makes the computer either store the input in memory, or begin computing with the input. Press ⸢ RETURN ⸣ . What happened?

4. Type: **PRINT "USE THE LEFT ARROW FOR CORRECTING."** Press ⸢ RETURN ⸣ .

5. Output appears on the screen after ⸢ RETURN ⸣ is pressed. Write the output.

B. PRINT is a BASIC command. It is used to make output appear on the screen. When used with quotation marks, what is inside the quotes is printed on the screen.

6. Type: **PRINT "THE CURSOR MOVES ON TO THE NEXT LINE WHEN IT REACHES THE END."**

   Do not press ⸢ RETURN ⸣ . Press the ⸢←⸣ key until the cursor is over the period. Now type: **OF THE SCREEN."** ⸢ RETURN ⸣ . Write the output.

C. The LET command makes the computer store input in its random access memory (RAM). The letters A to Z name 26 number-storage places; A$ to Z$ (read "Z string") name 26 word-storage places.

For example: **LET T = 12** ◀—— stores 12 in T.

But   **LET T$ = "TWELVE"** ◀—— stores TWELVE in T$.

Enter each pair of commands. Press [RETURN] after each one.
Write the output.

7. **LET T = 12**
   PRINT T

8. **LET F = 5**
   PRINT T" × "F
   " = "T∗F

9. **LET T$ =**
   "TWELVE"
   PRINT T$

D. The HOME command clears the screen of print, but does not erase the memory. Enter **HOME**. Now enter these and write the output.

10. **PRINT T**

11. **PRINT T" × "F**
    " = "T∗F

12. **PRINT T$**

E. The NEW command erases the computer's memory. When nothing else is stored in a storage place, a zero is held there. Enter **NEW**. Now enter these and write the output.

13. **PRINT T**

14. **PRINT T" × "F**
    " = "T∗F

15. **PRINT T$**

F. A colon is used to combine 2 or more program lines into a single line. <u>Type:</u> **PRINT "8":PRINT " ×9":PRINT "––":
PRINT 8 ∗ 9** [RETURN] .

The output is in 4 separate lines because the 4 separate commands are connected by colons. Write the output.

G. A semicolon is used between PRINT statements to make the computer print them on the same line. For these commands:

PRINT "8 × 9 = ";   The output is 8 × 9 = 72 because the
PRINT 8 ∗ 9   first command ends with a semicolon.

Try some of your own examples.

H. A comma is used with PRINT to make the computer print columns. Input these commands and write the output.

PRINT "8 × 9 = ",   72 is the output on top of the second
PRINT 8 ∗ 9   print column because the first
                         command ends with a comma.

Try some of your own examples.

THE COMPUTER AS A CALCULATOR

PRINT without quotation marks can make the computer work like a calculator. You may use a **?** in place of PRINT.

A. Like the calculator, the computer does not use commas. Enter these. (Remember to press | RETURN |.) Write the output.

   1. PRINT 338982 + 43876   **2. PRINT 382858 − 338982**

   3. PRINT 689591 / 79   **4. PRINT 79 * 689591**

B. The computer outputs a decimal when a quotient is not a whole number. If the decimal is longer than 9 places, the 9th digit is a rounded number. Enter these and write the output.

   5. PRINT 3/16   **6. PRINT 186/37**   **7. PRINT 158/19**

C. The computer follows the normal order of operations. Predict the output of these. Then enter each to check your answer.

   8. PRINT 3 + 8 * 4   **9. PRINT 20 − 4 / 2**   **10. PRINT 8 * 3 + 2**

D. The computer uses parentheses to change the order of operations. Predict the output of these; enter each to check your answer.

   11. PRINT (3 + 8) * 4   **12. PRINT (20 − 4) / 2**   **13. PRINT 8 * (3 + 2)**

E. The $\wedge$ is the computer's exponent key. 2^3 is entered as 2 $\wedge$ 3. Enter these and write the output.

   14. PRINT 5 $\wedge$ 3   **15. PRINT 13 $\wedge$ 2**   **16. PRINT 10 $\wedge$ 6**

F. The computer prints in scientific notation when an output is 1 billion or larger. 26,000,000,000 or 2.6×10^{10} is displayed as 2.6 E + 10. Enter each and write the output.

   17. PRINT 30000 * 400000   **18. PRINT 5 *10 $\wedge$ 11**

   19. Write standard scientific notation and a standard numeral for 1.2E + 10.

G. Round-off errors sometimes occur, especially with $\wedge$. This is because the computer must convert the base-ten input into base-two, do the computation, then convert the base-two answer back into base-ten for output display. The output in **18** might be 5.00000001E + 11.

404

Enter each input and write the output.

20. PRINT 7 ∧ 2   **21. PRINT 3 ∧ 6**   **22. PRINT 9 ∧ 3**

23. How large was the round-off error for each of these?

H. The integer function, INT, determines the integer portion of a decimal number in parentheses. INT also is used for rounding numbers. Enter these and write the output.

24. PRINT 89/7   **25. PRINT INT(89/7)**   **26. PRINT INT(89/7 + .5)**

I. The absolute value function, ABS, returns the absolute value of an expression in parentheses. Enter these; write the output.

27. PRINT ABS(⁻52)   **28. PRINT ABS (8 + ⁻13)**   **29. PRINT ABS (⁻12 − ⁻4)**

J. The square root function, SQR, returns the square root of a number in parentheses. Enter these and write the output.

30. PRINT SQR(225)   **31. PRINT SQR(.64)**   **32. PRINT SQR(2)**

Practice

Use the computer to answer these. Write decimal numbers for any that come out in scientific notation. Ignore round-off errors.

1. 8,396 + 21,342 + 178,112   **2.** 3,756,508 − 983,875

3. 589,860 × 52,000   **4.** 3,856,912 ÷ 5,824

5. 8 + 7 × 3 + 5   **6.** (8 + 7) × (3 + 5)

7. 3^4 **8.** 2^{10} **9.** 2.3×10^{11}

10. $(3 \times 10^3) \times (6 \times 10^6)$ **11.** $(8 \times 10^2) \div (4 \times 10^2)$

Write your input and the computer's output.

12. 864 ÷ 17 to the nearest whole number.

13. $|{}^-8 + {}^-3|$ **14.** $|{}^-18 - {}^-12|$ **15.** $|4 + {}^-7| \times |{}^-8|$

16. $\sqrt{625}$ **17.** $\sqrt{3.61}$ **18.** $\sqrt{10}$

THE COMPUTER PROGRAM

A. A program is a numbered list of command statements or program lines. The commands are carried out in the order of their line numbers. Here is an example of a computer program.

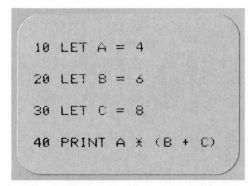

```
10 LET A = 4

20 LET B = 6

30 LET C = 8

40 PRINT A X (B + C)
```

1. To enter the program into the computer's memory (RAM), type each line including its line number. Press RETURN after each.

2. To list the program in the RAM, enter the command LIST. Are the program lines listed on the screen?

3. To run the program, enter the command RUN. Write the output.

B. A bug is an error that keeps the program from running the way you would like it to run. Removing the bugs is called debugging.

4. To replace a program line, enter it with the same line number. Change line 40 by typing:
40 PRINT A" × ("B" + "C") = "A" × "B" + "A" × "C" = "A * (B + C) RETURN .

5. Line numbers are skipped in writing a program in case you want to add lines later. Add this line 35 by typing:
35 PRINT "THE DISTRIBUTIVE PROPERTY:" RETURN .

6. To erase a program line, enter its line number. Enter line 50 by typing: 50 END RETURN . Then erase it by typing: 50 RETURN .

7. Enter LIST, then enter RUN. Write the output.

8. To erase the entire program from memory, enter NEW. Now enter LIST. Are the program lines on the screen?

★9. Write, enter, and run a program that stores 3 in D, 5 in E, and 7 in F. The output should demonstrate the associative property of addition. Use this lesson's program as a model.

THE INPUT COMMAND

The INPUT command makes the control unit stop and wait for input to be entered by the computer user.

```
10   PRINT "ENTER RADIUS OF CIRCLE
     ."
20   INPUT R
30   INPUT "ENTER UNIT OF MEASURE.
     ";U$
40   PRINT "CIRCUMFERENCE = "2 X 3
     .14 X R" "U$
50   PRINT "AREA = "3.14 X R ^ 2
     " SQUARE "U$
```

Empty quotes cause a space to be skipped in the output.

1. Enter NEW. Now enter and run this program.

2. INPUT in line 20 makes the computer wait for the measure of a radius to be entered. Type: 5 [RETURN] .

3. Line 30 combines PRINT and INPUT into a single line. Notice that a semicolon separates the message from U$. Enter FEET and write the output.

4. Erase line 20 and change line 10 to:
   10 INPUT "ENTER RADIUS OF CIRCLE. ";R
   Enter LIST. Have the changes been made?

5. Run the program again. Enter 12 for the radius and CM for the unit of measure. Write the output.

```
10 INPUT "ENTER RADIUS OF SPHERE
     ";R
20 INPUT "ENTER UNIT OF MEASURE
     ";U$
30 PRINT "VOLUME = "4 / 3 X 3.14
     X R ^ 3" CUBIC "U$
```

6. This program calculates the volume of a sphere. Enter NEW; enter and run this program twice, using different values for R and U$.

★ 7. Write, enter, and run a program that waits for the length, width, height, and unit of measure of a rectangular prism to be entered. It should output the surface area and volume of the prism.

BRANCHING IN A PROGRAM

A. The GOTO command makes control branch to the line number that follows it. GOTO creates an unconditional branch.

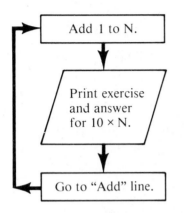

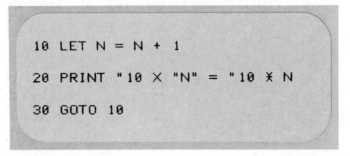

```
10 LET N = N + 1

20 PRINT "10 X "N" = "10 X N

30 GOTO 10
```

1. Enter NEW; enter and run this program. Describe the output.

2. When a loop is created by GOTO alone, the control unit does not have a way to exit. The looping continues until the run is broken by the user. To break the run, press [CONTROL] and [C] . Write the last line of output.

B. The IF/THEN command creates a conditional branch. Control branches to the line number that follows THEN only if the IF part is true.

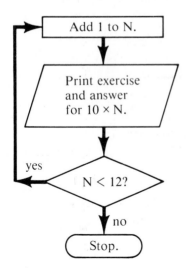

```
10 LET N = N + 1

20 PRINT "10 X "N" = "10 X N

30 IF N<12 THEN 10

40 END
```

3. Read the program and predict the output.

4. Enter NEW. Now enter and run the program. Describe the output. Was your prediction correct?

C. Line 10 contains a counter. With each loop, 1 is added to the number in N. The IF/THEN in line 30 checks to see if the number in N is less than 12. If it is, control branches back to line 10. When N reaches 12, the IF part is no longer true, control falls to line 40, and the program ends.

★**5.** Change lines 20 and 30 so the program outputs the multiples of 10 up to 10 × 25 in columns. Run the program. Did the changes work?

D. THEN may be followed by a command. The command is carried out only if the IF part is true.

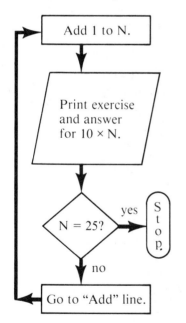

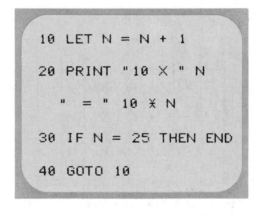

```
10 LET N = N + 1

20 PRINT "10 X " N
   " = " 10 * N

30 IF N = 25 THEN END

40 GOTO 10
```

6. Predict the output of this program.

7. Enter NEW. Now enter and run the program. Describe the output. Was your prediction correct?

★**8.** Write, enter, and run a program that makes the computer output the factors and products for the first 25 perfect square numbers (1 × 1 = 1 to 25 × 25 = 625) in columns. Use the program in B or D as a model.

THE FOR/STEP/NEXT COMMAND

A. FOR/STEP/NEXT is a 3-part command. It makes the computer count. FOR sets the lower and upper limits of the counting. STEP tells what to count by. NEXT checks to see if the upper limit has been reached.

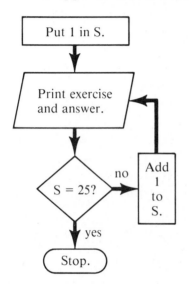

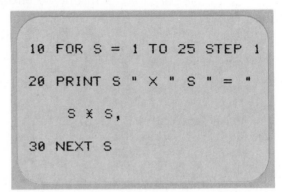

```
10 FOR S = 1 TO 25 STEP 1

20 PRINT S " X " S " = "

      S * S,

30 NEXT S
```

1. The comma at the end of line 20 makes the computer print the output in columns.

2. Predict the output of this program.

3. Enter NEW. Now enter and run the program. Describe the output. Was your prediction correct?

4. Change line 20 to: **20 PRINT S**. Run the program.
   10 FOR S = 1 TO 25 STEP 1 makes the computer count

        ▲   ▲   ▲

       from   1 to 25 by ones.

5. Complete.
   10 FOR S = 0 TO 104 STEP 13 makes the computer count
       from   _?_ to _?_ by _?_ .

   10 FOR S = 10 TO 250 STEP 10 makes the computer count
       from   _?_ to _?_ by _?_ .

6. Check your answers by changing line 10 to each line 10 in **5** and running the program. Were your answers correct?

B. FOR/STEP/NEXT may be used to make the control loop through one or more program lines as the counting takes place. We use STEP 1 to make it easy to keep track of the number of loops.

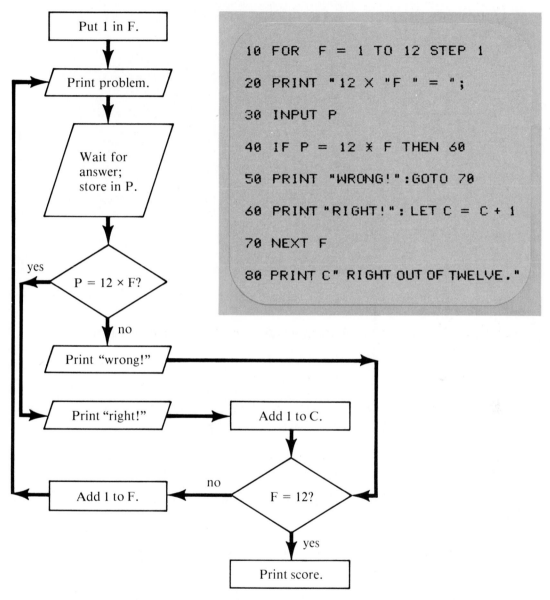

```
10 FOR  F = 1 TO 12 STEP 1

20 PRINT "12 X "F " = ";

30 INPUT P

40 IF P = 12 X F THEN 60

50 PRINT "WRONG!":GOTO 70

60 PRINT "RIGHT!": LET C = C + 1

70 NEXT F

80 PRINT C" RIGHT OUT OF TWELVE."
```

7. FOR/STEP/NEXT in lines 10 and 70 makes the control loop through lines 20 to 60 how many times?

8. Enter NEW. Enter and run the program. Take the quiz and write your score.

NESTED FOR/STEP/NEXT LOOP

One FOR/STEP/NEXT loop may be nested inside another.

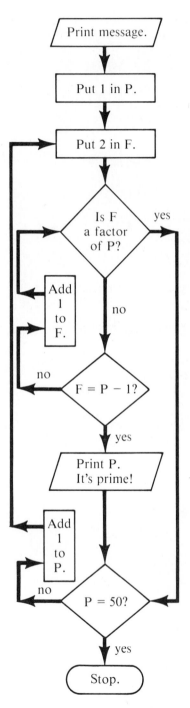

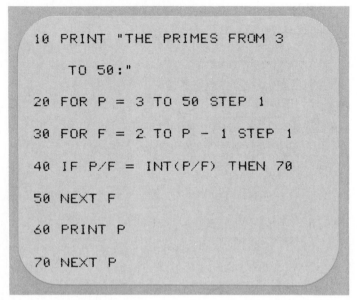

```
10 PRINT "THE PRIMES FROM 3
      TO 50:"
20 FOR P = 3 TO 50 STEP 1
30 FOR F = 2 TO P - 1 STEP 1
40 IF P/F = INT(P/F) THEN 70
50 NEXT F
60 PRINT P
70 NEXT P
```

1. Enter NEW; enter and run this program. Write the output.

2. Change lines 10 and 20 so that the primes from 51 to 101 are printed. Run the program and write the output.

★ 3. Write, enter, and run a program that gives the primes from 103 to 200. Use the program in section B as a model. Write the output.

THE RND FUNCTION

A. Random means without order. The random or RND function selects a 9 place decimal between 0 and 1.

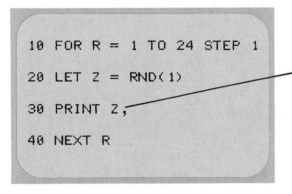

```
10 FOR R = 1 TO 24 STEP 1

20 LET Z = RND(1)

30 PRINT Z,

40 NEXT R
```

The comma makes the computer print in columns.

1. Enter NEW; enter and run this program. Describe the output.

2. Are any of the decimals on the screen repeated?

3. Change line 20 to: **LET Z = RND(1) * 5**
   Run the program. Do all the decimals have whole number parts from 0 to 4?

B. The integer function, INT, makes the computer print the closest lower integer.

4. Change line 20 to: **LET Z = INT(RND(1)*5)**
   Run the program. Are all the numbers integers from 0 to 4?

5. Change line 20 to: **LET Z = INT(RND(1)*5) + 1**
   Run the program. Are all the numbers integers from 1 to 5?

6. Change line 20 to: **LET Z = INT(RND(1)*10) + 3**
   Run the program. Are all the numbers integers from 3 to 12?

C. To get random integers from X to Y, use: **INT(RND(1)*(Y − X + 1)) + X**

★7. Change line 20 so that random integers from 5 to 10 are printed. Run the program. Did the change work?

★8. Write, enter, and run a program that gives a scored 10 problem multiplication quiz with random factors from 3 to 12. Use the program on page 411 as your model. (Hint: Make the computer store 2 random integers in each loop.)

READ/DATA

READ/DATA is a 2-part command. DATA allows you to store numbers or words in a line of a program. READ N makes the control unit look for a DATA line and put the first number it finds in N.

A. In this program, numbers are stored in a DATA line.

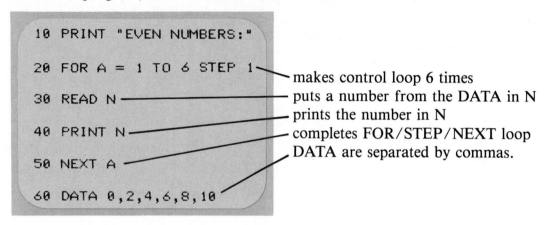

```
10 PRINT "EVEN NUMBERS:"

20 FOR A = 1 TO 6 STEP 1 ——— makes control loop 6 times

30 READ N ——————————————— puts a number from the DATA in N
                                   prints the number in N
40 PRINT N ——————————————— completes FOR/STEP/NEXT loop
                                   DATA are separated by commas.
50 NEXT A ———

60 DATA 0,2,4,6,8,10
```

1. Enter NEW; enter and run the program. Describe the output.

2. Add these lines 54 and 56.
   54 INPUT "LIKE TO SEE THEM AGAIN? (Y OR N)";A$
   56 IF A$ = "Y" THEN 10

3. You get an OUT OF DATA ERROR because the computer is reading at the end of the DATA line. Add this line 52: **52 RESTORE**

4. Run the program again. Did it work this time?

   ▶RESTORE resets the DATA pointer to the first DATA statement

5. Modify the program to print the even numbers from 0 to 20. You will have to change lines 20 and 60. How many times will control loop now? Run the program. Did you get the expected output?

6. Write, enter, and run a program to print the odd numbers from 21 to 29 using the READ/DATA command. Use the program above as a model.

B. This program stores words in DATA lines.

```
10  PRINT "COMPUTER HISTORY:"
20  FOR A = 1 TO 5 STEP 1
30  READ N$
40  PRINT N$
50  NEXT A
60  DATA    PASCAL,BABBAGE,BYRON
70  DATA    HOLLERITH,VON NEUMANN
```

a string variable is used to store words read from the DATA lines.

DATA may be stored in more than one line.

7. Enter NEW; enter and run this program. Describe the output.

★ 8. Add the lines necessary to give the computer user the option to have this list repeated. Run the program. Did your changes work?

★ 9. What was each of these people's contribution to the development of computers?

C. In this program, calculations are made with the numbers stored in data lines.

```
10 FOR A = 1 TO 9 STEP 1
20 READ N
30 PRINT 2 X  SQR (N)
40 NEXT A
50 DATA   9,25,49,81,121
60 DATA   169,225,289,361
```

prints the result of 2 times the square root of N with each loop

10. Read this program. Will the output consist of all odd or all even numbers? Why?
11. Enter NEW; enter and run this program. Was your prediction correct?

★12. Write, enter, and run a program to print the days of the week using the READ/DATA command.

Table of Trigonometric Ratios

Angle Measure	Sin	Cos	Tan	Angle Measure	Sin	Cos	Tan
0°	0.00	1.00	0.00	46°	.719	.695	1.04
1°	.017	1.00	.017	47°	.731	.682	1.07
2°	.035	.999	.035	48°	.743	.669	1.11
3°	.052	.999	.052	49°	.755	.656	1.15
4°	.070	.998	.070	50°	.766	.643	1.19
5°	.087	.996	.087	51°	.777	.629	1.23
6°	.105	.995	.105	52°	.788	.616	1.28
7°	.122	.993	.123	53°	.799	.602	1.33
8°	.139	.990	.141	54°	.809	.588	1.38
9°	.156	.988	.158	55°	.819	.574	1.43
10°	.174	.985	.176	56°	.829	.559	1.48
11°	.191	.982	.194	57°	.839	.545	1.54
12°	.208	.978	.213	58°	.848	.530	1.60
13°	.225	.974	.231	59°	.857	.515	1.66
14°	.242	.970	.249	60°	.866	.500	1.73
15°	.259	.966	.268	61°	.875	.485	1.80
16°	.276	.961	.287	62°	.883	.469	1.88
17°	.292	.956	.306	63°	.891	.454	1.96
18°	.309	.951	.325	64°	.899	.438	2.05
19°	.326	.946	.344	65°	.906	.423	2.15
20°	.342	.940	.364	66°	.914	.407	2.25
21°	.358	.934	.384	67°	.921	.391	2.36
22°	.375	.927	.404	68°	.927	.375	2.48
23°	.391	.921	.424	69°	.934	.358	2.61
24°	.407	.914	.445	70°	.940	.342	2.75
25°	.423	.906	.466	71°	.946	.326	2.90
26°	.438	.899	.488	72°	.951	.309	3.08
27°	.454	.891	.510	73°	.956	.292	3.27
28°	.469	.883	.532	74°	.961	.276	3.49
29°	.485	.875	.554	75°	.966	.259	3.73
30°	.500	.866	.577	76°	.970	.242	4.01
31°	.515	.857	.601	77°	.974	.225	4.33
32°	.530	.848	.625	78°	.978	.208	4.71
33°	.545	.839	.649	79°	.982	.191	5.15
34°	.559	.829	.675	80°	.985	.174	5.67
35°	.574	.819	.700	81°	.988	.156	6.31
36°	.588	.809	.727	82°	.990	.139	7.12
37°	.602	.799	.754	83°	.993	.122	8.14
38°	.616	.788	.781	84°	.995	.105	9.51
39°	.629	.777	.810	85°	.996	.087	11.4
40°	.643	.766	.839	86°	.998	.070	14.3
41°	.656	.755	.869	87°	.999	.052	19.1
42°	.669	.743	.900	88°	.999	.035	28.6
43°	.682	.731	.933	89°	1.00	.017	57.3
44°	.695	.719	.966	90°	1.00	0.00	
45°	.707	.707	1.00				

TABLE

EXTRA PRACTICE

Solve. *(Use with page 8.)*

1. Ed ran 35 minutes on Sunday, 25 minutes on Monday, and 48 minutes on Tuesday. How long did he run in all?

2. Jamie bought 4 ribbons, each 32 in. long. How much ribbon did she buy in all?

3. Joan bought shoes for $28.95, a sweater for $17.98, and a shirt for $17.50. How much did she spend in all?

4. Al handed out 158 flyers for a play. Pat handed out 63 more flyers than Al. How many flyers did Pat hand out?

5. Miss Marsh bought 4 rolls of paper. The paper cost $32.16. What was the cost of 1 roll of paper?

6. Tickets for a movie were $3.00 each. Mr. Lang bought $135 worth of tickets. How many tickets did he buy?

Add. *(Use with page 24.)*

1.	2.	3.	4.	5.
754 + 35	3,654 + 2,543	$ 5,189 + 2,675	36,136 + 52,375	47,967 + 31,358

6.	7.	8.	9.	10.
736,184 + 124,819	$ 359,487 + 427,891	4,189 7,156 + 5,398	7,157 3,984 5,619 + 3,487	$ 89,167 42,789 3,487 + 9,135

11.	12.	★ 13.	★ 14.
416,871 39,375 182,468 51,009 + 237,861	$ 2,386,143 1,194,865 544,328 167,246 + 4,891,387	15,489,163 27,564 39,986,115 656,789 + 8,132,108	31,376,181 497,275 73,891,476 65,971,183 + 1,246,971

Subtract. *(Use with page 26.)*

1.	2.	3.	4.
5,896 − 2,589	36,738 − 12,394	48,971 − 26,489	342,462 − 113,931

5.	6.	7.	8.
7,321 − 3,496	29,148 − 13,889	$ 56,782 − 24,395	563,436 − 351,189

9.	10.	11.	12.
4,618,497 − 2,948,699	$ 58,324.75 − 17,496.98	8,006 − 3,849	80,000 − 14,679

★ Find the missing numbers.

13. $358{,}647 - \square = 149{,}867$

14. $3{,}674{,}187 - \square = 1{,}897{,}379$

Estimate the sums. *(Use with page 28.)*

1. 634 + 798	**2.** 3,486 + 5,714	**3.** 71,863 + 13,974	**4.** 31,489 + 8,437	**5.** $ 67,874 + 1,396

6. 879
 7,253
 + 189

7. 13,875
 9,684
 + 57,183

8. $ 5,712
 6,410
 481
 + 759

★ **9.** 489,398
 + 217,167

★ **10.** 87,189
 2,788
 1,473
 8,689
 + 9,267

Estimate the differences.

11. 864
 − 395   **12.** 738
 − 53   **13.** 5,329
 − 1,875   **14.** 7,467
 − 89   **15.** 84,675
 − 32,198

16. 36,489
 − 8,114   **17.** 73,298
 − 1,640   **18.** 37,317
 − 3,843   ★ **19.** 467,483
 − 279,156   ★ **20.** 458,175
 − 37,137

Draw a diagram for each problem. Solve. *(Use with page 32.)*

1. A 65-ft television wire is cut into two pieces. The shorter piece is 27 ft long. How long is the longer piece?

2. The length of a garden is 72 ft longer than the width. The width is 94 ft. What is the length of the garden?

3. Ms. Gold bought a radio on sale for $98.98. She saved $50.95 off the original price. What was the original price of the radio?

4. Jan and Pietro drove toward each other from towns 146 km apart. When they met, Jan had driven 57 km. How far had Pietro driven?

5. Mr. Olin's fence is 240 ft long. He has painted 138 ft. How many feet does he have left to paint?

6. Sam walked south from a store for 460 m. Ilga walked north from the store 519 m. How far apart were they?

Multiply. *(Use with page 34.)*

1. 73
 × 9   **2.** 246
 × 5   **3.** $4,237
 × 7   **4.** 39,467
 × 6   **5.** $864.27
 × 8

6. 56
 × 28   **7.** 37
 × 89   **8.** 450
 × 25   **9.** $604
 × 34   **10.** 3,596
 × 72

11. 6,743
 × 67   **12.** $77.39
 × 41   **13.** 84,163
 × 74   **14.** 35,186
 × 29   **15.** $763.85
 × 37

16. 7 × 836   **17.** 85 × 31   **18.** 27 × 562

★ **19.** 36 × 4 × 93   ★ **20.** 56 × 8 × 37   ★ **21.** 37 × 6 × 82 × 4

Multiply. *(Use with page 36.)*

1. 346 × 121	**2.** 482 × 326	**3.** 507 × 827	**4.** 429 × 495	**5.** 360 × 128
6. 3,460 × 423	**7.** 8,067 × 212	**8.** 4,163 × 409	**9.** 3,170 × 604	**10.** 8,234 × 740
11. 42,316 × 819	**12.** 6,816 × 4,118	**13.** 3,939 × 2,617	**14.** 4,070 × 2,061	**15.** 14,152 × 2,020

16. 347 × 1,803   **17.** 1,830 × 4,072   ★ **18.** 31,463 × 2,004

★ **19.** 56 × 87 × 109   ★ **20.** 74 × 28 × 15 × 9   ★ **21.** 136 × 18 × 9 × 25

Divide. *(Use with page 38.)*

1. 4)96   **2.** 3)87   **3.** 6)83   **4.** 6)79   **5.** 5)86

6. 4)168   **7.** 3)159   **8.** 7)219   **9.** 6)$126   **10.** 5)183

11. 9)6,831   **12.** 8)7,580   **13.** 7)9,430   **14.** 6)7,594   **15.** 7)3,687

16. 6)8,172   **17.** 6)4,614   **18.** 8)$9,072   **19.** 4)5,073   **20.** 5)8,172

21. 6)24,738   **22.** 5)10,070   **23.** 8)26,183   **24.** 9)35,703   **25.** 8)26,924

★ Find the missing numbers.

26. □ ÷ 8 = 476 r 3   **27.** □ ÷ 7 = 4,587 r 6   **28.** □ ÷ 6 = 25,187 r 5

Divide. *(Use with page 40.)*

1. 25)75   **2.** 36)72   **3.** 34)67   **4.** 24)75

5. 21)756   **6.** 23)667   **7.** 33)$825   **8.** 27)189

9. 34)9,678   **10.** 44)9,618   **11.** 56)9,817   **12.** 25)6,718

13. 84)7,916   **14.** 73)5,629   **15.** 24)$1,584   **16.** 37)2,679

17. 382)9,168   **18.** 627)8,787   **19.** 148)8,436   **20.** 439)9,239

21. 875)55,175   **22.** 287)80,387   **23.** 676)13,008   **24.** 555)92,130

★ Find the missing numbers.

25. □ ÷ 37 = 29 r 36   **26.** □ ÷ 89 = 436 r 75   **27.** □ ÷ 189 = 374 r 59

Multiply. *(Use with page 44.)*

1. 50 × 7	**2.** 70 × 60	**3.** 300 × 50	**4.** 800 × 600	**5.** 50,000 × 2
6. 50,000 × 20	**7.** 50,000 × 200	**8.** 50,000 × 2,000	★ **9.** 60,000 × 50,000	★ **10.** 700,000 × 20,000

Divide.

11. 30)180   **12.** 40)1,600   **13.** 30)12,000   **14.** 60)24,000

15. 300)6,000   **16.** 400)28,000   **17.** 700)42,000   ★ **18.** 4,000)16,000

Estimate the products. *(Use with page 46.)*

1. 38 × 41	**2.** 350 × 29	**3.** $179 × 68	**4.** 479 × 881	**5.** 3,869 × 43
6. 8,704 × 56	**7.** 7,500 × 418	**8.** $8,164 × 893	★ **9.** 15,675 × 536	★ **10.** 38,471 × 986

Estimate the quotients.

11. 31)94   **12.** 36)418   **13.** 39)8,242   **14.** 51)$4,606

15. 42)29,416   **16.** 269)61,419   **17.** 289)16,681   ★ **18.** 837)334,198

Add or subtract. *(Use with page 51.)*

1. 8,734 + 3,978	**2.** 75,349 + 26,863	**3.** $ 526,679 + 29,187	**4.** 86,183 5,967 + 15,363	**5.** 418,379 67,186 + 153,208
6. 5,673 − 3,192	**7.** 6,837 − 2,988	**8.** 52,126 − 28,378	**9.** $ 7,003 − 5,394	**10.** 60,000 − 38,145

Multiply.

11. 3,918 × 7	**12.** $483 × 96	**13.** 5,670 × 72	**14.** $676 × 137	**15.** 5,379 × 802

Divide.

16. 5)975   **17.** 8)5,483   **18.** 27)189   **19.** 37)5,617   **20.** 64)$28,096

Estimate the answers.

21. 8,719 + 324	**22.** 63,495 − 8,961	**23.** 786 × 39	**24.** 6,917 × 257	**25.** 78)26,187

Select the equation to solve the problem. *(Use with page 88.)*

1. The regular price of a table is $269.50. On sale it is $199.25. How much is saved by buying the table on sale?

    a. $269.50 + 199.25 = c$
    b. $269.50 - 199.25 = c$
    c. $c - 269.50 = 199.25$
    d. $c - 199.25 = 269.50$

2. During a sale a person buying a plant for $7.50 received a second plant free. What is the average cost of each plant?

    a. $2 \times 7.50 = c$
    b. $7.50 + 7.50 = c$
    c. $7.50 + 7.50 = 2 \times c$
    d. $7.50 \div 2 = c$

3. Two cans of paint cost $13.96. How much do 12 cans of paint cost?

    a. $6 \times 13.96 = c$
    b. $12 \times 13.96 = c$
    c. $2 \times c = 13.96$
    d. $13.96 - 12 = c$

4. Baseballs cost $7.75 and footballs cost $9.98. How much more are footballs?

    a. $9.98 + 7.75 = c$
    b. $9.98 \times 7.75 = c$
    c. $c - 9.98 = 7.75$
    d. $9.98 - 7.75 = c$

5. Eighteen rulers cost $10.62. What is the cost of 1 ruler?

    a. $10.62 \div 18 = c$
    b. $18 \times 10.62 = c$
    c. $18 + 10.62 = c$
    d. $18 \div 10.62 = c$

6. Sara paid $11.98 for a book and $.59 for a card. How much did she pay in all?

    a. $11.98 - 0.59 = c$
    b. $c + 0.59 = 11.98$
    c. $0.59 \times 11.98 = c$
    d. $11.98 + 0.59 = c$

7. Five chairs, marked $36.75 a chair, were on sale at $5.00 off for each chair. How much did the five chairs cost on sale?

    a. $5 \times 36.75 = c$
    b. $36.75 - 5.00 = c$
    c. $5 \times 31.75 = c$
    d. $36.75 + 5.00 = c$

★ **8.** Roses are 6 for $1.98, with 50¢ off if you buy more than a dozen. How much would 3 dozen roses cost?

    a. $3 \times (1.98 - 0.50) = c$
    b. $6 \times (1.98 - 0.50) = c$
    c. $2 \times (1.98 - 0.50) = c$
    d. $18 \times (1.98 - 0.50) = c$

Which is the better buy? *(Use with page 96.)*

1. A package of 6 rolls is marked $1.05, and a package of 8 rolls is marked $1.30. Which is the better buy?

2. At store A, 3 cans of soup cost $0.59. At store B, the same soup is 2 for $0.43. Which is the better buy?

3. A store sold 6 pencils for $0.55 or 10 pencils for $0.85. Which is the better buy?

4. A box of 6 cupcakes costs $1.71, and a box of 10 cupcakes costs $2.90. Which is the better buy?

5. A box of 100 paper clips costs $0.39. A box of 1,000 costs $3.50. Which is the better buy?

6. Jill bought socks, 3 pairs for $2.98. Ann bought the same socks, 2 pairs for $1.89. Which was the better buy?

Add. *(Use with page 102.)*

1. 1.3 + 4.4	**2.** 7.5 + 8.9	**3.** 56.72 + 13.25	**4.** $ 8.72 + 3.19	**5.** 15.364 + 53.524

6. 0.5 0.8 + 0.4	**7.** 0.624 0.896 + 0.471	**8.** 0.1245 0.7862 + 0.1459	**9.** 8.6 3.4 + 3.6	**10.** $ 15.63 25.08 + 19.75

11. 3.6 8.9 7.4 + 6.7	**12.** $ 2.67 8.95 7.07 + 8.88	**13.** 3.561 1.089 7.364 + 8.516	**14.** 0.1465 0.3897 0.1642 + 0.3875	**15.** 35.4179 6.1175 9.2874 + 3.1579

16. 3.4 + 8 + 9.31   **17.** 9 + 1.5 + 0.011   ★ **18.** 0.6 + 0.095 + 0.48751

★ **19.** 8 + 1.606 + 9.5 + 7.0031   ★ **20.** 6.5 + 0.11316 + 8.145 + 6

Find the answers. *(Use with page 104.)*

1. 12.7 − 9.3	**2.** 358.7 − 149.8	**3.** $ 10.98 − 8.79	**4.** $ 45.73 − 24.89	**5.** 0.074 − 0.065

6. 0.897 − 0.648	**7.** 8.786 − 3.347	**8.** 0.0008 − 0.0005	**9.** 0.0074 − 0.0038	**10.** 8.1674 − 6.9848

11. 6.7 − 0.46   **12.** 0.7 − 0.56   **13.** 0.04 − 0.008   **14.** 0.83 − 0.079

15. 8.4 − 7.83   **16.** 7.3 − 2.186   **17.** 8 − 0.94   ★ **18.** 9 − 6.18765

★ **19.** 9.1 − 3.46153   ★ **20.** 8 − 2.6 + 0.451   ★ **21.** 7.6 − 0.43 + 9

Estimate to the nearest whole number or dollar. *(Use with page 106.)*

1. 0.98 + 3.063	**2.** 7.623 2.417 + 3.986	**3.** 3.1684 6.1875 + 9.7864	**4.** $ 6.75 2.09 + 3.86	**5.** $ 9.64 3.75 + 8.67

6. 8.5 − 0.76	**7.** 7.86 − 2.34	**8.** 8.3146 − 2.9108	**9.** $ 6.75 − 2.86	**10.** $ 37.89 − 5.98

Estimate the answers to the nearest tenth.

11. 0.75 + 0.81	**12.** 8.36 0.4 1.98 + 0.153	**13.** 8.167 9.4 3.046 + 6.13	**14.** 0.75 − 0.38	**15.** 7.496 − 2.1873

Multiply. *(Use with page 110.)*

1. 0.8
× 6

2. 4.7
× 5

3. 13.6
× 3

4. 256.7
× 8

5. 0.08
× 7

6. 0.37
× 4

7. 3.96
× 9

8. 4.08
× 7

9. 0.009
× 6

10. 0.034
× 2

11. 6.044
× 8

12. 3.497
× 3

13. 7.35
× 89

14. 6.103
× 48

15. 3.1463
× 32

★ **16.** 8.49
× 186

★ **17.** 0.1897
× 245

★ **18.** 7.18164
× 136

Multiply. *(Use with page 112.)*

1. 0.4
× 0.6

2. 3.4
× 0.8

3. 0.06
× 0.5

4. 3.86
× 0.4

5. 0.634
× 0.7

6. 0.83
× 0.45

7. 0.635
× 0.49

8. 7.865
× 0.32

9. 1.483
× 0.08

10. 2.59
× 0.006

11. 0.018
× 0.005

12. 6.508
× 0.038

13. 83.6
× 0.9

14. 0.8
× 0.65

15. 7.453
× 0.345

★ **16.** 3.03 × 0.7 × 7.14

★ **17.** 8.1 × 3.07 × 4.214

★ **18.** 0.08 × 3.45 × 1.111

★ **19.** 6.8 × 0.413 × 0.3

★ **20.** 0.61 × 7.3 × 0.043

★ **21.** 0.083 × 1.2 × 6.49

Solve. *(Use with page 114.)*

1. Ellen had 230 stamps. Bob had 4 times as many stamps as Ellen, but he gave 325 of his stamps away. How many stamps does he have now?

2. Mr. Howard bought 144 pens that cost $3 a dozen. How much did he pay in all?

3. Adam bought a shirt for $8.98, a suit for $87.50, and a tie for $4.75. He paid with 6 twenty-dollar bills. How much change did he receive?

4. Heather bought 225 m of movie film. The film is on 15 m rolls which cost $2.75 a roll. How much did Heather spend for the film?

5. Eric earns $2.35 an hour and averages $1.25 an hour in tips. He worked 4 hours on Saturday. How much did he earn?

6. Five pencils and 2 pens cost $3.73. Each pencil cost 7¢. How much does a pen cost?

7. At a sale, old records cost 24¢ each or 5 for $1.00. What is the greatest number of records that Hernandez can buy for $6.75?

8. Maria bought 3 kg of meat for $4.98 a kilogram and 5 kg of chicken for $3.79 a kilogram. How much change did she receive from a $50-bill?

Divide. *(Use with page 116.)*

1. $2\overline{)5.8}$ **2.** $4\overline{)1.2}$ **3.** $8\overline{)1.20}$ **4.** $7\overline{)3.57}$

5. $6\overline{)1.836}$ **6.** $5\overline{)47.855}$ **7.** $15\overline{)4.5}$ **8.** $23\overline{)29.9}$

9. $86\overline{)144.48}$ **10.** $72\overline{)169.20}$ **11.** $32\overline{)2.144}$ **12.** $39\overline{)2.184}$

★ **13.** $75\overline{)158.4225}$ ★ **14.** $253\overline{)89.056}$ ★ **15.** $418\overline{)883.9028}$ ★ **16.** $302\overline{)136.2322}$

Divide. *(Use with page 118.)*

1. $0.3\overline{)1.2}$ **2.** $0.3\overline{)0.06}$ **3.** $0.6\overline{)0.024}$ **4.** $0.4\overline{)8}$

5. $0.07\overline{)0.42}$ **6.** $0.09\overline{)0.036}$ **7.** $0.03\overline{)0.6}$ **8.** $0.008\overline{)0.016}$

9. $0.004\overline{)16}$ **10.** $0.003\overline{)1.503}$ **11.** $2.7\overline{)91.8}$ **12.** $2.8\overline{)169.12}$

★ **13.** $0.0005\overline{)10}$ ★ **14.** $0.0036\overline{)0.234}$ ★ **15.** $0.00025\overline{)0.375}$

Find each quotient to the nearest tenth. *(Use with page 120.)*

1. $6\overline{)1.9}$ **2.** $4\overline{)3.5}$ **3.** $8\overline{)6.2}$ **4.** $0.6\overline{)0.4}$

5. $0.8\overline{)0.5}$ **6.** $0.4\overline{)0.2}$ **7.** $2.3\overline{)0.9}$ **8.** $4.2\overline{)6.35}$

9. $6.8\overline{)5.96}$ **10.** $0.43\overline{)0.382}$ ★ **11.** $0.076\overline{)0.937}$ ★ **12.** $1.37\overline{)0.4364}$

Find each quotient to the nearest hundredth.

13. $0.04\overline{)0.2317}$ **14.** $0.32\overline{)0.397}$ **15.** $0.45\overline{)0.618}$ **16.** $32\overline{)0.469}$

17. $13\overline{)0.674}$ **18.** $0.23\overline{)0.5631}$ **19.** $0.26\overline{)0.7019}$ **20.** $1.9\overline{)0.2136}$

21. $4.2\overline{)3.0169}$ **22.** $3.8\overline{)6.452}$ ★ **23.** $0.026\overline{)0.1387}$ ★ **24.** $0.187\overline{)2.6179}$

Estimate the products. *(Use with page 121.)*

1. $\begin{array}{r} 7.6 \\ \times\ 2.3 \\ \hline \end{array}$ **2.** $\begin{array}{r} 4.019 \\ \times\ 8.35 \\ \hline \end{array}$ **3.** $\begin{array}{r} 0.491 \\ \times\ 6.8 \\ \hline \end{array}$ **4.** $\begin{array}{r} 8.634 \\ \times\ 2.9 \\ \hline \end{array}$ **5.** $\begin{array}{r} 0.876 \\ \times\ 0.735 \\ \hline \end{array}$

6. 2.8×3.7 **7.** 6.41×7.13 **8.** 8.34×0.913 ★ **9.** $6.45 \times 0.719 \times 3.6$

Estimate the quotients.

10. $1.6\overline{)6.116}$ **11.** $4.2\overline{)8.4132}$ **12.** $0.83\overline{)7.364}$ **13.** $6.7\overline{)0.3178}$ **14.** $0.23\overline{)6.251}$

15. $2.7\overline{)9.145}$ **16.** $0.41\overline{)0.863}$ **17.** $5.6\overline{)0.483}$ **18.** $0.39\overline{)8.67}$ ★ **19.** $4.19\overline{)8239.5}$

EXTRA PRACTICE

Solve for x and check. *(Use with page 122.)*

1. $x + 0.5 = 0.9$ **2.** $x - 2.4 = 6.7$ **3.** $0.3x + 0.4 = 3.1$ **4.** $0.5x - 2.4 = 0.6$

5. $4x - 3.1 = 4.1$ **6.** $3x + 5.6 = 9.5$ ★**7.** $3x + 9.5 = 3 + 4x$ ★**8.** $2.4x - 1.8 = 5.4 - 1.2x$

9. $3x = 0.6$ **10.** $0.9x = 7.2$ **11.** $\frac{x}{4} = 0.2$ **12.** $\frac{x}{3.1} = 0.5$

13. $\frac{x}{3} + 0.4 = 0.7$ **14.** $\frac{x}{0.5} - 0.8 = 0.2$ ★**15.** $\frac{x}{3} + 1.5 = 9.5 - x$

Add or subtract. *(Use with page 125.)*

1. 5.73	**2.** 7.437	**3.** $5 + 9.7 + 0.648$
2.89	-3.879	
$+1.57$		**4.** $7 + 0.486 + 9.4$

5. $0.8 - 0.57$

6. $9 - 2.74$

Multiply.

7. 4.56 **8.** 3.9 **9.** 0.37 **10.** 0.04×0.362 **11.** 0.7×0.005
$\underline{\times\ 8}$ $\underline{\times\ 0.4}$ $\underline{\times\ 0.09}$

Divide.

12. $3\overline{)0.9}$ **13.** $9\overline{)1.53}$ **14.** $0.8\overline{)0.32}$ **15.** $7.1\overline{)25.915}$ **16.** $0.25\overline{)6.25}$

Solve.

17. It snowed 8.56 cm on Monday and 12.85 cm on Tuesday. How much did it snow on both days?

18. Bill needs 5 pieces of rope each 24.5 cm long. How much will be left over from a piece 146.9 cm long?

Choose the best estimate. *(Use with page 138.)*

1. Ms. Kowalski drove 397 mi and used 21 gal of gas. How many miles per gallon (mpg) was this?
15 mpg   20 mpg   30 mpg

2. Mr. Stern drove 541.9 km in 10 h. What was his average kilometers per hour (km/h)?
50 km/h   60 km/h   70 km/h

3. Ms. Jackson used 114.9 gal of gasoline at a cost of $1.08 per gallon. What was the cost of the gasoline?
$50 $75 $100

4. Amy drove at an average speed of 78.7 km per hour for 12 hours. How far did she travel in that time?
600 km   700 km   800 km

5. The cost for food for a 5-day trip was $209.85. What was the average cost for food per day?
$4 $40 $400

6. Alex drove 213.9 km one day, 347.3 km the next day, and 186.4 km the third day. How far did he drive in all?
700 km   750 km   800 km

Solve. *(Use with page 152.)*

1. Mr. Garcia bought 1.8 kg of fish and 2.3 kg of meat. How much fish and meat did he buy in all?

2. A thickness of a piece of paper is 0.8 mm. How high is a pile of 2,000 sheets of this paper?

3. A box holds 7.6 kg of raisins. How many of these boxes are needed for 250 kg?

4. Jessie ran a race in 32.8 sec. Adam ran the same race in 28.9 sec. How much faster did Adam run the race?

5. Mrs. Kahn bought 11.5 yd of drapery material and 17.84 yd of dress fabric. How many yards did she buy in all?

6. Chad had a board 30.64 cm long. How many 7.66-cm pieces can be cut from it?

7. A plant was 4.8 ft high. The next year it was 5.03 ft high. How much did it grow?

8. A certain steel bar weighs 0.044 kg per centimeter. How much will a 28-cm length of this steel bar weigh?

Add and simplify. *(Use with page 168.)*

1. $\dfrac{2}{7}$ $+\dfrac{5}{7}$

2. $\dfrac{2}{9}$ $+\dfrac{4}{9}$

3. $\dfrac{3}{5}$ $+\dfrac{1}{5}$

4. $\dfrac{5}{8}$ $+\dfrac{3}{8}$

5. $\dfrac{3}{4}$ $+\dfrac{2}{4}$

6. $\dfrac{5}{8}$ $+\dfrac{3}{8}$

7. $\dfrac{1}{3}$ $+\dfrac{1}{2}$

8. $\dfrac{3}{5}$ $+\dfrac{1}{4}$

9. $\dfrac{1}{8}$ $+\dfrac{1}{5}$

10. $\dfrac{3}{4}$ $+\dfrac{5}{6}$

11. $\dfrac{1}{4}$ $+\dfrac{7}{8}$

12. $\dfrac{4}{5}$ $+\dfrac{3}{10}$

13. $\dfrac{3}{10} + \dfrac{4}{10} + \dfrac{2}{10}$

14. $\dfrac{3}{4} + \dfrac{1}{2} + \dfrac{1}{5}$

15. $\dfrac{1}{2} + \dfrac{3}{4} + \dfrac{5}{8}$

16. $\dfrac{3}{8} + \dfrac{1}{6} + \dfrac{7}{12}$

★17. $\dfrac{2}{3} + \dfrac{3}{4} + \dfrac{5}{6} + \dfrac{1}{2}$

★18. $\dfrac{5}{9} + \dfrac{4}{5} + \dfrac{5}{6}$

★19. $\dfrac{1}{2} + \dfrac{4}{7} + \dfrac{5}{9}$

★20. $\dfrac{2}{3} + \dfrac{1}{5} + \dfrac{5}{6} + \dfrac{7}{9}$

Add and simplify. *(Use with page 170.)*

1. $2\dfrac{2}{7}$ $+3\dfrac{3}{7}$

2. $2\dfrac{1}{6}$ $+5\dfrac{1}{6}$

3. $2\dfrac{1}{3}$ $+5\dfrac{2}{5}$

4. $6\dfrac{2}{3}$ $+5\dfrac{1}{4}$

5. $2\dfrac{1}{5}$ $+4\dfrac{3}{7}$

6. $4\dfrac{3}{8}$ $+3\dfrac{5}{8}$

7. $6\dfrac{4}{9}$ $+7\dfrac{8}{9}$

8. $5\dfrac{7}{12}$ $+7\dfrac{5}{8}$

9. $7\dfrac{2}{5}$ $+4\dfrac{3}{4}$

10. $3\dfrac{2}{3}$ $+4\dfrac{5}{6}$

11. $1\dfrac{1}{2}$ $+3\dfrac{2}{3}$

12. $3\dfrac{7}{8}$ $+4\dfrac{3}{4}$

13. $2\dfrac{3}{7} + 1\dfrac{2}{7} + 4\dfrac{1}{7}$

14. $5\dfrac{4}{9} + 1\dfrac{2}{9} + 1\dfrac{7}{9}$

15. $7\dfrac{2}{3} + 6\dfrac{5}{6} + 3\dfrac{4}{9}$

16. $2\dfrac{3}{5} + 4\dfrac{7}{10} + 2\dfrac{1}{2}$

★17. $1\dfrac{3}{4} + 3\dfrac{2}{3} + 1\dfrac{3}{5} + 2\dfrac{1}{2}$

★18. $4\dfrac{1}{3} + 2\dfrac{5}{14} + 3\dfrac{11}{21}$

★19. $2\dfrac{7}{24} + 1\dfrac{5}{18} + 3\dfrac{7}{9}$

★20. $3\dfrac{5}{12} + 6\dfrac{7}{18} + 1\dfrac{4}{15}$

Subtract and simplify. *(Use with page 173.)*

1. $\dfrac{7}{8}$ $-\dfrac{3}{8}$　　**2.** $\dfrac{5}{6}$ $-\dfrac{1}{6}$　　**3.** $\dfrac{5}{9}$ $-\dfrac{1}{3}$　　**4.** $\dfrac{7}{8}$ $-\dfrac{3}{4}$　　**5.** $\dfrac{1}{2}$ $-\dfrac{1}{3}$　　**6.** $\dfrac{5}{6}$ $-\dfrac{2}{3}$

7. $\dfrac{5}{8}$ $-\dfrac{1}{6}$　　**8.** $\dfrac{7}{10}$ $-\dfrac{2}{5}$　　**9.** $\dfrac{7}{9}$ $-\dfrac{1}{6}$　　**10.** $\dfrac{2}{3}$ $-\dfrac{1}{4}$　　**11.** $\dfrac{9}{10}$ $-\dfrac{1}{3}$　　**12.** $\dfrac{4}{5}$ $-\dfrac{1}{6}$

★**13.** $\dfrac{11}{15} - \dfrac{7}{18}$　　★**14.** $\left(\dfrac{11}{12} - \dfrac{2}{9}\right) - \dfrac{1}{2}$　　★**15.** $\left(\dfrac{3}{8} + \dfrac{2}{3}\right) - \dfrac{2}{3}$　　★**16.** $\left(\dfrac{7}{9} - \dfrac{1}{4}\right) - \left(\dfrac{2}{3} - \dfrac{5}{8}\right)$

Subtract and simplify. *(Use with page 174.)*

1. $6\dfrac{3}{8}$ $-2\dfrac{1}{8}$　　**2.** $7\dfrac{5}{12}$ $-2\dfrac{1}{12}$　　**3.** $9\dfrac{7}{8}$ $-2\dfrac{3}{4}$　　**4.** $8\dfrac{5}{6}$ $-2\dfrac{3}{4}$　　**5.** 5 $-1\dfrac{4}{5}$　　**6.** $5\dfrac{3}{8}$ -2

7. $7\dfrac{3}{8}$ $-2\dfrac{7}{8}$　　**8.** $8\dfrac{1}{12}$ $-3\dfrac{5}{12}$　　**9.** $7\dfrac{1}{2}$ $-3\dfrac{3}{4}$　　**10.** $8\dfrac{1}{3}$ $-2\dfrac{3}{4}$　　**11.** $3\dfrac{1}{4}$ $-1\dfrac{2}{5}$　　**12.** $4\dfrac{1}{9}$ $-3\dfrac{5}{6}$

★**13.** $7\dfrac{5}{24} - 3\dfrac{11}{18}$　　★**14.** $9\dfrac{7}{15} - 5\dfrac{9}{40}$　　★**15.** $7\dfrac{2}{9} - 2\dfrac{4}{5} + 4\dfrac{1}{2}$　　★**16.** $8\dfrac{3}{8} - 4\dfrac{11}{12} - 1\dfrac{5}{6}$

Identify only that information which is needed to solve each problem. Solve.
(Use with page 176.)

1. Francisca lost 3 kg during an illness. She now weighs 47 kg. Her sister weighs 53 kg. How much did Francisca weigh before she was ill?

2. Mr. Bryan spent $350 during his vacation for food and a motel. He stayed at the motel for 2 weeks for $220. How much did he spend for food?

3. A dozen eggs cost $0.98 and a liter of milk costs $0.48. How much do 7 dozen eggs cost?

4. Bob earns $18 a week. He saves $3 a week to buy a guitar. How long will it take him to save $45?

5. A regular box of cereal costs $0.96. The giant size costs $1.29. What is the cost of a dozen boxes of the regular size?

6. A bottle of shampoo costs $2.15. There are 24 bottles in a case. What is the cost of a dozen bottles of shampoo?

7. Cherries sell for $1.59 a kg and plums sell for $0.98 a kg. Andy bought 2.1 kg of cherries and 3.4 kg of plums. How much fruit did he buy in all?

8. Jodi ran 2.5 km on Sunday, 3.4 km on Monday, and 1.9 km on Tuesday. How much farther did she run on Sunday than on Tuesday?

Multiply and simplify. *(Use with page 178.)*

1. $\frac{1}{4} \times \frac{1}{5}$ **2.** $\frac{1}{8} \times \frac{1}{3}$ **3.** $\frac{2}{5} \times \frac{3}{7}$ **4.** $\frac{3}{4} \times \frac{1}{2}$ **5.** $\frac{1}{3} \times \frac{2}{5}$ **6.** $\frac{1}{2} \times 7$

7. $\frac{3}{4} \times 12$ **8.** $5 \times \frac{1}{4}$ **9.** $15 \times \frac{3}{5}$ **10.** $\frac{5}{8} \times \frac{6}{7}$ **11.** $\frac{5}{8} \times \frac{2}{5}$ **12.** $\frac{2}{3} \times \frac{9}{10}$

13. $\frac{1}{3} \times \frac{1}{2} \times \frac{1}{4}$ **14.** $\frac{2}{3} \times \frac{3}{4} \times \frac{1}{8}$ **15.** $\frac{3}{5} \times \frac{1}{4} \times \frac{5}{6}$

★ **16.** $\left(\frac{1}{2} \times \frac{1}{4}\right) + \left(\frac{1}{3} \times \frac{1}{3}\right)$ ★ **17.** $\left(\frac{3}{4} - \frac{1}{5}\right) \times \left(1 - \frac{2}{3}\right)$ ★ **18.** $\left(\frac{7}{8} - \frac{3}{4}\right) \times \left(\frac{4}{9} - \frac{1}{3}\right)$

Multiply and simplify. *(Use with page 180.)*

1. $8 \times 1\frac{1}{2}$ **2.** $12 \times 2\frac{3}{4}$ **3.** $10 \times 6\frac{1}{5}$ **4.** $4 \times 2\frac{1}{3}$ **5.** $2\frac{1}{2} \times 6$

6. $3\frac{1}{3} \times 6$ **7.** $1\frac{3}{5} \times 15$ **8.** $6\frac{3}{4} \times 5$ **9.** $\frac{1}{3} \times 1\frac{1}{2}$ **10.** $\frac{2}{5} \times 3\frac{1}{4}$

11. $\frac{3}{8} \times 2\frac{1}{3}$ **12.** $1\frac{1}{2} \times 2\frac{1}{3}$ **13.** $3\frac{1}{4} \times 2\frac{1}{5}$ **14.** $6\frac{2}{3} \times 1\frac{1}{2}$ **15.** $5\frac{2}{5} \times 2\frac{1}{4}$

16. $3 \times 1\frac{1}{2} \times 2\frac{1}{4}$ **17.** $4\frac{3}{4} \times 8 \times 1\frac{2}{5}$ ★ **18.** $3\frac{1}{6} \times 2\frac{1}{2} \times 1\frac{3}{4} \times 2\frac{2}{3}$

★ **19.** $\left(\frac{3}{5} + \frac{1}{2}\right) \times \left(\frac{3}{8} + \frac{1}{4}\right)$ ★ **20.** $\left(3\frac{1}{2} - 1\frac{7}{8}\right) \times \left(4\frac{3}{4} - 2\frac{5}{6}\right)$ ★ **21.** $\left(\frac{7}{8} + \frac{4}{5}\right) \times \left(4 - 3\frac{1}{5}\right)$

Divide. *(Use with page 182.)*

1. $\frac{1}{5} \div \frac{1}{3}$ **2.** $\frac{3}{4} \div \frac{2}{5}$ **3.** $\frac{5}{6} \div \frac{1}{8}$ **4.** $\frac{7}{8} \div \frac{1}{2}$ **5.** $\frac{3}{4} \div \frac{2}{3}$

6. $\frac{3}{8} \div \frac{9}{10}$ **7.** $1\frac{3}{5} \div \frac{2}{5}$ **8.** $2\frac{5}{6} \div \frac{2}{6}$ **9.** $\frac{5}{8} \div 3$ **10.** $\frac{3}{4} \div 3$

11. $4 \div \frac{3}{4}$ **12.** $3 \div \frac{6}{7}$ **13.** $3\frac{4}{5} \div 2$ **14.** $8 \div 2\frac{5}{6}$ **15.** $\frac{7}{8} \div 1\frac{1}{2}$

16. $\frac{8}{9} \div 2\frac{2}{3}$ **17.** $1\frac{1}{2} \div 1\frac{3}{4}$ **18.** $2\frac{1}{3} \div 3\frac{1}{2}$ **19.** $6\frac{2}{3} \div 3\frac{1}{6}$ ★ **20.** $\left(4\frac{5}{6} \div 1\frac{7}{8}\right) \div \frac{4}{5}$

★ **21.** $\left(\frac{4}{5} \times \frac{7}{8}\right) \div \frac{5}{6}$ ★ **22.** $\left(4\frac{1}{4} \div 3\frac{2}{3}\right) \div \left(5\frac{2}{7} \div \frac{4}{7}\right)$ ★ **23.** $\left(\frac{7}{9} + \frac{5}{6}\right) \div \left(\frac{2}{3} - \frac{3}{8}\right)$

Write decimals. *(Use with page 184.)*

1. $\frac{9}{2}$ **2.** $\frac{9}{8}$ **3.** $\frac{3}{10}$ **4.** $\frac{6}{25}$ **5.** $\frac{17}{50}$ **6.** $\frac{11}{40}$ **7.** $\frac{23}{125}$

8. $3\frac{3}{5}$ **9.** $5\frac{7}{50}$ **10.** $4\frac{9}{40}$ **11.** $6\frac{7}{10}$ ★ **12.** $\frac{3}{80}$ ★ **13.** $4\frac{9}{80}$ ★ **14.** $\frac{1}{160}$

Add or subtract. Simplify. *(Use with page 187.)*

1. $\frac{3}{7}$
$+\frac{2}{7}$

2. $\frac{3}{4}$
$+\frac{1}{3}$

3. $3\frac{4}{5}$
$+4\frac{5}{6}$

4. $\frac{7}{8}$
$-\frac{3}{8}$

5. $5\frac{7}{9}$
$-3\frac{2}{3}$

6. $4\frac{2}{3}$
$-1\frac{4}{5}$

7. $\frac{3}{4} + \frac{1}{8} + \frac{5}{6}$

8. $2\frac{2}{3} + 4\frac{1}{4} + 3\frac{3}{5}$

Multiply or divide. Simplify.

9. $\frac{3}{5} \times \frac{2}{7}$

10. $\frac{2}{3} \times 6$

11. $\frac{4}{5} \div \frac{4}{9}$

12. $6 \div 3\frac{1}{6}$

13. $4\frac{3}{4} \div 2\frac{3}{8}$

Solve.

14. Sara bought 3 tennis balls for $2.98 and 3 baseballs for $11.98. How much did she spend in all?

15. Sal had $\frac{3}{4}$ of a pizza. He ate $\frac{1}{2}$ of it. How much of the pizza did he eat?

Solve, if possible. *(Use with page 206.)*

1. The perimeter of a rectangle is 60 mm. What is the length?

2. The perimeter of a square is 40 mm. What is the length of a side?

3. In right triangle *ABC*, *C* is a right angle. $m \angle A = 34°$. What is $m \angle B$?

4. The perimeter of a triangle is 36 cm. How long is each side?

5. The width of a garden is 6 ft. At $3.59 a foot, what is the cost of fencing the garden?

6. A yard is in the shape of a scalene triangle. One side is 86 ft. What is the perimeter of the yard?

Solve. *(Use with page 218.)*

1. Ms. Best spent $167.98 for a car radio, $37.75 for carpet protectors, and $486.95 for air-conditioning. How much did she spend in all?

2. The average gas mileage for a certain car is 23 miles per gallon. The car has an 18-gal gas tank. How far can it go on a tank of gas?

3. Mr. Joseph spent $\frac{1}{2}$ h with one car dealer and $1\frac{1}{4}$ h with a second dealer. How much more time did he spend with the second dealer?

4. One dealer charged $7,186.78 for a certain car. Another dealer charged $6,794.89 for the same car. How much more did the first dealer charge?

5. Mrs. Genzoff will pay $8,256 in 12 equal payments for a new car. How much will each payment be?

6. Bill's new car had 4.8 mi on the odometer when he bought it. A week later the odometer read 142.6 mi. How far had he gone that week?

Compute. *(Use with page 226.)*

1. $7 \times (5 + 8)$　　　**2.** $9 \times (7 - 3)$　　　**3.** $(8 + 16) - (2 \times 6)$　　　**4.** $8 \times 6 + 5$

5. $25 - 3 \times 6$　　　**6.** $6 \times (3 \times 4 - 8)$　　★**7.** $\left(\frac{8+7}{3}\right) \times \left(\frac{9+9}{9}\right)$　　★**8.** $\left[(6 + 8) - \left(\frac{7-3}{4}\right)\right]$

Evaluate.

9. $2a + 7$ if $a = 3$

10. $6x - 7$ if $x = 9$

11. $5a + b$ if $a = 3$, $b = 6$

12. $\frac{a+7}{b}$ if $a = 3$, $b = 2$

13. $\frac{n}{2} \cdot (a + b)$ if $n = 8$, $a = 1$, $b = 13$

14. $\frac{n}{2} \cdot (a + l)$ if $n = 100$, $a = 1$, $l = 20$

15. $\frac{a(2+b)}{c} \cdot ab$ if $a = 2$, $b = 3$, $c = 5$

★**16.** $abc \cdot \frac{(2a + 3b - c)}{d}$ if $a = 3$, $b = 4$, $c = 6$, $d = 2$

Write equations. Solve. *(Use with page 230.)*

1. Tickets for a concert are $3 each. Jeff's club paid $69 for tickets. How many members bought tickets?

2. Sam bought twice as many stamps as Charlie. Sam bought 54 stamps. How many stamps did Charlie buy?

3. Maria bought 4 greeting cards, each the same price, and a book for a total of $6.58. The book cost $3.98. What was the cost of each card?

4. Mrs. Hill bought a chair and a desk. The desk was $34.98 more than the chair. The chair was $148.79. How much was the desk?

5. Danny's mother is 38. She is 2 years more than 4 times Danny's age. How old is Danny?

6. A string was 45 cm long. It was cut into 6 pieces of the same length and a piece 3 cm long. How long was each of the 6 pieces?

Solve. Use the distance formula. *(Use with page 236.)*

1. Mrs. Alms averaged 80 km/h during a 4-hour trip. How far did she drive?

2. Amanda drove 135 mi in 3 hours. What was her average speed?

3. Ms. Ruff drove 840 km at an average speed of 70 km/h. How long did it take her?

4. A plane flew 3,840 km at an average speed of 640 km/h. How long was the trip?

5. Juan rode a bicycle $18\frac{3}{4}$ mi. It took him $2\frac{1}{2}$ hours. What was his average speed?

6. A plane flew $2\frac{1}{2}$ hours at an average speed of 450 mph. How far did the plane fly?

Find the ratios. *(Use with page 239.)*

Jim planted 12 tulip bulbs. Seven of them sprouted.

1. What is the ratio of the number of bulbs planted to the number which sprouted?

2. What is the ratio of the number of bulbs which sprouted to the number of bulbs planted?

Eight girls and 10 boys joined a club.

3. What is the ratio of the number of girls to the number of boys.

4. What is the ratio of the number of boys to the total number in the club?

Which ratios are equal to 6 is to 18?

5. 1 is to 3   **6.** 5 is to 15   **7.** 8 is to 16

Solve the proportions. *(Use with page 240.)*

1. $1:3 = 3:x$ **2.** $2:3 = x:9$ **3.** $2:x = 6:12$ **4.** $x:3 = 8:12$

5. $5:3 = x:9$ **6.** $4:1 = x:5$ **7.** $3:x = 4:8$ **8.** $x:6 = 4:3$

9. $\frac{9}{12} = \frac{6}{x}$ **10.** $\frac{2}{3} = \frac{8}{x}$ **11.** $\frac{8}{x} = \frac{12}{3}$ **12.** $\frac{2}{9} = \frac{6}{x}$ **13.** $\frac{x}{3} = \frac{12}{9}$

14. $\frac{x}{12} = \frac{10}{5}$ **15.** $\frac{x}{12} = \frac{8}{6}$ **16.** $\frac{10}{x} = \frac{5}{4}$ **17.** $\frac{9}{54} = \frac{2}{x}$ **18.** $\frac{14}{16} = \frac{x}{24}$

19. $\frac{x}{18} = \frac{2}{3}$ **20.** $\frac{5}{9} = \frac{x}{27}$ **21.** $\frac{6}{x} = \frac{9}{15}$ **22.** $\frac{4}{7} = \frac{20}{x}$ **23.** $\frac{12}{15} = \frac{x}{5}$

★ **24.** $\frac{4}{3} = \frac{10}{x}$ ★ **25.** $\frac{x}{9} = \frac{4}{5}$ ★ **26.** $\frac{x}{4} = \frac{16}{x}$ ★ **27.** $\frac{x}{0.6} = \frac{1.5}{0.2}$ ★ **28.** $\frac{1.8}{0.2} = \frac{x}{2.3}$

Solve. Use the lever formula. *(Use with page 242.)*

1. A 9-g mass is 8 cm from the fulcrum. What mass 12 cm from the fulcrum will balance it?

2. A 33-lb weight is 3 ft from the fulcrum. What weight 9 feet from the fulcrum will balance it?

3. A 6-g mass is 28 cm from the fulcrum. How far from the fulcrum is a 21-g mass which balances the 6-gram mass?

4. Val has a mass of 36 kg. She is sitting 3 m from the fulcrum. Mike has a mass of 54 kg. The seesaw is balanced. How far is Mike from the fulcrum?

5. A 24-lb weight 14 inches from the fulcrum balances a 16-lb weight. How far from the fulcrum is the 16-lb weight?

6. Josh has a mass of 42 kg. He is sitting 2 m from the fulcrum. He is balanced by his sister who is 3 m from the fulcrum. What is the mass of his sister?

Solve. *(Use with page 244.)*

1. Shirts are selling at 3 for $42. What is the cost of 1 dozen shirts?

2. Ann saves $4 of each $15 she earns. If she earned $75, how much did she save?

3. Mrs. Gleason earned $80 by selling 2 television sets. How many sets would she have to sell to earn $200?

4. The ratio of the width of a rectangle to its length is 3 to 5. The length is 30 in. What is the width?

5. Tires are selling at 2 for $75. How much will 8 cost?

6. Mr. Hu's gas bill was $45 for 2 months. What will his gas bill be for 1 year?

7. At a party there were 4 boys to every 5 girls. There were 320 boys. How many girls were there?

8. At a school there were 3 bicycles for every 5 children. There were 45 children. How many bicycles were there?

Write the simplest fractions, whole numbers, or mixed numbers. *(Use with page 250.)*

1. 60%   2. 28%   3. 65%   4. 17%   5. 800%   6. 500%

7. $66\frac{2}{3}\%$ 8. $37\frac{1}{2}\%$ 9. $\frac{2}{5}\%$ 10. 350%   11. $\frac{1}{4}\%$ 12. 175%

13. $2\frac{1}{5}\%$ 14. $3\frac{1}{2}\%$ ★ 15. $\frac{3}{8}\%$ ★ 16. $2\frac{5}{8}\%$ ★ 17. 1,040%   ★ 18. $1,250\frac{3}{5}\%$

Write percents.

19. $\frac{3}{4}$ 20. $\frac{4}{5}$ 21. $\frac{7}{25}$ 22. $\frac{3}{10}$ 23. $\frac{2}{3}$ 24. $\frac{7}{200}$

25. $\frac{300}{100}$ 26. $\frac{13}{50}$ 27. $\frac{3}{2}$ 28. $\frac{8}{5}$ 29. $\frac{17}{20}$ 30. $\frac{29}{400}$

31. $\frac{5}{12}$ 32. $\frac{7}{8}$ ★ 33. $\frac{23}{1,000}$ ★ 34. $\frac{15}{10,000}$ ★ 35. $\frac{382}{600}$ ★ 36. $\frac{240}{400}$

Change to decimals. *(Use with page 252.)*

1. 4%   2. 8%   3. 15%   4. 36%   5. 150%   6. 180%

7. 200%   8. 340%   9. 0.4%   10. 2.4%   11. 0.08%   12. 6.14%

13. $\frac{7}{10}\%$ 14. $3\frac{1}{4}\%$ ★ 15. $3\frac{7}{8}\%$ ★ 16. 1,350%   ★ 17. $9\frac{5}{6}\%$ ★ 18. $1,000\frac{1}{8}\%$

Change to percents.

19. 0.06   20. 0.53   21. 0.2   22. 4   23. 3.7   24. 0.031

25. 0.004   26. 0.0086   ★ 27. $0.05\frac{1}{2}$ ★ 28. $0.08\frac{3}{4}$ ★ 29. $0.37\frac{1}{2}$ ★ 30. $0.33\frac{1}{3}$

Compute. *(Use with page 254.)*

1. 8% of 20
2. 36% of 100
3. 100% of 40

4. 150% of 90
5. 300% of 500
6. 0.7% of 240

7. $2\frac{1}{2}$% of 60
8. $33\frac{1}{3}$% of 90
9. $\frac{1}{4}$% of 1,200

10. 1.8% of $4,000
11. 0.05% of 200
12. 0.08% of 350

★ 13. 40% of 0.8
★ 14. $66\frac{2}{3}$% of $5\frac{5}{6}$
★ 15. 300% of $\frac{1}{2}$

Compute. *(Use with page 256.)*

1. What percent of 18 is 12?
2. 6 is what percent of 12?

3. What percent of 10 is 6?
4. 7 is what percent of 4?

5. What percent of 6 is 10?
6. 8 is what percent of 8?

7. What percent of 3 is 4?
8. 0.3 is what percent of 10?

9. What percent of 100 is 0.5?
10. 0.08 is what percent of 24?

★ 11. What percent of 20 is 200?
★ 12. 2,500 is what percent of 200?

★ 13. What percent of $\frac{1}{2}$ is $\frac{1}{6}$?
★ 14. $\frac{1}{5}$ is what percent of $\frac{1}{10}$?

Compute. *(Use with page 258.)*

1. 75% of what number is 150?
2. 12 is 6% of what number?

3. $66\frac{2}{3}$% of what number is 64?
4. 56 is $87\frac{1}{2}$% of what number?

5. 35% of what number is 24.5?
6. 18 is 120% of what number?

7. 300% of what number is 36?
8. 64 is 3.2% of what number?

9. $1\frac{1}{2}$% of what number is 72?
10. 8 is 0.1% of what number?

★ 11. 25% of what number is $\frac{1}{16}$?
★ 12. 2% of what number is $1\frac{3}{5}$?

★ 13. $66\frac{2}{3}$% of what number is 1?
★ 14. $16\frac{1}{2}$% of what number is $\frac{5}{24}$?

Write mini-problems. Solve. *(Use with page 262.)*

1. Betsy worked 6 hours on Saturday and 8 hours on Sunday. She earns $2.90 an hour. How much did she earn in all?

2. Ms. Frankel bought 3 dozen roses at $2.95 a dozen. How much change did she receive from a $10-bill?

3. Joe bought 4 shirts for $63.80. If he exchanges one for a tie costing $8.79, how much change will he receive?

4. Kim has 18 records more than Beth, who has 78 records. If Kim stacks them in piles of eight, how many piles will she have?

5. A theater has 18 rows with 32 seats in a row. For one movie all the seats were taken except the first two rows. How many people saw the movie?

6. Jim typed 34 pages of 70 pages on Sunday and finished the rest in 4 hours on Monday. How many pages did he type per hour on Monday?

Find the discount and the sale price. *(Use with page 264.)*

The rate of discount is 25%. The regular prices are given.

1. Radio: $40   **2.** Television: $150   **3.** Stereo: $200

Find the sale prices. The regular prices are given.

4. Hat marked $15 at a 10%-off sale

5. Gloves marked $8 at a 30%-off sale

6. Shoes marked $32 at a 25%-off sale

7. Tie marked $12 at a 20%-off sale

Find the rate of discount as a percent.

8. Bat: was $9, now $6   **9.** Ball: was $4, now $3   **10.** Mitt: was $10, now $8

Solve.

11. A tire selling for $45 was on sale at 20% off. What was the sale price?

12. An $8 book was sold for $5.60 during a sale. What was the rate of discount?

Find the interest. *(Use with page 266.)*

1. $120 at 8% for 2 years

2. $5,000 at 6.5% for 3 years

3. $600 at 6% for 4 months

4. $600 at $8\frac{1}{2}$% for 6 months

Find the total amount to be paid back.

5. $6,000 at 9% for 5 years

6. $2,000 at 9.5% for 4 years

7. $2,000 at 10% for 10 years

8. $400 at 8% for 6 months

9. $200 at $5\frac{1}{2}$% for 3 months

10. $5,000 at 11.25% for 3 years

Find the percent increase. *(Use with page 270.)*

1. Last year: 40 games won
This year: 52 games won

2. Last week: 16 dolls sold
This week: 28 dolls sold

3. 1st math test: 75
2nd math test: 100

4. Last year: gloves cost $8
This year: gloves cost $10

Find the percent decrease.

5. Sold 250 records last week
Sold 125 records this week

6. Last month's telephone bill: $20
This month's telephone bill: $15

7. Population 10 years ago: 2,000
Population now: 800

8. 600 customers last year
450 customers this year

Write percents. *(Use with page 273.)*

1. $\frac{8}{100}$ **2.** 0.06   **3.** 0.58   **4.** $\frac{3}{4}$ **5.** $2\frac{1}{2}$ **6.** 0.089   **7.** 2.75

Write simplest fractions, whole numbers, or mixed numbers.

8. 2%   **9.** 35%   **10.** 200%   **11.** 325%   **12.** $16\frac{1}{2}$%   **13.** $\frac{3}{5}$%

Write decimals.

14. 5%   **15.** 275%   **16.** 8.6%   **17.** $\frac{3}{4}$%   **18.** 0.8%   **19.** $\frac{6}{10}$%

Compute.

20. 8% of 70   **21.** $\frac{1}{4}$% of 20   **22.** 0.6% of 200

23. What percent of 20 is 5?   **24.** 0.4 is what percent of 80?

25. 50% of what number is 16?   **26.** 9% of what number is 40.5?

Solve.

27. A radio selling for $85 was on sale at 20% off. What was the sale price?

28. Al borrowed $1,200 at 8% interest for 4 years. How much interest did he pay?

29. A $150 stereo was sold for $105 during a sale. What was the rate of discount?

30. Jill borrowed $500 at 7.5% for 6 months. How much did she pay back at the end of 6 months?

31. Last year a hat cost $12. This year the hat cost $16. What is the percent increase?

32. Last week a store sold 200 books. This week the store sold 150 books. What is the percent decrease?

Add. *(Use with page 280.)*

1. $^+4 + {}^+7$ **2.** $^+8 + {}^+6$ **3.** $^+15 + {}^+16$ **4.** $^+56 + {}^+12$

5. $^-3 + {}^-7$ **6.** $^-8 + {}^-7$ **7.** $^-31 + {}^-14$ **8.** $^-23 + {}^-47$

9. $\begin{array}{r} ^+9 \\ + \ ^+3 \\ \hline \end{array}$ **10.** $\begin{array}{r} ^+17 \\ + \ ^+19 \\ \hline \end{array}$ **11.** $\begin{array}{r} ^-6 \\ + \ ^-7 \\ \hline \end{array}$ **12.** $\begin{array}{r} ^-81 \\ + \ ^-10 \\ \hline \end{array}$

13. $^+4 + {}^+8 + {}^+3$ **14.** $^+35 + {}^+18 + {}^+11$ **15.** $^-5 + {}^-1 + {}^-3$

16. $^-13 + {}^-6 + {}^-19$ ★ **17.** $^+4 + {}^+9 + {}^+13 + {}^+8$ ★ **18.** $^+35 + {}^+16 + {}^+7 + {}^+1$

★ **19.** $^-8 + {}^-6 + {}^-2 + {}^-1$ ★ **20.** $^-11 + {}^-24 + {}^-6 + {}^-3$ ★ **21.** $^-15 + {}^-31 + {}^-64 + {}^-32$

Add. *(Use with page 282.)*

1. $^+3 + {}^-1$ **2.** $^+6 + {}^-5$ **3.** $^+12 + {}^-18$ **4.** $^+63 + {}^-49$

5. $^-3 + {}^+5$ **6.** $^-34 + {}^+25$ **7.** $^-23 + {}^+47$ **8.** $^-46 + {}^+31$

9. $\begin{array}{r} ^+8 \\ + \ ^-5 \\ \hline \end{array}$ **10.** $\begin{array}{r} ^-9 \\ + \ ^+6 \\ \hline \end{array}$ **11.** $\begin{array}{r} ^-5 \\ + \ ^+7 \\ \hline \end{array}$ **12.** $\begin{array}{r} ^+6 \\ + \ ^-4 \\ \hline \end{array}$ **13.** $\begin{array}{r} ^+10 \\ + \ ^-8 \\ \hline \end{array}$

14. $\begin{array}{r} ^-15 \\ + \ ^+38 \\ \hline \end{array}$ **15.** $\begin{array}{r} ^-35 \\ + \ ^+14 \\ \hline \end{array}$ **16.** $\begin{array}{r} ^+46 \\ + \ ^-29 \\ \hline \end{array}$ **17.** $\begin{array}{r} ^+83 \\ + \ ^-96 \\ \hline \end{array}$ **18.** $\begin{array}{r} ^-57 \\ + \ ^+48 \\ \hline \end{array}$

19. $^+3 + {}^-4 + {}^+8$ **20.** $^-8 + {}^+4 + {}^+5$ **21.** $^+6 + {}^+3 + {}^-9$

22. $^-6 + {}^+3 + {}^-5$ ★ **23.** $^-6 + {}^-8 + {}^+4 + {}^-2$ ★ **24.** $^+3 + {}^-4 + {}^-3 + {}^+6$

★ **25.** $^+8 + {}^-6 + {}^+1 + {}^+3$ ★ **26.** $^+4 + {}^-3 + {}^+1 + {}^+2$ ★ **27.** $^-3 + {}^+4 + {}^+5 + {}^-2$

Subtract. *(Use with page 284.)*

1. $^+9 - {}^+4$ **2.** $^+4 - {}^+6$ **3.** $^+23 - {}^+47$ **4.** $^+53 - {}^+38$

5. $^+3 - {}^-1$ **6.** $^+6 - {}^-9$ **7.** $^+16 - {}^-12$ **8.** $^+13 - {}^-15$

9. $^-3 - {}^+7$ **10.** $^-12 - {}^+3$ **11.** $^-16 - {}^+4$ **12.** $^-24 - {}^+36$

13. $^-6 - {}^-9$ **14.** $^-10 - {}^-8$ **15.** $^-25 - {}^-14$ **16.** $^-47 - {}^-56$

★ Simplify.

17. $^+8 + {}^-3 + {}^+6 - {}^-9$ **18.** $^+3 - {}^+8 - {}^-4 - {}^-9$ **19.** $^-6 - {}^-3 + {}^-9 + {}^+4$

20. $^-1 + {}^+2 - {}^-3 + {}^+4$ **21.** $^-9 + {}^+7 + {}^+3 - {}^-6$ **22.** $^+8 + {}^+4 - {}^-2 - {}^-5$

Multiply. *(Use with page 290.)*

1. $^+3 \cdot {}^+6$ **2.** $^+9 \cdot {}^+8$ **3.** $^+5 \cdot {}^+6$ **4.** $^-3 \cdot {}^-4$ **5.** $^-6 \cdot {}^-2$

6. $^-7 \cdot {}^-8$ **7.** $^+6 \cdot {}^-3$ **8.** $^+2 \cdot {}^-4$ **9.** $^+8 \cdot {}^-7$ **10.** $^-9 \cdot {}^+4$

11. $^-8 \cdot {}^+3$ **12.** $^-6 \cdot {}^+9$ **13.** $0 \cdot {}^+4$ **14.** $^-9 \cdot 0$ **15.** $^+1 \cdot {}^-6$

16. $^+9 \cdot {}^-1$ **17.** $^+9 \cdot {}^-13$ **18.** $^-11 \cdot {}^-6$ **19.** $^-13 \cdot {}^+18$ **20.** $^+23 \cdot {}^+17$

★ **21.** $^+5 \cdot {}^-6 \cdot {}^+4$ ★ **22.** $^-4 \cdot {}^+6 \cdot {}^-5$ ★ **23.** $^-3 \cdot {}^-6 \cdot {}^-4$

★ **24.** $^-8 \cdot {}^+5 \cdot {}^+9$ ★ **25.** $^+6 \cdot {}^+3 \cdot {}^-4$ ★ **26.** $^-7 \cdot {}^+6 \cdot {}^-3$

Divide. *(Use with page 294.)*

1. $^+10 \div {}^+2$ **2.** $^-16 \div {}^-8$ **3.** $^-64 \div {}^-8$ **4.** $^+48 \div {}^-8$

5. $^+32 \div {}^-4$ **6.** $^-45 \div {}^+9$ **7.** $^-72 \div {}^+9$ **8.** $0 \div {}^-6$

9. $\frac{^+36}{^+6}$ **10.** $\frac{^-40}{^-5}$ **11.** $\frac{^-16}{^-4}$ **12.** $\frac{^+56}{^-8}$ **13.** $\frac{^+35}{^-7}$ **14.** $\frac{^-64}{^+8}$

15. $\frac{^-54}{^+9}$ **16.** $\frac{0}{^+5}$ **17.** $\frac{^+15}{^-15}$ **18.** $\frac{^-32}{^-1}$ **19.** $\frac{^+70}{^-10}$ **20.** $\frac{^-42}{^+21}$

★ **21.** $(^+81 \div {}^-9) \div {}^+3$ ★ **22.** $^+92 \div (^-16 \div {}^+4)$

★ **23.** $(^-64 \div {}^+8) \div (^-12 \div {}^-6)$ ★ **24.** $(^+28 \div {}^-7) \div (^-24 \div {}^-12)$

Solve and check. *(Use with page 296.)*

1. $x + {}^+6 = {}^+8$ **2.** $x + {}^-6 = {}^-4$ **3.** $x + {}^-3 = {}^+9$ **4.** $x - {}^+3 = {}^+5$

5. $x - {}^-5 = {}^-8$ **6.** $x - {}^-9 = {}^-4$ **7.** $^+3x + {}^+5 = {}^+14$ **8.** $^+3x + {}^-7 = {}^-19$

9. $^+6x - {}^+4 = {}^+8$ **10.** $\frac{x}{^+2} + {}^+2 = {}^+5$ **11.** $\frac{x}{^-3} - {}^+4 = {}^-3$ **12.** $\frac{x}{^+9} + {}^-3 = {}^-2$

13. $^+4x = {}^+28$ **14.** $^+9x = {}^-72$ **15.** $^-3x = {}^-27$ **16.** $^-3x = {}^+12$ **17.** $^+5x = {}^-25$

18. $\frac{x}{^+4} = {}^-3$ **19.** $\frac{x}{^-3} = {}^+7$ **20.** $\frac{x}{^+6} = {}^+6$ **21.** $\frac{x}{^-4} = {}^-12$ **22.** $\frac{x}{^-1} = {}^+1$

★ Solve. Replacements for x: $^-8, {}^-7, \ldots, {}^+7, {}^+8$

23. $x + {}^+2 > {}^+5$ **24.** $x + {}^-5 > {}^-2$ **25.** $^+2x > {}^+8$

26. $^-5x > {}^-26$ **27.** $^+2x + {}^+3 > {}^+2$ **28.** $^+7x + {}^-3 > {}^+2$

Add. *(Use with page 301.)*

1. $^+3 + {}^+9$ **2.** $^-8 + {}^-6$ **3.** $^-9 + {}^+4$ **4.** $^-6 + {}^+8$

Subtract.

5. $^+6 - {}^-3$ **6.** $^+7 - {}^+5$ **7.** $^-7 - {}^-3$ **8.** $^-4 - {}^+7$

Multiply.

9. $^+8 \cdot {}^+3$ **10.** $^-9 \cdot {}^-5$ **11.** $^+6 \cdot {}^-5$ **12.** $^-3 \cdot {}^+6$

Divide.

13. $^+32 \div {}^-4$ **14.** $^+64 \div {}^+8$ **15.** $\dfrac{^-63}{^-7}$ **16.** $\dfrac{^-20}{^+5}$

Solve.

17. $x - {}^+3 = {}^-7$ **18.** $x + {}^-8 = {}^-4$ **19.** $\dfrac{x}{^-2} = {}^+9$ **20.** $^+3x + {}^+5 = {}^+14$

Write in the form $\frac{a}{b}$. *(Use with page 306.)*

1. 7   **2.** $^-6$ **3.** 0   **4.** 12   **5.** 0.7   **6.** $^-0.3$

7. $^-6.93$ **8.** 4.013   **9.** $^-4.9$ **10.** $^-3.0001$ ★ **11.** $\dfrac{\frac{1}{3}}{9}$ ★ **12.** $\dfrac{\frac{1}{4}}{20}$

Compare. Replace ☰ with >, <, or =

13. $^-7 \equiv 2\frac{3}{4}$ **14.** $5 \equiv {}^-2.7$ **15.** $3 \equiv {}^-4.7$ **16.** $^-1.8 \equiv 2.4$ **17.** $\frac{1}{6} \equiv 0$

18. $\frac{^-1}{8} \equiv 0$ **19.** $\frac{^-3}{4} \equiv \frac{^-6}{8}$ **20.** $\frac{3}{4} \equiv \frac{^-1}{2}$ **21.** $^-1 \equiv \frac{^-3}{4}$ **22.** $^-1\frac{2}{5} \equiv {}^-1\frac{3}{4}$

23. $^-1.4 \equiv {}^-8.0$ **24.** $^-3\frac{1}{2} \equiv {}^-1\frac{1}{4}$ ★ **25.** $\dfrac{\frac{1}{2}}{5} \equiv \dfrac{\frac{2}{3}}{8}$ ★ **26.** $\dfrac{\frac{3}{4}}{5} \equiv \dfrac{\frac{2}{5}}{7}$ ★ **27.** $\dfrac{\frac{5}{6}}{3} \equiv \dfrac{\frac{3}{4}}{2}$

Write terminating decimals. *(Use with page 306.)*

1. $\frac{4}{5}$ **2.** $\frac{5}{8}$ **3.** $\frac{^-9}{10}$ **4.** $\frac{^-13}{25}$ **5.** $\frac{9}{40}$ **6.** $\frac{^-7}{50}$

7. $\frac{9}{4}$ **8.** $\frac{^-15}{8}$ **9.** $2\frac{3}{20}$ **10.** $^-4\frac{7}{10}$ ★ **11.** $\frac{^-3}{125}$ ★ **12.** $\frac{119}{200}$

Write repeating decimals.

13. $\frac{7}{9}$ **14.** $\frac{^-2}{7}$ **15.** $\frac{5}{11}$ **16.** $\frac{^-8}{9}$ **17.** $^-2\frac{5}{6}$ **18.** $4\frac{4}{11}$

19. $\frac{4}{3}$ **20.** $\frac{11}{6}$ **21.** $\frac{^-9}{7}$ ★ **22.** $\frac{7}{33}$ ★ **23.** $^-1\frac{1}{99}$ ★ **24.** $^-2\frac{13}{45}$

Add and simplify. *(Use with page 308.)*

1. $\frac{1}{5} + \frac{2}{5}$ 　　　　 **2.** $\frac{^-7}{10} + \frac{^-3}{10}$ 　　　　 **3.** $\frac{^-5}{6} + \frac{^-3}{6}$ 　　　　 **4.** $\frac{3}{4} + \frac{^-3}{4}$

5. $\frac{5}{6} + \frac{3}{8}$ 　　　　 **6.** $\frac{^-3}{5} + \frac{^-3}{4}$ 　　　　 **7.** $\frac{^-2}{3} + \frac{1}{2}$ 　　　　 **8.** $\frac{^-3}{4} + \frac{5}{8}$

9. $\frac{^-7}{9} + \frac{1}{9}$ 　　　　 **10.** $\frac{5}{9} + \frac{^-2}{3}$ 　　　　 **11.** $\frac{^-3}{10} + \frac{^-7}{10}$ 　　　　 **12.** $\frac{^-1}{4} + \frac{7}{8}$

13. $0.8 + {}^-0.4$ 　　　　 **14.** $^-0.09 + {}^-0.06$ 　　　　 **15.** $7.3 + {}^-8.1$ 　　　　 **16.** $^-3.7 + 6.2$

★ **17.** $\frac{^-2}{3} + \frac{4}{5} + \frac{^-6}{7}$ 　　 ★ **18.** $\frac{3}{4} + \frac{^-1}{2} + \frac{3}{5}$ 　　 ★ **19.** $\frac{^-7}{9} + \frac{^-5}{6} + \frac{^-1}{3}$ 　　 ★ **20.** $\frac{1}{4} + \frac{^-3}{7} + \frac{5}{14}$

Subtract and simplify. *(Use with page 310.)*

1. $\frac{3}{4} - \frac{1}{4}$ 　　　　 **2.** $\frac{5}{6} - \frac{^-1}{6}$ 　　　　 **3.** $\frac{^-5}{9} - \frac{1}{9}$ 　　　　 **4.** $\frac{^-5}{7} - \frac{^-2}{7}$

5. $\frac{^-3}{5} - \frac{1}{2}$ 　　　　 **6.** $\frac{1}{2} - \frac{^-3}{4}$ 　　　　 **7.** $\frac{5}{6} - \frac{3}{8}$ 　　　　 **8.** $\frac{^-7}{12} - \frac{^-3}{4}$

9. $\frac{7}{8} - \frac{^-3}{8}$ 　　　　 **10.** $\frac{^-7}{12} - \frac{5}{12}$ 　　　　 **11.** $\frac{^-7}{8} - 0$ 　　　　 **12.** $0 - \frac{^-3}{5}$

13. $0.9 - {}^-0.5$ 　　　　 **14.** $^-1.07 - {}^-1.06$ 　　　　 **15.** $3.4 - {}^-1.7$ 　　　　 **16.** $^-4.6 - 8.0$

★ **17.** $\left(\frac{^-2}{5} - \frac{^-4}{5}\right) - \frac{1}{5}$ 　 ★ **18.** $\left(\frac{^-3}{4} - \frac{1}{2}\right) - \frac{^-5}{8}$ 　 ★ **19.** $\frac{2}{3} - \left(\frac{^-7}{8} - \frac{3}{4}\right)$ 　 ★ **20.** $\frac{^-3}{5} - \left(\frac{7}{9} - \frac{^-2}{3}\right)$

Multiply and simplify. *(Use with page 312.)*

1. $\frac{2}{3} \cdot \frac{3}{4}$ 　　　　 **2.** $\frac{^-5}{6} \cdot \frac{^-1}{2}$ 　　　　 **3.** $\frac{^-3}{4} \cdot \frac{2}{7}$ 　　　　 **4.** $\frac{5}{8} \cdot \frac{^-1}{2}$

5. $1\frac{3}{8} \cdot \frac{^-8}{3}$ 　　　　 **6.** $^-1\frac{1}{2} \cdot \frac{^-3}{8}$ 　　　　 **7.** $^-7.8 \cdot 0.4$ 　　　　 **8.** $0.6 \cdot {}^-0.1$

9. $^-1.6 \cdot {}^-2.3$ 　　 ★ **10.** $\frac{^-3}{4} \cdot \frac{4}{5} \cdot \frac{^-2}{3}$ 　　 ★ **11.** $8 \cdot \frac{^-5}{6} \cdot \frac{4}{5}$ 　　 ★ **12.** $^-0.6 \cdot {}^-0.4 \cdot 0.3$

Divide and simplify.

13. $0.16 \div {}^-0.4$ 　　　　 **14.** $^-0.32 \div {}^-0.4$ 　　　 **15.** $\frac{5}{6} \div \frac{1}{2}$ 　　　　 **16.** $\frac{^-3}{4} \div \frac{2}{3}$

17. $\frac{1}{3} \div \frac{^-2}{3}$ 　　　　 **18.** $\frac{^-7}{8} \div \frac{^-3}{4}$ 　　　　 **19.** $8 \div \frac{^-4}{7}$ 　　　　 **20.** $^-1\frac{3}{4} \div \frac{^-2}{5}$

21. $^-1\frac{1}{2} \div 1\frac{5}{6}$ 　 ★ **22.** $\left(\frac{^-2}{3} \div \frac{4}{9}\right) - \frac{^-1}{8}$ 　 ★ **23.** $\left(1\frac{3}{7} \div \frac{^-3}{5}\right) \div 1\frac{2}{3}$ 　 ★ **24.** $1\frac{3}{4} \div \left(\frac{^-5}{8} \div \frac{^-5}{16}\right)$

Compute. *(Use with page 319.)*

1. 3^2 **2.** 11^2 **3.** $(^-1)^2$ **4.** $(^-17)^2$ **5.** $\left(\frac{1}{4}\right)^2$ **6.** $\left(\frac{^-5}{8}\right)^2$

7. $(0.9)^2$ **8.** $(1.6)^2$ **9.** $(^-2.3)^2$ ★ **10.** $(1.05)^2$ ★ **11.** $(^-0.003)^2$ ★ **12.** $(0.0004)^2$

Give 2 numbers whose square is the given number.

13. 36   **14.** 64   **15.** 81   **16.** 144   ★ **17.** 10,000   ★ **18.** 625

Give 2 square roots. *(Use with page 320.)*

1. 121   **2.** 400   **3.** $\frac{9}{4}$ **4.** $\frac{16}{49}$ **5.** 0.81   **6.** 0.0004

Find square roots.

7. $\sqrt{25}$ **8.** $^-\sqrt{64}$ **9.** $\sqrt{0.25}$ **10.** $^-\sqrt{0.09}$ **11.** $^-\sqrt{0.0049}$ **12.** $\sqrt{8,100}$

13. $^-\sqrt{1,600}$ **14.** $^-\sqrt{\frac{1}{4}}$ **15.** $^-\sqrt{\frac{9}{64}}$ **16.** $\sqrt{\frac{49}{100}}$ ★ **17.** $\sqrt{0.000009}$ ★ **18.** $^-\sqrt{\frac{900}{10,000}}$

Estimate, then find the exact square root.

19. $\sqrt{324}$ **20.** $\sqrt{729}$ **21.** $\sqrt{1,156}$ **22.** $\sqrt{2,116}$ **23.** $\sqrt{5,184}$ **24.** $\sqrt{6,889}$

Estimate, then find the square root to the nearest tenth.

25. $\sqrt{3}$ **26.** $\sqrt{6}$ **27.** $\sqrt{17}$ **28.** $\sqrt{35}$ **29.** $\sqrt{68}$ **30.** $\sqrt{85}$

Solve. First use rounded numbers. *(Use with page 348.)*

1. A set of golf clubs sells for $297.75. During a sale it sold for $189.50. How much was saved during the sale?

2. Mr. Aarons bought 5 cans of tennis balls. He paid $14.90 for them. How much is one can of balls?

3. A baseball club bought 18 baseballs that cost $7.98 each. How much did they spend on all the balls?

4. A golf umbrella is marked $6.95 during a sale. This is $2.50 less than the regular price. What is the regular price?

5. Joanna bought a fishing rod for $31.98, fishing boots for $19.75, and a tackle box for $8.79. How much did she spend in all?

6. Jay paid $28.98 for a warm-up suit. Pete paid $36.89 for the same suit. How much more did Pete pay?

7. Ms. Sherman paid $5.96 for 4 pairs of tennis socks. Each pair was the same price. How much did each pair cost?

8. Mr. Edwards bought 8 rubber balls for his school. Each ball was $4.89. How much did he pay in all?

TABLE OF MEASURES

Metric | Customary

Length

1 kilometer (km) = 1,000 meters
1 hectometer (hm) = 100 meters
1 dekameter (dam) = 10 meters
1 meter (m)
1 decimeter (dm) = 0.1 meter
1 centimeter (cm) = 0.01 meter
1 millimeter (mm) = 0.001 meter

1 foot (ft) = 12 inches (in.)

$1 \text{ yard (yd)} = \begin{cases} 3 \text{ feet} \\ 36 \text{ inches} \end{cases}$

$1 \text{ mile (mi)} = \begin{cases} 5,280 \text{ feet} \\ 1,760 \text{ yards} \end{cases}$

Mass/Weight

1 kilogram (kg) = 1,000 grams
1 hectogram (hg) = 100 grams
1 dekagram (dag) = 10 grams
1 gram (g)
1 decigram (dg) = 0.1 gram
1 centigram (cg) = 0.01 gram
1 milligram (mg) = 0.001 gram
1 metric ton (t) = 1,000 kilograms

1 pound (lb) = 16 ounces (oz)
1 ton (T) = 2,000 pounds

Capacity

1 kiloliter (kL) = 1,000 liters
1 hectoliter (hL) = 100 liters
1 dekaliter (daL) = 10 liters
1 liter (L)
1 deciliter (dL) = 0.1 liter
1 centiliter (cL) = 0.01 liter
1 milliliter (mL) = 0.001 liter

1 teaspoon = 5 milliliters
1 tablespoon = 12.5 milliliters

1 liter = 1,000 cubic centimeters (cm^3)
1 milliliter = 1 cubic centimeter

1 cup (c) = 8 fluid ounces (fl oz)
1 pint (pt) = 2 cups
1 quart (qt) = 2 pints
1 gallon (gal) = 4 quarts

Time

1 minute (min) = 60 seconds (s)
1 hour (h) = 60 minutes
1 day (d) = 24 hours
1 week = 7 days
$1 \text{ year (y)} = \begin{cases} 12 \text{ months} \\ 365 \text{ days} \end{cases}$
1 decade = 10 years
1 century = 100 years

GLOSSARY

This glossary contains an example, an illustration, or a brief description of important terms used in this book.

Absolute value of a number The number or its opposite, whichever is positive.
Examples $|{}^+5| = {}^+5$
$|{}^-5| = {}^+5$

Acute angle An angle whose measure is less than $90°$.

Addition property for equations The same number may be added to each side of an equation to form an equivalent equation.
If $a = b$, then $a + c = b + c$.

Alternate interior angles In the figure below, $\angle a$ and $\angle d$ are alternate interior angles, and so are $\angle b$ and $\angle c$.

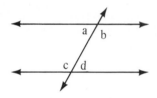

Associative property of addition For all numbers a, b, and c, $(a + b) + c = a + (b + c)$.
Example $(3 + 4) + 5 = 3 + (4 + 5)$

Associative property of multiplication For all numbers a, b, and c, $(a \cdot b) \cdot c = a \cdot (b \cdot c)$.
Example $(3 \cdot 2) \cdot 5 = 3 \cdot (2 \cdot 5)$

BASIC A computer language, BASIC stands for Beginner's All-purpose Symbolic Instructional Code.

Bisect To divide into two congruent parts.

Central angle An angle whose vertex is at the center of a circle.

Circumference Distance around a circle.

Common factor A factor of two or more numbers.
Example 2 is a common factor of 8 and 10.

Common multiple A multiple of two or more numbers.
Example 12 is a common multiple of 3 and 4.

Commutative property of addition For all numbers a and b, $a + b = b + a$.
Example $6 + 4 = 4 + 6$

Commutative property of multiplication For all numbers a and b, $a \cdot b = b \cdot a$.
Example $5 \cdot 3 = 3 \cdot 5$

Complementary angles Two angles, the sum of whose measures is $90°$.

Composite number A number that has more than two factors.
Example 10 is a composite number.
Factors 1, 2, 5, 10

Congruent figures Figures that have the same size and shape.

Coordinates Numbers matched with points on a line. Number pairs matched with points on a plane.

Corresponding angles In the figure below, pairs of corresponding angles are $\angle a$ and $\angle e$, $\angle b$ and $\angle f$, $\angle c$ and $\angle g$, $\angle d$ and $\angle h$.

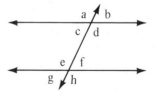

442

Cosine A trigonometric ratio. $\cos A = \frac{b}{c}$

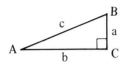

Decimal A number shown with a decimal point.
Examples   0.84   0.12$\overline{12}$

Denominator In $\frac{5}{8}$, 8 is the denominator.

Diagonal A line segment joining two nonconsecutive vertices of a polygon. $\overline{AC}$ is a diagonal.

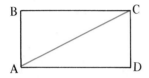

Discount Amount deducted from the marked price.

Distributive property For all numbers a, b, and c, $a \cdot (b + c) = (a \cdot b) + (a \cdot c)$.
Example $3 \cdot (2 + 4) = (3 \cdot 2) + (3 \cdot 4)$

Divisible If a number is divided by a second number and the remainder is zero, the number is divisible by the second number.
Example   18 is divisible by 2, but not by 4.

Division property for equations Each side of an equation may be divided by the same number and an equivalent equation is formed.
If $a = b$, $\frac{a}{c} = \frac{b}{c}$, $c \neq 0$.

Equilateral triangle A triangle with the three sides the same length.

Equivalent equations Two equations with the same solution.
Example $2x = 6$ $x = 3$

Equivalent fractions Fractional numerals for the same number.
Example $\frac{1}{2} = \frac{4}{8}$

Exponent In 10^4, 4 is the exponent. It means that 10 is used as a factor 4 times: $10^4 = 10 \cdot 10 \cdot 10 \cdot 10$.

Extremes In a proportion, the first and fourth terms are the extremes.
Example $\frac{1}{2} = \frac{5}{10}$
           1 and 10 are the extremes.

Factor A number to be multiplied.
Example $2 \times 4 = 8$ 2 and 4 are factors.

FOR/STEP/NEXT A three part command that makes a computer count and loop.

Graph of an equation A picture of all solutions to an equation.

Greatest common factor The largest common factor of two or more numbers.
Example   For 9 and 15,
           3 is the greatest common factor.

Greatest possible error One half the smallest unit of measurement used.

Hypotenuse The longest side of a right triangle. The side opposite the right angle. $\overline{AB}$ is the hypotenuse.

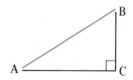

IF/THEN A command that tells a computer to make a decision.
Example   IF N < 100 THEN 50 tells a computer to go to line 50 if the number in storage place N is less than 100; otherwise it is to go on to the next line.

Infinite Continues without end; endless.

INPUT A command that tells a computer to wait for input and then to store the input in its memory.

INT In a computer program, INT(X) makes a computer cut off all the digits of a number X to the right of the decimal point.

Integer Any of these numbers: ..., $^-2$, $^-1$, 0, 1, 2, 3, 4, ...

Interest Payment for use of money.

Irrational number A number named by a non-repeating and nonterminating decimal.
Examples 0.123123312333... $\sqrt{28}$

Isosceles triangle A triangle with two or more sides the same length.

Least common multiple The smallest common multiple of two or more numbers.
Example For 6 and 9, 18 is the least common multiple.

LET A command that tells a computer to store information in its memory.

Mean, in a proportion The means are the second and the third terms of a proportion.
Example $\frac{1}{2} = \frac{3}{6}$
2 and 3 are the means.

Mean, in statistics The average of a set of numbers.
Example The mean for 13, 16, 26, 33 is 22.

Median When numbers are arranged in order, the middle number or the average of the middle two numbers.
Examples The median for 17, 19, 23 is 19.
The median for 8, 11, 15, 19 is 13.

Mixed number A number such as $3\frac{1}{5}$ or $9\frac{2}{3}$.

Mode The number occurring most often in a set of numbers.

Multiple A number that is the product of the given number and another factor.
Example A multiple of 2 is 12.

Multiplication property for equations If each side of an equation is multiplied by the same number, the result is an equivalent equation.
If $a = b$, $ac = bc$, $c \neq 0$.

NEW A command that tells a computer to erase any programs or information stored in its memory.

Numerator In $\frac{3}{4}$, 3 is the numerator.

Obtuse angle An angle whose measure is more than 90° and less than 180°.

Opposite The sum of a number and its opposite is zero.
Example $^-12 + {}^+12 = 0$
$^-12$ is the opposite of $^+12$.
$^+12$ is the opposite of $^-12$.

Origin The point assigned to 0 on the number line or the point where the *x*- and *y*-axes intersect.

Parallel lines Two or more lines in a plane that do not intersect.

Parallelogram A quadrilateral with both pairs of opposite sides parallel.

Percent Ratio of a number to 100, using the % sign.
Example 8% means 8 out of 100.

Perfect square A number that can be named as the product of two equal factors.
Example 16 is a perfect square because $16 = 4 \cdot 4$.

Perimeter Sum of the lengths of the sides of a polygon.

Periods in numerals The groups of three digits set off by a comma in a numeral.

Perpendicular lines Two lines that intersect so that each angle they form measures 90°.

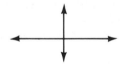

Polygon A simple closed curve made up of line segments.

Precision The smaller the unit with which a measurement is made, the more precise the measurement.

Prime factorization A factorization in which all factors are prime numbers.
Example $30 = 2 \cdot 3 \cdot 5$

Prime number A natural number with exactly two factors, itself and one.

Principal The amount of money on which interest is paid.

Prism A solid with two parallel and congruent bases that are polygons.

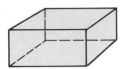

Probability The number of favorable outcomes divided by the number of all possible outcomes. A number from 0 to 1.

Proportion A statement of equality for two ratios.

Example $\frac{3}{6} = \frac{1}{2}$

Pyramid A solid figure with triangular regions for faces and a polygonal region for a base.

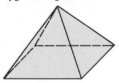

Pythagorean relationship A relationship between the measures of the sides of a right triangle: $a^2 + b^2 = c^2$.

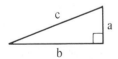

Quadrants The *x*-axis and the *y*-axis separate the plane into four parts called quadrants.

Quadrilateral A polygon with 4 sides.

Radius A line segment from a point on a circle to the center of the circle.

RAM The part of the memory of a computer where the user can store information and programs. RAM stands for Random Access Memory. This part of the memory is erased when NEW is entered.

Range The difference between the largest and the smallest number in a set of data. Sometimes, the range is given as an interval.
Example The range for 3, 8, 11, and 15 is 3 to 15.

Ratio Comparison of two numbers by division.

Rational number A number that can be expressed in the form $\frac{a}{b}$ where *a* and *b* are integers and $b \neq 0$.

READ/DATA A two part command for a computer. DATA allows you to store numbers or words in a line of a program. READ N tells the computer to look for a DATA line and put the first item of information it finds in storage place N.

Real numbers The set of rational and irrational numbers.

Reciprocals Two numbers whose product is 1.
Example $\frac{3}{4}$ and $\frac{4}{3}$ are reciprocals of each other because $\frac{3}{4} \cdot \frac{4}{3} = 1$.

Regular polygon A polygon with all sides the same length and all angles the same size.

Relatively prime Two numbers whose only common factor is 1.
Example 3 and 8 are relatively prime.

Repeating decimal A decimal with one or more digits repeating endlessly.
Examples $0.3\overline{3}$ $0.09\overline{09}...$

Rhombus A parallelogram with all sides congruent.

Right angle An angle whose measure is 90°.

RND(1) In a computer program, RND(1) makes a computer select a 9-place decimal between 0 and 1.

ROM The part of the memory of a computer where the meanings of the BASIC commands are stored. ROM stands for Read Only Memory. This part of the memory is not erased when NEW is entered.

RUN A command that tells a computer to carry out the program commands in its memory.

Sample space The set of possible outcomes of an experiment.
Example If a die is tossed, the sample space is {1, 2, 3 4, 5, 6}.

Scale drawing A drawing that has the same shape as an object, but that can be larger, the same size, or smaller than the object.

Scalene triangle A triangle with no two sides the same length.

Scientific notation Expressing a number as a product of two factors. One factor is a power of 10. The other factor is from 1 to 10.
Example 2.3×10^4 is scientific notation for 23,000.

Significant digits Those digits used to express the number of units of measurement in a measurement.
Example In 0.061 meter, the unit of measurement is 0.001 meter, and there are 61 of the units. 6 and 1 are significant digits, but 0 is not.

Similar triangles Triangles with the same shape, but not necessarily the same size.

Simplest form A fraction is in simplest form when its numerator and denominator are relatively prime.
Examples $\frac{4}{5}$ $\frac{9}{11}$

Sine A trigonometric ratio. $\sin A = \frac{a}{c}$

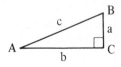

Solution(s) A replacement that makes a number sentence true.
Examples $2x + 8 = 14$ 3 is the solution.
   $x < 4$ 0, 1, 2, 3 are solutions.

Square root A number that when multiplied by itself give the original number.
Example 5 is the square root of 25.

Straight angle An angle whose measure is 180°.

Subtraction property for equations The same number may be subtracted from each side of an equation to form an equivalent equation.
If $a = b$, then $a - c = b - c$.

Supplementary angles Two angles, the sum of whose measures is 180°.

Surface area The total area of the surface of a solid.

Symmetry The correspondence of parts on opposite sides of a point, line, or plane.

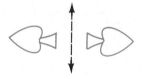

Tangent A trigonometric ratio. $\tan A = \frac{a}{b}$

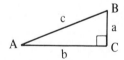

Terminating decimal A decimal that does not repeat.
Example 0.75

Transversal A line intersecting two or more lines at a different point on each line. $\overleftrightarrow{AB}$ is a transversal.

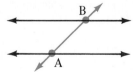

Trapezoid A quadrilateral with exactly one pair of parallel sides.

Variable A letter or other symbol that represents a number.

Vertex A point common to two rays of an angle or two sides of a polygon.

Vertical angles Angles formed by two intersecting lines. ∠ *a* and ∠ *b* form a pair of vertical angles.

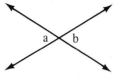

Whole number Any of these numbers: 0, 1, 2, 3, 4, 5, . . .

LIST OF SYMBOLS

		Page
≐	is approximately equal to	4
<	is less than	72
>	is greater than	72
10^5	ten to the fifth power	90
≅	is congruent to	192
$\overrightarrow{AB}$	ray AB	192
∠ ABC	angle ABC	192
$m\angle A$	measure of angle A	192
$\overline{AB}$	line segment AB	200
△ABC	triangle ABC	200
π	pi (about 3.14)	208
$\overleftrightarrow{AB}$	line AB	214
‖	is parallel to	212
⊥	is perpendicular to	214
3:4	three to four	239
10%	ten percent	250
⁻3	negative three	278
⁺3	positive three	278
|⁻2|	absolute value of negative 2	278
$0.3\overline{3}$	repeating decimal	306
√	square root	320
~	is similar to	336

ANSWERS TO THE LEARNING STAGE

CHAPTER 1

PAGES 2–3
A. 1. eighty-seven thousand, nine hundred sixty-four
    2. five hundred sixty-four thousand, one hundred ninety-two
    3. three hundred fourteen million, sixteen thousand, nine hundred twelve
    4. nine hundred seventy-four billion, one hundred eighty-four million, three hundred twelve thousand, one hundred forty-seven
B. 5. 23,000 6. 4,167,000
    7. 6,700,425 8. 45,600,024,000
C. 9. 81,341,007; 81,341,070
       81,341,700; 82,373,710
D. 10. 14,500 11. $500,000
   12. 23,500,000
   13. 500,000,000

PAGE 4
A. 1. 6,000 2. 38,000
    3. 531,000 4. 1,790,000
B. 5. 50,000 6. 630,000
    7. 2,640,000 8. 7,000,000
C. 9. 500,000 10. 800,000
   11. 5,800,000 12. 555,600,000
D. 13. 9,000,000 14. 35,000,000
   15. 489,000,000

PAGE 8
A. 1. How many hours were worked in all to prepare the scenery?
    2. Twelve students; each worked 16 hours on the scenery.
    3. Multiplication
    4. $16 \times 12 = n$
    5. 192 6. 12 7. 16
    8. 192 hours 9. yes
B. 10. 14 rows

PAGE 10
A. 1. yes 2. no 3. yes 4. yes
B. 5. 2, 5, 10 6. 2 7. 5
    8. 2, 5, 10 9. 2 10. 5
   11. 2 12. 2, 5, 10 13. 2 14. 5
C. 15. neither 16. 3 17. 3
   18. 3, 9 19. 3, 9 20. neither
   21. 3, 9 22. 3, 9
   23. 3 24. 3, 9

D. 25. 6 26. no 27. no
   28. 6 29. 6

PAGES 12–13
A. 1. 1, 2, 3, 4, 6, 12
    2. 1, 2, 4, 5, 10, 20
    3. 1, 2, 3, 6, 9, 18
    4. 1, 23
    5. 1, 2, 3, 4, 6, 9, 12, 18, 36
B. 6. 2, 3, 5, 7
    7. 4, 6
C. 8. no 9. yes 10. no
   11. no 12. yes 13. yes
D. 14. yes 15. no 16. yes
   17. no 18. yes 19. yes
E. 20. $2 \times 2 \times 2$ 21. $3 \times 3 \times 3$
   22. $2 \times 2 \times 2 \times 5$ 23. $2 \times 3 \times 3 \times 3$
   24. $3 \times 3 \times 3 \times 3$

PAGE 14
A. 1. Add 3 to each number.
    2. Multiply each number by 3.
B. 3. 243 4. 65
C. 5. 21, 28, 36
    6. 240, 1,440, 10,080
D. 7. 22, 23, 28
    8. 22, 23, 46

CHAPTER 2

PAGE 22
A. 1. 27 2. 90 3. 51 4. 72
B. 5. 70 6. 295 7. 40 8. 45 9. 160

PAGE 24
A. 1. 452 2. 3,643
    3. 945 4. $73.14
    5. 132,752
B. 6. 2,920 7. $551.55
C. 8. 15,360 9. $252.95
   10. 59,330 11. 2,572,158
   12. 16,844 13. 86,239

PAGE 26
A. 1. 516 2. 5,374
    3. 25,819 4. $611,117
B. 5. 355 6. 2,229
    7. 31,604 8. $828,889
C. 9. 363 10. 1,406
   11. 1,864 12. 46,241

PAGE 28
- **A.** 1. 1,100 2. 8,000
  - 3. 560 4. 7,400
  - 5. 72,400
- **B.** 6. 400 7. 3,000
  - 8. 560 9. 2,700
  - 10. 14,400
- **C.** 11. 1,600 12. 13,700
  - 13. 7,900 14. 85,000
  - 15. 12,060

PAGE 32
- **A.** 1. $499.99 2. 720 m 3. 350 cm
- **B.** 4. 3 km 5. 115 cm

PAGE 34
- **A.** 1. 1,872 2. 9,184
  - 3. 10,530 4. 112,648
  - 5. $1,707.75
- **B.** 6. 144 7. 720 8. 864
- **C.** 9. 1,645 10. 30,958
  - 11. 410,976 12. 583,398
  - 13. $11,450.65 14. 2,968
  - 15. 14,030 16. $32.52
  - 17. 420 18. 3,910
  - 19. 62,816

PAGE 36
- **A.** 1. 158,912
  - 2. 995,320
  - 3. 15,897,867
  - 4. 8,071,656
  - 5. 30,135,571
  - 6. 255,750,768
- **B.** 7. 169,728
  - 8. 991,420
  - 9. 5,296,100
  - 10. 58,201,500

PAGE 38
- **A.** 1. 143 2. $474
  - 3. 149 4. 408
  - 5. 32 6. $86
  - 7. 344 8. 432
- **B.** 9. 806 10. 403
  - 11. 1,707 12. 1,203
  - 13. 54r2 14. 163r1
  - 15. 78r2 16. 395r5
- **C.** 17. 212 18. $432
  - 19. 3,323 20. 2,034
  - 21. 317 22. 952
  - 23. 812r2 24. 3,020r1

PAGE 40
- **A.** 1. 24 2. 54r2
  - 3. $234 4. 43

- **B.** 5. 181 6. 256
  - 7. 63r3 8. 1,689r32
  - 9. 47 10. 697
  - 11. 144r583

PAGE 44
- **A.** 1. 240 2. 350
  - 3. 600 4. 4,500
  - 5. 32,000 6. 72,000
  - 7. 420,000
  - 8. 150,000
  - 9. 320,000
  - 10. 3,200,000
  - 11. 32,000,000
  - 12. 45,000,000
  - 13. 3,200
  - 14. 18,000
  - 15. 280,000
  - 16. 2,800,000
- **B.** 17. 4 18. 6 19. 2
  - 20. 8 21. 40
  - 22. 40 23. 60
  - 24. 60 25. 20
  - 26. 30 27. 300

PAGE 46
- **A.** 1. 1,800 2. $1,200
  - 3. 42,000 4. 4,000,000
- **B.** 5. 4,000 6. $32,000
  - 7. 280,000
  - 8. $5,400,000
- **C.** 9. 3 10. $20
  - 11. 80 12. 200
- **D.** 13. 2 14. 6
  - 15. 300 16. 700

PAGE 48
- **A.** 1. associative property of addition
  - 2. property of zero for multiplication
  - 3. distributive property
  - 4. commutative property of addition
  - 5. commutative property of multiplication
  - 6. associative property of multiplication
  - 7. property of zero for addition
  - 8. property of one for multiplication
- **B.** 9. $n = 7$ 10. $n = 8$
  - 11. $n = 0$ 12. $n = 8$
  - 13. $n = 4$ 14. $n = 42$

CHAPTER 3

PAGE 56
- **A.** 1. $x + 5$
  - 2. $x + 5 - 5$
  - 3. x

Continued on page 450.

Page 56 (continued)

B. **4.** 7   **5.** 4   **6.** $\frac{3}{4}$ **7.** 0.4

C. **8.** $x - 3$
9. $x - 3 + 3$
10. x

D. **11.** 2   **12.** 9   **13.** $\frac{1}{2}$ **14.** 0.4

E. **15.** divide by 3   **16.** divide by 4
17. multiply by 2   **18.** multiply by 5

PAGE 58

A. **1.** $x = 2$; $x + 3 = 2 + 3$
2. $y = 3$; $y + 4 = 7$
B. Answers may vary. Examples:
3. $x = 7$; $x + 4 = 7 + 4$
4. $b = 12$; $b + 1 = 12 + 1$
5. $c = 4$; $c + 9 = 4 + 9$
6. $d = 24$; $d + 3 = 24 + 3$
C. Answers may vary. Examples:
7. $x + 7 = 12$; $x = 5$
8. $d + 8 = 14$; $d = 6$
9. $n + 4 = 15$; $n = 11$
10. $r + 6 = 10$; $r = 4$
D. **11.** yes   **12.** no

PAGE 60

A. **1.** 4   **2.** 5   **3.** 8
B. **4.** 3   **5.** 7   **6.** 7
C. **7.** 4   **8.** 5   **9.** 8
D. **10.** 11   **11.** 17   **12.** 23
E. **13.** 5   **14.** 9   **15.** 16
16. 14   **17.** 46   **18.** 39

PAGE 64

A. **1.** $x = 2$; $3 \cdot x = 3 \cdot 2$
2. $n = 3$; $4 \cdot n = 12$; $2 \cdot n = 6$
B. Answers may vary. Examples:
3. $x = 4$; $3x = 12$
4. $x = 6$; $2 \cdot x = 2 \cdot 6$
5. $x = 8$; $5x = 40$
6. $x = 0.7$; $3x = 2.1$
C. Answers may vary. Examples:
7. $4x = 8$; $x = 2$
8. $6x = 30$; $x = 5$
9. $8x = 24$; $x = 3$
10. $9x = 18$; $x = 2$
D. **11.** yes   **12.** yes

PAGE 66

A. **1.** divide by 3   **2.** multiply by 3
3. divide by 12
B. **4.** multiply by 4
5. $\frac{x}{4} \cdot 4 = 10 \cdot 4$
6. $x = 40$ **7.** 40
C. **8.** 16   **9.** 72
10. 13   **11.** 34

D. **12.** 28   **13.** 7
14. 6   **15.** 312

PAGE 68

A. **1.** about 230   **2.** 100   **3.** about 130
4. more people over 21   **5.** up
B. **6.** $1,250   **7.** 1, 2, 3, 5
8. about $400   **9.** generally up

PAGE 70

A. **1.** add 4 to both sides of equation
2. $\frac{n}{2} - 4 + 4 = 7 + 4$
3. $\frac{n}{2} = 11$ **4.** $n = 22$
5. 22   **6.** $\frac{22}{2} - 4 = 7$
B. **7.** 6   **8.** 23   **9.** 10
10. 16   **11.** 64   **12.** 42

PAGE 72

A. **1.** $x - 4 = 3$ **2.** 7   **3.** yes
4. no   **5.** 8, 9, 10
B. **6.** $2x = 8$ **7.** 4   **8.** no
9. yes   **10.** 5, 6, 7, 8, 9, 10
C. **11.** 3, 4, 5, 6, 7, 8, 9, 10
12. 0, 1, 2
13. 5, 6, 7, 8, 9, 10

CHAPTER 4

PAGES 80–81

A. **1.** tenths
2. thousandths
3. millionths
4. hundredths
B. **5.** four tenths
6. seventy-five thousandths
7. four hundred thirteen thousand
one hundred twenty-eight millionths
8. six and nine thousand four hundred
twelve ten-thousandths
C. **9.** 0.065
10. 0.0065
11. 2.300
12. 21.000005
D. **13.** 0.009   **14.** 0.0003   **15.** 0.000007
16. 8,000
E. **17.** $(3 \times 1{,}000) + (8 \times 100) + (7 \times 10) +$
(6×1)
18. $(1 \times 100{,}000) + (4 \times 10{,}000) +$
$(3 \times 1{,}000) + (7 \times 100) + (6 \times 10) +$
(2×1)
19. $(2 \times 0.1) + (4 \times 0.01) + (1 \times 0.001)$
20. $(7 \times 0.1) + (8 \times 0.01) + (6 \times 0.001)$
$+ (0 \times 0.0001) + (2 \times 0.00001)$
F. **21.** 960   **22.** 0.90468   **23.** 20.52

PAGE 82
A. 1. 0.70   2. 0.60
   3. 0.10   4. 3.00
   5. 8.10
B. 6. 0.500   7. 0.800
   8. 0.320
   9. 3.000
   10. 4.000
C. 11. >   12. <
   13. >   14. =
D. 15. >   16. >
   17. <   18. <
E. 19. >   20. >
   21. <   22. =
F. 23. $0.9 > 0.134 > 0.06$
   24. $0.37 > 0.037 > 0.0037$
   25. $0.201 > 0.13 > 0.07$
   26. $0.4 > 0.104 > 0.03006$

PAGE 84
A. 1. 0.435   2. 0.742
   3. 2.315   4. 3.006
   5. 0.169   6. 0.170
   7. 6.099   8. 3.000
B. 9. 0.05   10. 0.05
   11. 0.01   12. 0.61
   13. 6.42   14. 7.01
   15. 9.00   16. 3.00
C. 17. 9.7   18. 0.3
   19. 1.1   20. 3.4
   21. 2.5   22. 6.9
   23. 2.9   24. 4.0
D. 25. 6   26. 7
   27. 3   28. 7
   29. 1   30. 0
   31. 1   32. 0

PAGE 88
A. 1. c

PAGES 90–91
A. 1. 10^7 2. 10^9 3. 10^8
B. 4. 10^6 5. 10^6
   6. 10^7 7. 10^5
C. 8. 10^3 9. 10^2
   10. 10^4 11. 10^5
   12. 10^5 13. 10^1
   14. 10^8 15. 10^7 16. 10^0
D. 17. $4 \times 10^2 + 8 \times 10^1 + 7 \times 10^0$
   18. $5 \times 10^3 + 6 \times 10^2 + 8 \times 10^1 +$
       5×10^0
   19. $9 \times 10^4 + 3 \times 10^3 + 4 \times 10^2 +$
       $0 \times 10^1 + 6 \times 10^0$
   20. $4 \times 10^5 + 0 \times 10^4 + 3 \times 10^3 +$
       $6 \times 10^2 + 1 \times 10^1 + 7 \times 10^0$

PAGE 92
A. 1. yes   2. no   3. no
   4. yes   5. yes
B. 6. 5×10^2 7. 4×10^4
   8. 6×10^5 9. 7×10^6
   10. 8×10^7
C. 11. 7.4×10^3 12. 8.6×10^5
   13. 3.8×10^6 14. 4.61×10^{11}
D. 15. 31,000   16. 6,200,000
   17. 600,000,000
   18. 7,310,000,000

PAGE 94
A. 1. 9.3×10^{13} 2. 4.27×10^{16}
B. 3. 6×10^{16} 4. 9×10^{15}
   5. 8×10^{13} 6. 4×10^{16}
C. 7. 2.8×10^{10} 8. 1.8×10^{14}
   9. 2.1×10^{13} 10. 1.6×10^{19}
D. 11. 2×10^6 12. 3.6×10^3

CHAPTER 5
PAGE 102
A. 1. 1.8   2. 11.85
   3. 8.930   4. 48.0573
B. 5. 12.708   6. 1.556   7. 11.016
C. 8. $33.34   9. $75.97
   10. $2.54   11. $946.41

PAGE 104
A. 1. 0.17   2. 0.186
   3. 0.7   4. 1.788
   5. 6.1529
B. 6. 0.33   7. 0.084
   8. 0.178   9. 2.21
C. 10. 2.9   11. 7.3
   12. 54.53   13. 92.581
D. 14. $5.91   15. $2.86
   16. $11.04   17. $34.69
   18. $137.74

PAGE 106
A. 1. 6   2. 15
   3. 3   4. 3
B. 5. 1.5   6. 16.2
   7. 0.7   8. 2.4
C. 9. $31   10. $29
   11. $4   12. $13

PAGE 110
A. 1. 0.6   2. 2.8
   3. 12.8   4. 4.68
   5. 1.56
B. 6. 13.86   7. 138.6
   8. 1,386   9. 234
C. 10 90   11. 900
   12. 80   13. 3,100

PAGE 112
- A. 1. 0.48 2. 0.936
     3. 4.3491 4. 0.4182
- B. 5. 0.0012 6. 0.00952
     7. 0.000008 8. 0.016024
     9. 1.206 10. 1.7572
     11. 0.6042 12. 0.044356

PAGE 114
- A. 1. $6.21
     2. $3.36

PAGE 116
- A. 1. 0.2 2. 0.3
     3. 6.1 4. 1.9
- B. 5. 0.03 6. 0.07
     7. 0.21 8. 2.88
     9. 0.003 10. 0.030
     11. 0.100 12. 4.332
     13. 4.8 14. 2.45
     15. 0.023 16. 2.01
     17. 0.507 18. 3.52
     19. 0.451 20. 0.25

PAGE 118
- A. 1. 4 2. 4 3. 5 4. 6
- B. 5. 6 6. 4 7. 3 8. 21
- C. 9. 0.12 10. 0.23
     11. 4,000 12. 34

PAGE 120
- A. 1. 1.7 2. 1.6
     3. 1.8 4. 2.1
- B. 5. 0.43 6. 2.98
     7. 0.07 8. 0.43

PAGE 121
- A. 1. 18 2. 24 3. 2.1
- B. 4. 2 5. 0.6
     6. 4 7. 60

PAGE 122
- A. 1. 2.8 2. 2.2 3. 0.4
     4. 2.6 5. 0.05 6. 6
     7. 3.6 8. 0.04 9. 0.72
- B. 10. 0.3, 0.3
     11. 1.2
     12. 2,2 13. 0.6
- C. 14. 0.4 15. 0.5 16. 3
     17. 5 18. 1.2 19. 0.16

CHAPTER 6

PAGE 130
- A. 1. 7,000 2. 600
     3. 500 4. 4
     5. 2,000 6. 30,000

- B. 7. 5 8. 0.7 9. 6
     10. 2.5 11. 9 12. 0.8

PAGE 132
- A. 1. 67.8 2. 0.000678
     3. 678 4. 6.78
- B. 5. 521 6. 3,200
     7. 2,400 8. 2.4
     9. 0.432 10. 0.037

PAGE 134
- A. 1. AB = 3 cm, BC = 5 cm, AC = 5 cm
     2. AB = 32 mm, BC = 54 mm, AC = 49 mm
     3. Measurements to the nearest millimeter are more precise.
- B. 4. 8 mm 5. 14 m 6. 7.4 mm
     7. 0.3 cm 8. 9 in. 9. 0.04 mi
- C. 10. 0.1 cm
     11. 0.05 cm
     12. 0.05 cm
- D. 13. 0.05 cm 14. 0.5 m
     15. 0.5 mm 16. 0.005 km

PAGE 138
- A 1. 400 2. 80 3. 5 km/L
- B. 4. 400 km 5. 300 km
     6. 100 km 7. 800; 800 km
- C. 8. $0.30 9. 80 L 10. $24.00; $24

PAGE 140
- A. 1. 34; 2 2. 12; 2
     3. 3; 1 4. 14; 2
- B. 5. 17 m²
     6. 308 cm²

PAGE 142
- A. 1. 3,500 2. 42,000
     3. 5,600 4. 25,000
- B. 5. 0.125 6. 0.050
     7. 0.225 8. 0.005
- C. 9. 7,000 10. 4.5

PAGE 144
- A. 1. 3,000 2. 8,000
     3. 0.002 4. 0.015
     5. 5,600 6. 12,800
     7. 0.017 8. 0.384

PAGE 146
- A. 1. 24 2. 72
     3. 108 4. 9
     5. 10,560 6. 5,280
     7. 4 8. 6
     9. 2 10. 2
     11. 4 12. 3
     13. 6 14. 12
     15. 13,200 16. 40
     17. 10 18. 5, 4
     19. 1, 5

PAGE 148

A. 1. 48 2. 4,000
    3. 1,000 4. 4
    5. 2 6. 5
B. 7. 6 8. 32 9. 6
    10. 8 11. 3 12. 6

PAGE 149

A. 1. 100°C 2. 0°C 3. 100°
B. 4. cold 5. hot
    6. comfortable
C. 7. about 30°C 8. about 10°C

PAGES 150–151

A. 1. 3, later
    2. 2 pm 3. 2 pm
B. 4. 11:30 am
    5. 12:30 am
C. 6. no
    7. one day earlier
    8. 12:00 noon, June 8th

CHAPTER 7

PAGE 158

A. 1. 3, 6, 9, 12, 15
    2. 8, 16, 24, 32, 40
    3. 12, 24, 36, 48, 60
B. 4. 8, 16, 24, 32, 40, 48, 56, . . .
    5. 12, 24, 36, 48, 60, 72, 84, . . .
    6. 24
C. 7. 24 8. 20
    9. 45 10. 12
    11. 18 12. 25
    13. 36 14. 24

PAGE 160

A. 1. $\frac{2}{6}, \frac{3}{9}, \frac{4}{12}$ 2. $\frac{4}{10}, \frac{6}{15}, \frac{8}{20}$
B. 3. 2 4. 12 5. 8 6. 15

PAGE 162

A. 1. 3 2. 2 3. 1 4. 6
B. 5. yes 6. no 7. yes 8. yes
C. 9. yes 10. yes 11. no 12. no
    13. no
D. 14. $\frac{2}{5}$ 15. yes
E. 16. $\frac{1}{2}$ 17. $\frac{1}{3}$ 18. $\frac{1}{2}$ 19. $\frac{3}{10}$
    20. $\frac{2}{3}$

PAGE 164

A. 1. > 2. < 3. > 4. =
B. 5. 24 6. $\frac{9}{24}, \frac{4}{24}$
    7. > 8. >
C. 9. > 10. < 11. = 12. <

PAGE 166

A. 1. $1\frac{1}{2}$ 2. $1\frac{1}{2}$ 3. $2\frac{1}{3}$
    4. $1\frac{1}{2}$ 5. $2\frac{3}{5}$
B. 6. $\frac{19}{6}$ 7. $\frac{17}{3}$ 8. $\frac{37}{5}$
    9. $\frac{11}{6}$ 10. $\frac{15}{4}$ 11. $\frac{13}{2}$
C. 12. $\frac{13}{4}$ 13. $\frac{47}{8}$
    14. $\frac{25}{6}$ 15. $\frac{5}{2}$

PAGE 168

A. 1. $\frac{5}{8}$ 2. 1 3. $\frac{1}{3}$ 4. $1\frac{3}{10}$
    5. $1\frac{1}{3}$ 6. $\frac{1}{2}$ 7. $\frac{9}{10}$ 8. $2\frac{1}{4}$
B. 9. 10 10. 5 11. 4
    12. $\frac{9}{10}$ 13. $\frac{9}{10}$
C. 14. $\frac{7}{8}$ 15. $\frac{13}{24}$ 16. $1\frac{1}{6}$
    17. $1\frac{9}{20}$ 18. $1\frac{1}{12}$

PAGE 170

A. 1. $5\frac{7}{8}$ 2. $3\frac{3}{4}$ 3. $5\frac{3}{4}$
    4. $6\frac{13}{24}$ 5. $7\frac{2}{3}$
B. 6. $4\frac{1}{4}$ 7. $6\frac{1}{2}$ 8. $8\frac{1}{8}$
    9. $13\frac{7}{12}$ 10. $11\frac{7}{8}$
C. 11. $6\frac{1}{5}$ 12. $8\frac{1}{4}$ 13. $10\frac{1}{4}$

PAGE 173

A. 1. $\frac{2}{3}$ 2. $\frac{3}{5}$ 3. $\frac{1}{2}$
    4. $\frac{2}{5}$ 5. $\frac{2}{3}$
B. 6. $\frac{7}{20}$ 7. $\frac{1}{2}$ 8. $\frac{5}{24}$
    9. $\frac{7}{15}$ 10. $\frac{7}{12}$

PAGE 174

A. 1. $1\frac{1}{2}$ 2. $3\frac{1}{6}$ 3. $3\frac{5}{24}$
    4. $3\frac{1}{3}$ 5. $2\frac{7}{12}$
B. 6. $5\frac{3}{8}$ 7. $5\frac{1}{6}$ 8. $2\frac{1}{2}$
    9. $1\frac{3}{8}$ 10. $4\frac{1}{5}$
C. 11. $4\frac{4}{5}$ 12. $3\frac{5}{8}$ 13. $1\frac{11}{15}$
    14. $4\frac{19}{24}$ 15. $2\frac{11}{12}$

PAGE 176

A. 1. the cost of 10 cans
    2. The cost of 2 cans is $0.79.
    3. the number of cans in a case (24)
    4. The cost of 10 cans is $3.95.

Continued on page 454.

Page 176 (continued)

B. 5. the cost of a dozen large size bottles
6. the cost of one large size bottle
7. the cost of the next smaller size
8. The cost of a dozen bottles is $11.88.

PAGE 178

A. 1. $\frac{2}{15}$ 2. $\frac{3}{40}$ 3. $\frac{1}{12}$ 4. $\frac{1}{20}$

B. 5. 6 6. 8 7. 1 8. $4\frac{1}{2}$

C. 9. $\frac{5}{24}$ 10. $\frac{21}{40}$ 11. $\frac{3}{5}$

D. 12. $\frac{5}{8}$ 13. $\frac{2}{3}$ 14. $\frac{4}{15}$ 15. 4

PAGE 180

A. 1. 28 2. $27\frac{1}{5}$
3. $26\frac{1}{4}$ 4. $1\frac{2}{3}$
5. $1\frac{5}{8}$ 6. 10
7. $4\frac{7}{12}$ 8. $6\frac{3}{5}$

PAGE 181

A. 1. 1 2. 1 3. 1 4. 1

B. 5. $\frac{3}{2}$ 6. $\frac{7}{5}$ 7. $\frac{1}{6}$ 8. $\frac{4}{13}$

C. 9. $\frac{4}{3}$ 10. $\frac{21}{1}$ 11. $\frac{8}{21}$

D. 12. $\frac{2}{3}$ 13. $\frac{1}{2}$ 14. $\frac{1}{6}$ 15. $\frac{1}{6}$

PAGE 182

A. 1. 6 2. $\frac{6}{1}$ 3. 4 4. 4

B. 5. 6 6. $\frac{3}{20}$ 7. 6 8. $\frac{9}{10}$

C. 9. $8\frac{5}{8}$ 10. $\frac{5}{18}$ 11. $1\frac{5}{6}$ 12. $2\frac{3}{16}$

PAGES 184–185

A. 1. 0.2 2. 0.4
3. 1.5 4. 1.2
5. 3.5

B. 6. 0.25 7. 0.04
8. 0.04 9. 0.16
10. 1.25

C. 11. 0.375 12. 0.625
13. 0.025 14. 0.075
15. 0.008

D. 16. 2.5 17. 3.75
18. 4.2 19. 4.04
20. 3.125

CHAPTER 8

PAGES 192–193

A. 1. E
2. $\overrightarrow{ED}, \overrightarrow{EC}$
3. $\angle CED, \angle DEC, \angle E$

C. 9. 30° 10. 60° 11. 90°
12. 120° 13. 30° 14. 60°
15. 120° 16. 90°

D. 17. $\angle STU \cong \angle PQR$

PAGES 194–195

A. 1. right 2. obtuse
3. obtuse 4. acute

B. 5. acute 6. right 7. obtuse

C. 8. isosceles 9. scalene
10. equilateral 11. scalene

PAGES 198–199

A. 1. quadrilateral
2. pentagon
3. octagon

B. 4. 1 5. 2 6. They are congruent.
7. It is a parallelogram with all four sides congruent.
8. It is a parallelogram with four right angles.
9. It is a rectangle with all four sides congruent.

C. 10. 7 mm 11. 60° 12. 5 cm

PAGES 200–201

A. 1. 180° 2. 122° 3. 58°

B. 4. A, B, C, D 5. $\overline{BD}$

C. 6. 180° 7. 360°

D. 8. 60° 9. 90° 10. 108°

PAGE 202

A. 1. 94 m 2. 18 ft 3. 145 mm

B. 4. 100 mm 5. 10.2 cm 6. 32 cm

C. 7. $p = 3s$ 8. 72 cm

PAGE 206

A. 1. the measure of $\angle T$
2. the measure of $\angle R$
3. no

B. 4. $m\angle U$ and $m\angle V$
5. $m\angle U$ and $m\angle V$
6. yes
7. $m\angle W = 60°$

C. 8. no
9. the length of the rectangle

PAGES 208–209

A. 1. 124 mm 2. 1 m
3. 4.2 cm 4. 19.2 in.

B. 5. 16 m 6. 3 cm
7. 7.5 mm 8. 1.9 in.

C. 9. 10 π m 10. 3.5 π cm
11. 8 π mm 12. 10.8 π m

D. 13. 34.54 cm 14. 65.94 mm
15. 56.52 mm 16. 9.42 in.

PAGE 210
- **A.**
  1. complementary
  2. supplementary
  3. supplementary
  4. complementary
- **B.** 5. $60°$ **6.** $66°$ **7.** $23°$
  8. $36°$ **9.** $1°$
- **C.** 10. $150°$ **11.** $60°$ **12.** $126°$
  13. $63°$ **14.** $1°$
- **D.** 15. $90°$ **16.** $180°$ **17.** $90°$
- **E.** 18. $63°$ **19.** $48°$ **20.** $50°$

PAGE 212
- **A.**
  1. $\angle w$ and $\angle z$
  2. $\angle a$ and $\angle w$, $\angle x$ and $\angle c$, $\angle z$ and $\angle d$
- **B.** 4. They are congruent.
  6. They are congruent.
- **C.** 7. $50°$ **8.** $130°$
  9. $120°$ **10.** $60°$

PAGE 214
- **A.** 1. $90°$ **2.** $90°$

PAGE 216
- **A.** 1. 1

CHAPTER 9

PAGE 226
- **A.** 1. $n - 4$ **2.** $n + 3$ **3.** $\frac{n}{3}$
  4. $2n + 3$ **5.** $\frac{1}{2}n$ **6.** n^2
- **B.** 7. 18 **8.** 48 **9.** 24 **10.** 0
- **C.** 11. 64 **12.** 17

PAGE 228
- **A.** 1. $n + 6 = 47$
  2. $n - 8 = 63$
  3. $\frac{n}{8} = 18$
  4. $3n = 129$
- **B.** 5. $2n$
  6. $2n - 4$
  7. $2n - 4 = 18$
- **C.** 8. $2x - 9 = 47$
  9. $3x + 7 = 43$
  10. $7x - 5 = 23$

PAGE 230
- **A.** 1. How many hours did Abe work?
  2. $4x$
  3. $4x = 92$; $x = 23$; Abe worked 23 hours.
- **B.** 4. What was the cost of each shirt?
  5. $3n$
  6. $3n - 5$
  7. $3n - 5 = 43$; $n = 16$; Each shirt costs $16.

PAGE 234
- **A.** 1. yes **2.** yes **3.** no
- **B.** 4. no

PAGE 236
- **A.** 1. $d = rt$ **2.** $180 = r \cdot 4$
  3. $r = 45$ mph
- **B.** 4. $d = rt$ **5.** $260 = 65 \cdot t$
  6. $t = 4$ hours

PAGE 239
- **A.** 1. 9:3 or $\frac{9}{3}$ **2.** 9:12 or $\frac{9}{12}$
- **B.** 3. yes **4.** no **5.** yes

PAGE 240
- **A.** 1. no **2.** yes **3.** yes
- **B.** 4. 2 and 2 **5.** 3 and 2
  6. 8 and 2 **7.** 5 and 40
- **C.** 8. 16 **9.** 16 **10.** yes
- **D.** 11. 20 **12.** 4 **13.** 9

PAGE 242
- **A.** 1. $w_1 \cdot d_1 = w_2 \cdot d_2$
  2. $10 \cdot 6 = x \cdot 2$
  3. $x = 30$; The mass of the rock is 30 kg.
- **B.** 4. $w_1 \cdot d_1 = w_2 \cdot d_2$
  5. $16 \cdot 5 = x \cdot 4$
  6. $x = 20$; A mass of 20 kg will balance the 16 kg mass.

CHAPTER 10

PAGE 250
- **A.** 1. $\frac{1}{10}$ **2.** $\frac{1}{4}$ **3.** $\frac{2}{5}$
  4. 1 **5.** 2
- **B.** 6. $\frac{1}{8}$ **7.** $\frac{1}{6}$ **8.** $\frac{9}{200}$
  9. $\frac{1}{400}$ **10.** $\frac{1}{200}$
- **C.** 11. 75%
- **D.** 12. 50% **13.** $12\frac{1}{2}\%$ **14.** $\frac{1}{2}\%$
  15. 300% **16.** 175%

PAGES 252–253
- **A.** 1. 0.07 **2.** 0.81
  3. 1.30 **4.** 0.004
  5. 0.026
- **B.** 6. 0.63 **7.** 1.50
  8. 0.005 **9.** 0.023
  10. 0.0304
- **C.** 11. 0.0025 **12.** 0.002
  13. 0.0075 **14.** 0.025
  15. 0.0525
- **D.** 16. 78% **17.** 50% **18.** 100%
  19. 240% **20.** 7.5%
- **E.** 21. $\frac{1}{5}$, 0.20 **22.** 50%, 0.50 **23.** 25%, $\frac{1}{4}$

PAGE 254
- **A.** 1. 5.2 2. 18 3. 18
- **B.** 4. $140\% \times 65 = n$
  5. $1.40 \times 65 = n$
  6. $n = 91$
  7. 91
- **C.** 8. 2 9. 4 10. 15.75
- **D.** 11. 30 12. 9 13. 8

PAGE 256
- **A.** 1. $n \cdot 30 = 12$
  2. $n = 0.40$
  3. 40%
- **B.** 4. 12% 5. $62\frac{1}{2}\%$
- **C.** 6. 300% 7. 525%
- **D.** 8. $n \cdot 10 = 0.05$
  9. $n = 0.005$
  10. 0.5%
- **E.** 11. $2\frac{1}{2}\%$ 12. 0.25%

PAGE 258
- **A.** 1. $48 = 40\% \cdot n$
  2. 0.40
  3. $n = 120$
  4. $48 = 0.40 \times 120$
- **B.** 5. 24 6. 27
  7. 42 8. 54
- **C.** 9. 4 10. 4
  11. 800 12. 2,000

PAGE 262
- **A.** 1. Hammer cost $5.95; screwdriver
  cost $3.75
  How much money was returned?
  $2.20
- **B.** 2. Mini-problem 1
  Second chair cost $149.50
  First chair cost $125.00
  Additional cost? $24.50
  Mini-problem 2
  Gave $50.00
  Additional cost $24.50
  Change? $25.50

PAGE 264
- **A.** 1. $12, $36
  2. $7.75, $23.25
  3. $4.45, $13.35
- **B.** 4. $14.40 5. $16.80 6. $19.20
- **C.** 7. $4 8. $4 = $n \cdot 16$
  9. $n = 0.25$ 10. 25%

PAGE 266
- **A.** 1. $1,008 2. $330

PAGE 266 (cont.)
- **B.** 3. $3,815 4. $4,936.75
  5. $1,380 6. $517.50
  7. $6,695 8. $2,463

PAGE 268
- **A.** 1. $70 2. $1,070 3. $74.90
  4. $1,144.90 5. $1,225.04
- **B.** 6. $848.72 7. $265.34

PAGE 270
- **A.** 1. 4 2. $4 = r \cdot 32$
  3. $r = \frac{1}{8}$ 4. 12.5%
- **B.** 5. 20% 6. 25%

CHAPTER 11

PAGE 278
- **A.** 1. > 2. < 3. > 4. <
- **B.** 5. ⁻6 6. ⁺7 7. ⁻15
  8. ⁺11 9. ⁺43 10. ⁻73
- **C.** 11. ⁺5 12. ⁺5
- **D.** 13. ⁺9 14. ⁺7 15. ⁺21
  16. ⁺16 17. ⁺37 18. ⁺86
- **E.** 19. < 20. > 21. =

PAGE 280
- **A.** 1. ⁺4 2. ⁺7
  3. ⁺3 4. ⁺6
- **B.** 5. ⁺87 6. ⁺41
  7. ⁺72 8. ⁺31
- **C.** 9. ⁻4 10. ⁻6
  11. ⁻4 12. ⁻6
- **D.** 13. ⁻29 14. ⁺82
  15. ⁻56 16. ⁺73
  17. ⁺24 18. ⁺44 19. ⁻79

PAGES 282–283
- **A.** 4. ⁻3
- **B.** 5. ⁺1 6. ⁻3 7. ⁺2
- **C.** 8. positive 9. 8 10. ⁺8
- **D.** 11. ⁺4 12. ⁻3 13. ⁻15 14. ⁻6

PAGE 284
- **A.** 1. ⁻6 2. ⁺14 3. ⁺3
- **B.** 4. ⁻3 5. ⁻3 6. ⁺4 7. ⁺4
  8. ⁺8 9. ⁻5 10. ⁻8 11. ⁺7
- **C.** 12. ⁺2 13. ⁺2
  14. ⁺11 15. ⁺11
  16. ⁺11 17. ⁻4
  18. ⁺7 19. ⁻15

PAGE 286
- **A.** 1. ⁺11 2. ⁺13
  3. ⁻13 4. ⁻10
- **B.** 5. ⁺8 6. ⁻3
  7. ⁺4 8. ⁻7

PAGE 288

A. 1. Program C 2. 25%
   3. 15 4. 18 5. 27

B. 6. 100% 7. non-union
   8. 138 9. 262

PAGE 290

A. 1. $^+$56 2. $^+$6 3. $^+$63 4. $^+$35

B. 5. $^-$6 6. $^-$12
   7. 0 8. $^-$18
   9. $^-$28 10. $^-$54

C. 11. $^-$18 12. $^-$20
   13. $^-$8 14. $^-$8
   15. $^-$42 16. $^-$54

D. 17. $^+$5 18. $^+$10
   19. $^+$27 20. $^+$32
   21. $^+$63 22. $^+$56

PAGE 292

A. 1. yes 2. yes 3. yes
   4. yes 5. yes 6. yes
   7. yes 8. yes 9. yes

PAGE 294

A. 1. $^+$4 2. $^+$4 3. $^+$3
   4. $^+$8 5. $^+$8 6. $^+$2
   7. $^-$6 8. $^-$6 9. $^-$6
   10. $^-$3 11. $^-$17 12. $^-$6

B. 13. 0 14. 0
   15. no answer possible
   16. no answer possible

C. 17. 0 18. $^-$1
   19. no answer possible 20. 0

PAGE 296

A. 1. $^-$7 2. $^+$11 3. $^-$12

B. 4. $^+$8 5. $^+$8 6. $^-$5

C. 7. $^-$21 8. $^+$32 9. $^-$54

D. 10. $^+$4 11. $^-$4 12. $^+$3

PAGE 298

A. 1. $^-$1 2. $^-$2
   3. $^-$4 4. 10
   5. 100 6. 100,000

B. 7. 100 8. 0.01
   9. 10 10. 0.1
   11. 1,000,000 12. 0.000001

C. 13. 10^1 14. 10^{-1} 15. 10^3
   16. 10^{-3} 17. 10^2 18. 10^{-2}

PAGE 299

A. 1. 2 2. $^-$1 3. $^-$3 4. $^-$5

B. 5. $(3 \times 10^0) + (4 \times 10^{-1})$
   $+ (6 \times 10^{-2}) + (7 \times 10^{-3})$
   6. $(2 \times 10^1) + (8 \times 10^0) + (5 \times 10^{-1})$
   $+ (1 \times 10^{-2}) + (0 \times 10^{-3}) + (6 \times 10^{-4})$

7. $(3 \times 10^2) + (7 \times 10^1) + (4 \times 10^0)$
   $+ (6 \times 10^{-1}) + (9 \times 10^{-2}) + (1 \times 10^{-3})$
   $+ (8 \times 10^{-4}) + (0 \times 10^{-5}) + (2 \times 10^{-6})$

C. 8. 346.878 9. 4,200.634

CHAPTER 12

PAGE 306

A. 1. $\frac{8}{1}$ 2. $\frac{-8}{1}$ 3. $\frac{0}{1}$ 4. $\frac{-1}{10}$ 5. $\frac{21}{10}$

B. 6. < 7. < 8. < 9. >

C. 10. $^-$6, $^-$1.1, $\frac{2}{5}$, $\frac{7}{2}$
   11. $^-$4, $^-$1.4, $\frac{-3}{4}$, $\frac{3}{5}$

D. 12. $^-$0.5 13. 0.12 14. $^-$0.125
   15. 0.4 16. $^-$1.75

E. 17. $0.6\bar{6}$ 18. $0.\bar{1}$ 19. $^-1.\bar{3}$
   20. $^-3.\overline{142857}$ 21. $0.\bar{4}$

PAGE 308

A. 1. $\frac{-4}{7}$ 2. $\frac{2}{5}$
   3. $^-1\frac{1}{4}$ 4. $^-$5.8

B. 5. $\frac{-1}{3}$ 6. $\frac{-1}{2}$
   7. $\frac{-1}{2}$ 8. 1.4
   9. $\frac{1}{4}$ 10. $\frac{1}{2}$
   11. $\frac{-1}{12}$ 12. $^-$2.5

C. 13. $\frac{1}{2}$ 14. $\frac{-3}{4}$
   15. $\frac{5}{6}$ 16. $^-$1.2

D. 17. 0 18. 0 19. 0 20. 0

PAGE 310

A. 1. $\frac{3}{8}$ 2. $\frac{1}{2}$

B. 3. $^-$1 4. $\frac{-3}{5}$ 5. $^-$0.4 6. 5.7

C. 7. 2 8. $\frac{3}{4}$ 9. $1\frac{1}{4}$

D. 10. $1\frac{5}{8}$ 11. $^-1\frac{5}{12}$ 12. $\frac{-11}{24}$ 13. $^-9\frac{1}{6}$

E. 14. $\frac{-1}{2}$ 15. $\frac{2}{3}$ 16. $\frac{-1}{3}$ 17. $^-$0.4
   18. 0 19. 0 20. 0 21. 0

PAGE 312

A. 1. $\frac{-2}{5}$ 2. $\frac{3}{8}$ 3. $^-$2 4. $^-$3.22

B. 5. $\frac{1}{2}$ 6. $\frac{-7}{30}$ 7. $\frac{-2}{3}$ 8. $1\frac{1}{4}$

C. 9. $\frac{-4}{3}$ 10. $\frac{-8}{5}$ 11. $\frac{4}{3}$ 12. $\frac{-2}{3}$

D. 13. $\frac{15}{16}$ 14. $^-$4 15. $\frac{-1}{4}$ 16. 2

E. 17. 4

PAGE 314

   A. **1.** rational   **2.** irrational
        3. rational   **4.** irrational

PAGE 316

   A. **1.** 35
        2. 55
        3. October
        4. November
        5. 20
   B. **6.** 15 fiction, 20 biographies
        7. October
        8. November

PAGE 318

   A. **1.** $\frac{1}{3}$ **2.** $\frac{5}{9}$ **3.** $\frac{2}{3}$ **4.** $\frac{7}{9}$ **5.** $\frac{8}{9}$
   B. **6.** $\frac{4}{11}$ **7.** $\frac{5}{11}$ **8.** $\frac{2}{33}$ **9.** $\frac{26}{33}$ **10.** $\frac{10}{33}$

PAGE 319

   A. **1.** 16 **2.** 64 **3.** 9
        4. $\frac{4}{9}$ **5.** $\frac{4}{9}$ **6.** 0.16
        7. 5.29 **8.** 3.61
        9. 9.61 **10.** 17.64
   B. **11.** ⁻5 **12.** 10, ⁻10
   C. **13.** no **14.** yes **15.** yes
        16. yes **17.** no

PAGE 320

   A. **1.** 5, ⁻5 **2.** 10, ⁻10 **3.** $\frac{2}{3}$, $\frac{⁻2}{3}$
        4. 0.3, ⁻0.3 **5.** 0.8, ⁻0.8
   B. **6.** 6 **7.** 9 **8.** $\frac{1}{6}$
        9. ⁻0.4 **10.** ⁻20
   C. **11.** 17
   D. **12.** 15 **13.** 23 **14.** 31
        15. 53 **16.** 92
   E. **17.** 2.2
   F. **18.** 2.6 **19.** 3.3 **20.** 4.9
        21. 8.5 **22.** 9.7

PAGES 322–323

   A. **1.** 2.646 **2.** 3.162
        3. 5.385 **4.** 6.403
   B. **5.** rational **6.** irrational
        7. irrational **8.** rational

PAGE 324

   A. **1.** 13 **2.** 15
   B. **3.** 9 **4.** 8

PAGE 326

   A. **5.** $\sqrt{2}$
   B. **7.** ⁻2, ⁻1.6̄6̄, ⁻1, ⁻$\frac{1}{2}$, 0, 0.3̄3̄, 0.764, 1, 2
        8. ⁻$\sqrt{3}$, ⁻0.909009…, $\sqrt{2}$, 1.5678…

CHAPTER 13

PAGES 334–335

   A. **1.** $\overline{LM} \cong \overline{PQ}$; $\overline{MN} \cong \overline{QR}$; $\overline{NL} \cong \overline{RP}$
        2. $\angle L \cong \angle P$; $\angle N \cong \angle R$; $\angle M \cong \angle Q$
   B. **8.** yes
   C. **14.** yes
   D. **15.** 11 cm **16.** 60°

PAGE 336

   A. **1.** yes **2.** yes **3.** yes
   B. **4.** PR **5.** 6 **6.** 12 **7.** $x = 3$
   C. **8.** $x = 50$

PAGE 338

   A. **1.** 3 cm
        2. $0.5x = 3$; $x = 6$ m
        3. 6 m
   B. **4.** approximately 75 mi
   C. **5.** 2 m
        6. $2x = 4.8$; $x = 2.4$ cm
        7. 3.3 cm
        8. 2.4 cm × 3.3 cm

PAGES 340–341

   A. **1.** 5 **2.** 3 **3.** 4
   B. **4.** $\frac{3}{4}$ **5.** $\frac{3}{5}$ **6.** $\frac{4}{5}$ **7.** $\frac{4}{3}$
   C. **8.** 0.616
        9. 0.934
      10. 3.49
      11. adjacent
      12. opposite
      13. tangent
      14. 72.7
      15. 72.7 m

PAGES 342–343

   A. **1.** 12 m² **2.** 132 cm²
        3. 60 mm² **4.** 6 m²
   B. **5.** 16 cm² **6.** 121 m²
   C. **9.** parallelogram
      10. 8
      11. 16 square units
      12. The area of a trapezoid is one half the area of a parallelogram.
      13. 8 square units
   D. **14.** 66 mm²
      15. 24 cm²

PAGE 344–345

   A. **1.** 100 **2.** 1,000,000
        3. 20,000 **4.** 400
        5. 3,000,000 **6.** 1,000,000
   B. **7.** 0.01 **8.** 0.000001
        9. 0.000001 **10.** 0.0025
      11. 0.0005 **12.** 4
   C. **13.** 20,000 **14.** 4 **15.** 400
   D. **16.** km² **17.** m² **18.** ha
      19. cm² **20.** mm² **21.** m²

PAGE 346

A. 1. 314 cm² 2. 50.2 m²
3. 78.5 cm² 4. 254.3 cm²
B. 5. 452.16 cm²
6. 113.04 cm²
8. 339.12 cm²

PAGE 348

A. 1. $2.00 2. 20 3. multiplication
4. $40.00 5. $47.76 6. yes

PAGE 350

A. 1. 432 2. 36
3. 8 4. 2
B. 5. 12 yd² 6. 12 yd²
C. 7. 87,120 8. 1,280

PAGES 352–353

A. 1. 528 cm² 2. 864 cm²
B. 3. circle 4. 12.56 cm²
5. 25.12 cm² 6. rectangle
7. 10 cm 8. 12.56 cm
9. 125.6 cm² 10. 150.72 cm²
C. 11. 64 cm²
12. 48 cm², 192 cm²
13. 256 cm²
D. 14. 78.5 cm²
15. 314 cm²
16. 392.5 cm²
E. 17. 200.96 cm²

PAGE 354

A. 1. 4,000,000
2. 7,000
3. 1,000,000,000
B. 4. 0.000008
5. 0.0005
6. 0.009
7. 5
8. 0.007

PAGES 356–357

A. 1. 240 m³
2. 27 cm³
B. 3. triangle
4. 30 cm²
5. 540 cm³
C. 6. circle
7. 314 cm²
8. 9,420 cm³
D. 9. 48,000 mm³
10. 100,480 cm³
E. 11. 267.95 cm³

PAGE 358

A. 1. 5 2. 2.5
3. 10 4. 0.65
B. 5. 8,000 6. 7,300
7. 3,000 8. 2,500

C. 9. 3 10. 400
11. 240 12. 7.4
D. 13. 7 g 14. 34.1 g
15. 2,300 g 16. 3,400 g
E. 17. 3,000 cm³ 18. 8 cm³
19. 341.7 cm³ 20. 4,600 cm³

PAGES 360–361

A. 5. $\frac{1}{3}$ full 6. $\frac{1}{3}$
B. 7. 70 m³ 8. 192 cm³
C. 9. $\frac{1}{3}$ 10. $\frac{1}{3}$
D. 11. 50.24 cm³ 12. 628 m³

PAGE 362

A. 1. 54 2. 270
3. 5,184 4. 17,280
5. 2 6. 5

CHAPTER 14

PAGES 372–373

A. 1. (3, 2) 2. (2, 3) 3. (⁻2, 2)
4. (⁻3, ⁻1) 5. (1, ⁻2) 6. (⁻2, ⁻3)
7. (⁻3, 1) 8. (4, ⁻2) 9. (1, 0)
B. 10. – 14.

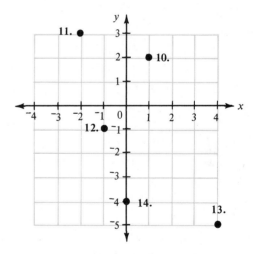

C. 15. first 16. fourth
17. second 18. third
D. 19. (1, ⁻3), (3, ⁻1), (6, ⁻3)

PAGE 374

A. 1. $y = 4$ 2. $y = 6$
B. 3. (0, 4) 4. (1, 5)
5. (2, 6) 6. (3, 7)
C. 7. $y = $ ⁻3, ⁻1, 1, 3

Continued on page 460.

Page 374 (continued)

C. **8.** ($^-$2, $^-$3); ($^-$1, $^-$1); (0, 1); (1, 3)

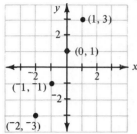

PAGE 376

A. **1.** yes **2.** no

B. **3.**

x	$^-$2	$^-$1	0	1	2
y	$^-$4	$^-$2	0	2	4

4. ($^-$2, $^-$4); ($^-$1, $^-$2); (0, 0); (1, 2); (2, 4)

5. − **6.**

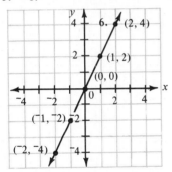

C. **7.** (2, 3); (1, 1); (0, $^-$1); ($^-$1, $^-$3); ($^-$2, $^-$5)

8. (2, 1); (0, 0); ($^-$2, $^-$1); ($^-$4, $^-$2); ($^-$6, $^-$3)

9. (2, $^-$3); (1, $^-$1); (0, 1); ($^-$1, 3); ($^-$2, 5)

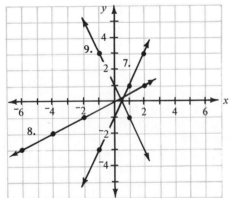

D. **10.** (3,2), (2,1), (1,0), (0,$^-$1)

($^-$1, $^-$2), ($^-$2, $^-$3)

11. $y = x - 1$

PAGE 378

A. **1.**

x	y
$^-$2	$^-$5
$^-$1	$^-$3
0	$^-$1
1	1
2	3

2.

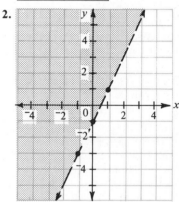

3. no **4.** yes **5.** See graph above.

B. **6.** − **7.**

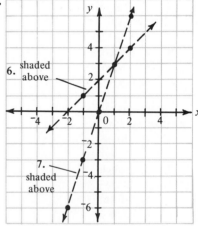

8. − **9.**

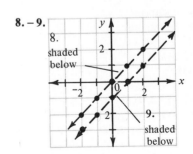

PAGE 382

   A. **4.** The lines intersect at noon.

        5. The son will overtake his father at noon.

   B. **6.** In twelve weeks both José and Susan will have 48 records.

PAGE 384

   A. **1.** 100, 95, 90, 85, 85, 85, 80, 75, 70

        2. 70 to 100

        3. 85

        4. 85

        5. 85

   B. **6.** 108

        7. 784

   C. **8.** 6 salaries; $246, $182, $156, $105, $98, $75

        9. no

      10. $156 and $105

      11. $130.50

PAGE 389

   A. **1.** 82 **2.** yes

   B. **3.** 100×3 (300); 95; 91×2 (182); 90; 89; 87×2 (174); 85×2 (170); 83×2 (166); 82×4 (328); 81; 77×2 (154); 76; 73; 72; 70×3 (210); 66; 64; 50

        4. 2,440 **5.** $81\frac{1}{3}$

PAGE 390

   A. **1.** 4 **2.** 360°

   B. **3.** $\frac{4}{30}$ or $\frac{2}{15}$; 48°

        4. $\frac{8}{30}$ or $\frac{4}{15}$; 96°

        5. $\frac{5}{30}$ or $\frac{1}{6}$; 60°

        6. $\frac{3}{30}$ or $\frac{1}{10}$; 36°

PAGE 392

   A. **1.** 4 **2.** 8

        3. $\frac{4}{8}$ **4.** $\frac{1}{2}$

   B. **5.** $\frac{1}{5}$ **6.** $\frac{4}{5}$

   C. **7.** $\frac{4}{4}$ or 1 **8.** $\frac{0}{4}$ or 0

PAGE 394

   A. **1.** $\frac{1}{6}$ **2.** $\frac{1}{2}$ **3.** $\frac{1}{2}$

   B. **4.** (H, T) **5.** (T, T) **6.** (T, H)

   C. **7.** $\frac{1}{4}$ **8.** 25

PAGE 396

   A. **1.** $\frac{7}{12}$ **2.** $\frac{1}{12}$ **3.** $\frac{2}{3}$

   B. **4.** $\frac{1}{2}$ **5.** $\frac{1}{2}$

        6. (H, H); (H, T); (T, T); (T, H)

        7. $\frac{1}{4}$ **8.** true

   C. **9.** 25 times; Answers will vary.

INDEX

ART CREDITS

PHOTO CREDITS